Methods in Molecular Biology

Series Editor
John M. Walker
School of Life and Medical Sciences
University of Hertfordshire
Hatfield, Hertfordshire, UK

For further volumes:
http://www.springer.com/series/7651

For over 35 years, biological scientists have come to rely on the research protocols and methodologies in the critically acclaimed Methods in Molecular Biology series. The series was the first to introduce the step-by-step protocols approach that has become the standard in all biomedical protocol publishing. Each protocol is provided in readily-reproducible step-by step fashion, opening with an introductory overview, a list of the materials and reagents needed to complete the experiment, and followed by a detailed procedure that is supported with a helpful notes section offering tips and tricks of the trade as well as troubleshooting advice. These hallmark features were introduced by series editor Dr. John Walker and constitute the key ingredient in each and every volume of the Methods in Molecular Biology series. Tested and trusted, comprehensive and reliable, all protocols from the series are indexed in PubMed.

Manuscript Instructions https://drive.google.com/file/d/1K4Az5vpAFVzYadJ3jKx2-QSo7Dqa_SNA/view?usp=sharing

MicroRNA Target Identification

Methods and Protocols

Second Edition

Edited by

Alessandro Laganà

Department of Oncological Sciences, Icahn School of Medicine at Mount Sinai, New York, NY, USA

Giovanni Nigita

Division of Medical Oncology, Department of Internal Medicine, College of Medicine, The Arthur G. James Comprehensive Cancer Center, The Ohio State University, Columbus, OH, USA

Humana Press

Editors
Alessandro Laganà
Department of Oncological Sciences
Icahn School of Medicine at Mount Sinai
New York, NY, USA

Giovanni Nigita
Division of Medical Oncology
Department of Internal Medicine
College of Medicine
The Arthur G. James Comprehensive Cancer Center
The Ohio State University
Columbus, OH, USA

ISSN 1064-3745 ISSN 1940-6029 (electronic)
Methods in Molecular Biology
ISBN 978-1-0716-5391-3 ISBN 978-1-0716-5392-0 (eBook)
https://doi.org/10.1007/978-1-0716-5392-0

This Humana imprint is published by the registered company Springer Science+Business Media, LLC, part of Springer Nature.
The registered company address is: 1 New York Plaza, New York, NY 10004, U.S.A.

Preface

MicroRNAs (miRNAs) are small endogenous RNA molecules that play a central role in post-transcriptional gene regulation. Since their initial characterization, research on miRNAs has expanded dramatically, driven by continuous advances in molecular biology, high-throughput sequencing, and computational analysis. These efforts have established miRNAs as key regulators of complex biological programs and as critical contributors to human disease, while also highlighting their potential as biomarkers and therapeutic targets. In recent years, the field has undergone a further conceptual shift, moving beyond canonical targeting models toward a more nuanced view that encompasses molecular diversity, regulatory context, and multi-layered control.

MiRNAs exert their regulatory activity primarily through sequence-dependent interactions with target messenger RNAs (mRNAs), leading to translational repression or destabilization of the target transcript. Despite decades of methodological development, the accurate identification of miRNA targets remains a challenging task. This difficulty reflects both biological complexity and technical limitations, including incomplete knowledge of non-canonical binding modes, the functional relevance of isomiRs and RNA editing, and the influence of competing endogenous RNAs and cellular context. As a result, miRNA target prediction and validation increasingly rely on integrative strategies that combine computational modeling, network-based analysis, and experimental approaches.

This second edition aims to provide an updated, comprehensive, state-of-the-art overview of miRNA target identification, spanning miRNA binding site prediction, integrative and network-based analyses, emerging technologies, and experimental validation strategies. Building on the foundation established in the first edition, this volume includes updated versions of several core chapters as well as entirely new contributions that reflect major conceptual and technological advances in the field. As in the first edition, some chapters review methodologies and computational resources, while others provide step-by-step protocols and practical tutorials to guide readers through specific analytical and experimental tasks.

Chapters 1–6 focus on the principles and evolving features of miRNA target recognition and prediction. Chapter 1 provides an updated overview of past and current computational approaches for miRNA target prediction, highlighting both their achievements and remaining limitations. Chapter 2 introduces the emerging concept of target-directed miRNA degradation, discussing its molecular mechanisms and biological significance. Chapters 3–5 present updated treatments of established computational approaches, covering RNA secondary structure–based prediction methods, miRNA targeting at the isomiR level, and consensus-based prediction strategies that integrate canonical and modified miRNAs. Chapter 6 introduces MiREDiBase, a comprehensive resource dedicated to miRNA editing events, underscoring the growing importance of RNA modifications in miRNA biology.

Chapters 7–12 are devoted to systems-level and integrative analyses of miRNAs and their targets. Chapter 7 describes methodologies for constructing and analyzing miRNA regulatory networks, while Chap. 8 focuses on computational approaches for pathway and network analysis that incorporate miRNAs. Chapter 9 reviews computational resources for studying miRNA–long non-coding RNA interactions, with an emphasis on ceRNA mechanisms.

Chapter 10, newly added in this edition, extends this perspective to circRNA–miRNA–mRNA regulatory networks, reflecting the rapid expansion of circRNA research. Chapter 11 presents updated network-based approaches for identifying disease–miRNA associations across multiple human complex diseases, and Chap. 12 discusses integrative multi-omics strategies to identify cancer-related miRNAs.

Chapters 13 and 14 address emerging technologies and translational applications. Chapter 13 introduces microRNA profiling at single-cell resolution, highlighting recent advances in single-cell sequencing technologies and their implications for studying cellular heterogeneity and miRNA regulation. Chapter 14 presents an innovative nanopore-based approach for microRNA quantification in liquid biopsies, demonstrating how technological advances are enabling highly sensitive, clinically relevant miRNA measurements.

The final three chapters, Chaps. 15–17, are dedicated to the experimental validation of predicted miRNA targets. These chapters provide detailed laboratory protocols for the most widely used validation techniques, including luciferase reporter assays, mutagenesis of miRNA binding regions, and western blot analysis. Together, they offer practical guidance for translating computational predictions into experimentally supported biological insights.

We are pleased to present this second edition, which reflects the continued evolution of the field and the increasing integration of computational and experimental perspectives in miRNA research. We hope that this volume will serve as a helpful reference for researchers working in computational biology, molecular biology, and translational medicine.

We thank all contributing authors for their expertise and commitment to this project. Our sincere gratitude also goes to the series editor, John Walker, and the Springer editorial and production teams for their professional guidance and support throughout the preparation of this volume.

New York, NY, USA *Alessandro Laganà*
Columbus, OH, USA *Giovanni Nigita*

Contents

Contributors

MARIO ACUNZO • *Department of Medicine and Surgery, LUM "Giuseppe Degennaro", Casamassima, Bari, Italy; Division of Pulmonary Diseases and Critical Care Medicine, Virginia Commonwealth University, Richmond, VA, USA*
SALVATORE ALAIMO • *Department of Clinical and Experimental Medicine, University of Catania, Catania, Italy*
MICHAEL ATTATHIKHUN • *Department of Translational Molecular Pathology, The University of Texas MD Anderson Cancer Center, Houston, TX, USA*
SIMONE AVESANI • *Department of Computer Science, University of Verona, Verona, Italy*
ROSARIO NICOLA BRANCACCIO • *Department of Precision Medicine in Medical, Surgical and Critical Care, University of Palermo, Palermo, Italy*
SIMONE CALIGOLA • *Veneto Institute of Oncology IOV-IRCCS, Padova, Italy*
GEORGE A. CALIN • *Department of Translational Molecular Pathology, The University of Texas MD Anderson Cancer Center, Houston, TX, USA; Department of Cancer Biology, The University of Texas MD Anderson Cancer Center, Houston, TX, USA; Center for RNA Interference and Non-Coding RNAs, The University of Texas MD Anderson Cancer Center, Houston, TX, USA*
FEDERICA CALORE • *Department of Cancer Biology and Genetics, The James Comprehensive Cancer Center, The Ohio State University Wexner Medical Center, Columbus, OH, USA*
DAVIDE CAPOZZI • *Laboratory of Tumor Immunology and Cell Therapy, Department of Experimental Medicine, Sapienza University of Rome, Rome, Italy*
LUCIANO CASCIONE • *Institute of Oncology Research, Faculty of Biomedical Sciences, Università della Svizzera italiana (USI), Bellinzona, Switzerland; Swiss Institute of Bioinformatics (SIB), Lausanne, Switzerland*
FEDERICA CONTE • *Institute for Systems Analysis and Computer Science Antonio Ruberti, National Research Council, Rome, Italy*
ILARIA COSENTINI • *Institute for Biomedical Research and Innovation (IRIB), National Research Council of Italy, Palermo, Italy*
FEDERICA D'ANNUNZIO • *Department of Translational and Precision Medicine, Sapienza University of Rome, Rome, Italy*
SEBASTIANO DI BELLA • *Department of Precision Medicine in Medical, Surgical and Critical Care, University of Palermo, Palermo, Italy*
YE DING • *Wadsworth Center, New York State Department of Health, Albany, NY, USA*
ROSARIO DISTEFANO • *Division of Hematology, Department of Internal Medicine, The Ohio State University, Columbus, OH, USA*
ADRIENNE M. DORRANCE • *Division of Oncological Sciences, Huntsman Cancer Institute, University of Utah, Salt Lake City, UT, USA*
SHAOHONG FENG • *Department of Biomedical Informatics, The Ohio State University, Columbus, OH, USA*
GIULIA FISCON • *Institute for Systems Analysis and Computer Science Antonio Ruberti, National Research Council, Rome, Italy; Department for the Promotion of Human Sciences and Quality of life, San Raffaele Roma Open University, Rome, Italy*

Miriam Gaggianesi • *Department of Precision Medicine in Medical, Surgical and Critical Care, University of Palermo, Palermo, Italy*
Ramiro Garzon • *Division of Hematology and Hematological Malignancies, Huntsman Cancer Institute, University of Utah, Salt Lake City, UT, USA*
Rosalba Giugno • *Department of Computer Science, University of Verona, Verona, Italy*
Chinmayee Goda • *Division of Oncological Sciences, Huntsman Cancer Institute, University of Utah, Salt Lake City, UT, USA*
Shaopeng Gu • *Department of Biomedical Informatics, The James Comprehensive Cancer Center, College of Medicine, The Ohio State University, Columbus, OH, USA; Pelotonia Institute for Immuno-Oncology, The Arthur G. James Comprehensive Cancer Center, The Ohio State University, Columbus, OH, USA; Department of Biomedical Informatics, The Ohio State University, Columbus, OH, USA*
Haiyan Hu • *Department of Computer Science, University of Central Florida, Orlando, FL, USA*
Ancuta Jurj • *Department of Translational Molecular Pathology, The University of Texas MD Anderson Cancer Center, Houston, TX, USA*
Anastassia Kanavarioti • *Yenos Analytical LLC, El Dorado Hills, CA, USA*
Shaveta Kanoria • *Wadsworth Center, New York State Department of Health, Albany, NY, USA*
Roma Karna • *Department of Surgery, Division of Surgical Oncology, The James Comprehensive Cancer Center, The Ohio State University Wexner Medical Center, Columbus, OH, USA*
Alessandro Laganà • *The Tisch Cancer Center, Icahn School of Medicine at Mount Sinai, New York, NY, USA; Department of Oncological Sciences, Icahn School of Medicine at Mount Sinai, New York, NY, USA; Department of Genetics and Genomic Sciences, Icahn School of Medicine at Mount Sinai, New York, NY, USA*
Xiaoman Li • *Burnett School of Biomedical Sciences, College of Medicine, University of Central Florida, Orlando, FL, USA*
Junhao Liu • *Computer Science and Engineering, College of Engineering, The Ohio State University, Columbus, OH, USA*
Jun Lu • *Department of Genetics and Yale Stem Cell Center, Yale University, New Haven, CT, USA*
Anjun Ma • *Department of Biomedical Informatics, The James Comprehensive Cancer Center, College of Medicine, The Ohio State University, Columbus, OH, USA; Pelotonia Institute for Immuno-Oncology, The Arthur G. James Comprehensive Cancer Center, The Ohio State University, Columbus, OH, USA*
Qin Ma • *Department of Biomedical Informatics, The James Comprehensive Cancer Center, College of Medicine, The Ohio State University, Columbus, OH, USA; Pelotonia Institute for Immuno-Oncology, The Arthur G. James Comprehensive Cancer Center, The Ohio State University, Columbus, OH, USA; Department of Biomedical Informatics, The Ohio State University, Columbus, OH, USA*
Gioacchino P. Marceca • *Independent Researcher, Vittoria, RG, Italy*
Giovanni Nigita • *Pelotonia Institute for Immuno-Oncology, The Arthur G. James Comprehensive Cancer Center, The Ohio State University, Columbus, OH, USA; Center for RNA Biology, The Ohio State University, Columbus, OH, USA; Division of Medical Oncology, Department of Internal Medicine, College of Medicine, The Arthur G. James Comprehensive Cancer Center, The Ohio State University, Columbus, OH, USA*

FRANCESCO ORILIO • *Department of Precision Medicine in Medical, Surgical and Critical Care, University of Palermo, Palermo, Italy*
PAOLA PACI • *Institute for Systems Analysis and Computer Science Antonio Ruberti, National Research Council, Rome, Italy; Department of Computer, Control, and Management Engineering Antonio Ruberti, Sapienza University of Rome, Rome, Italy*
ERNESTO PICARDI • *Department of Bioscience, Biotechnology and Environment, University of Bari, Bari, Italy; Institute of Biomembranes, Bioenergetics and Molecular Biotechnologies (IBIOM) of the National Research Council (CNR), Bari, Italy*
ALFREDO PULVIRENTI • *Department of Clinical and Experimental Medicine, University of Catania, Catania, Italy*
WILLIAM RENNIE • *Wadsworth Center, New York State Department of Health, Albany, NY, USA*
CHRISTIAN ROLFO • *Division of Medical Oncology, Department of Internal Medicine, College of Medicine, The Arthur G. James Comprehensive Cancer Center, The Ohio State University, Columbus, OH, USA*
GIULIA ROMANO • *Division of Pulmonary Diseases and Critical Care Medicine, Virginia Commonwealth University, Richmond, VA, USA*
ESWAR SHANKAR • *Division of Medical Oncology, Department of Internal Medicine, College of Medicine, The Arthur G. James Comprehensive Cancer Center, The Ohio State University, Columbus, OH, USA*
SOUMYA SHARMA • *College of Medicine and Life Sciences, The University of Toledo, Toledo, OH, USA*
GIORGIO STASSI • *Department of Precision Medicine in Medical, Surgical and Critical Care, University of Palermo, Palermo, Italy*
MATILDE TODARO • *Department of Health Promotion, Mother and Child Care, Internal Medicine and Medical Specialities (PROMISE), University of Palermo, Palermo, Italy*
FRANCESCO VERONA • *Department of Precision Medicine in Medical, Surgical and Critical Care, University of Palermo, Palermo, Italy*

Check for updates

Chapter 1

miRNA Target Prediction: An Overview of the Past and Current Tools

Sebastiano Di Bella, Francesco Orilio, Giovanni Nigita, Matilde Todaro and Giorgio Stassi

Abstract

MicroRNAs (miRNAs) are among the most studied molecules in recent years, and since their discovery, many miRNAs have been identified across various species. As members of the non-coding RNA family, miRNAs are key players in post-transcriptional gene regulation. These molecules can inhibit translation or promote degradation of messenger RNA (mRNA) by binding to the 3′ untranslated region (UTR) of mRNA, thereby influencing almost all biological processes. To identify a miRNA's biological role, it is essential to predict the target sites to which it binds, a goal made possible through bioinformatics tools. This chapter discusses the bioinformatics tools commonly used for this purpose. Also, it analyzes the main factors considered in target prediction, such as seed match, free energy, conservation, site accessibility, multiple binding site contribution, and machine learning and deep learning approaches. Understanding the principles underlying these predictive methodologies is crucial for advancing one's biological research on miRNAs.

Keywords MicroRNA, Target prediction, Seed match, Conservation, Free energy, Site accessibility, Bioinformatic tools, Machine learning

1 Introduction

The existence of miRNAs was first reported in 1993 [1], and subsequent investigations of their biological roles have revolutionized our understanding of gene regulatory networks. These molecules, composed of approximately 22 nucleotides, have been found in various organisms, including animals, plants, and viruses [2–4]. Experimentally discovered miRNAs are deposited in the public database "miRBase." The latest version of the miRBase database (v22) collects miRNA sequences from 271 organisms, including 38,589 hairpin precursors and 48,860 mature miRNAs. In the most recent updates, miRBase has

Sebastiano Di Bella, Francesco Orilio. These authors equally contributed.

Alessandro Laganà and Giovanni Nigita (eds.), *MicroRNA Target Identification*, Methods in Molecular Biology, vol. 3056, https://doi.org/10.1007/978-1-0716-5392-0_1,

improved annotation quality and functional information by integrating 1493 small RNA deep-sequencing datasets and mapping a total of 5.5 billion reads to miRNAs. These data support the validity of 20%–65% of miRNA annotations in well-studied species, after removing over 200 unreliable sequences. Functional information has been further enriched with Gene Ontology terms and over 500,000 sentences extracted from 18,542 scientific articles mentioning miRNAs [5]. miRBase is freely accessible at http://mirbase.org/. The currently available version, miRBase 22.1, is a minor update that includes modifications to deep-sequencing datasets mapped to miRBase sequences, leading to recalibration of the high-confidence miRNA set.

The functions of miRNAs are closely associated with the RNA-induced (RNAi) silencing complex (RISC). After transcription and cleavage, these molecules undergo nuclear maturation and, once released into the cytoplasm, can bind target messenger RNAs and load them onto the silencing complex [6]. The pairing between miRNA and mRNA is facilitated by the interaction of the 3′ UTR of the messenger with the region defined as the "seed" of the miRNA [7]. miRNAs can reduce gene expression by promoting mRNA degradation or inhibiting translation [8]. It is estimated that a single miRNA can regulate hundreds, if not thousands, of target genes and that multiple miRNAs can regulate a single gene [9]. Various biological processes are regulated by miRNA activity, including cell differentiation [10], cell division [11], cell proliferation [12], apoptosis [13], metabolism [14], and tissue development [15]. Given these molecules' significant role in gene regulation, miRNAs are often associated with diseases, where they are recognized as biomarkers [16] and therapeutic targets [17]. Multiple databases have been created to identify the target genes of miRNAs and thus understand their biological functions [18]. At the same time, various computational algorithms have been developed to efficiently predict these targets.

This chapter describes the features considered in computational tools for miRNA target prediction analyses. Additionally, it provides an overview of the most commonly used bioinformatics tools for predicting miRNA target sites. We also discuss the strengths and limitations of these tools, as this information will be valuable for users in assessing and selecting the most suitable option for their specific needs. Furthermore, we will focus on the latest deep-learning approaches in this field.

2 Common Characteristics of miRNA Target Prediction Tools

Several factors can be considered when developing a miRNA target prediction algorithm. First and foremost, it is essential to pay attention to the interactions between the "seed" component of the miRNA and the 3′ UTR of the messenger RNA. Sequence comparisons across species are then performed to determine whether the target sequence is conserved. Moreover, the number of target sites for a given miRNA can be quantified, as multiple binding sites may regulate mRNAs. Some tools can conduct thermodynamic analysis of the defined miRNA–mRNA duplexes, including calculating the free energy of duplex formation and evaluating binding thermodynamics.

Additionally, the structural accessibility of the 3′ UTR is examined to ensure that miRNAs can effectively bind to their targets. These features can be utilized stepwise. The following sections will describe each feature. Recent advancements in machine learning and deep learning have improved miRNA target prediction algorithms by analyzing complex relationships between miRNAs and their target genes. These methods utilize large datasets, thermodynamic models, and structural features to enhance prediction accuracy and adapt to the specific characteristics of different species (Fig. 1).

2.1 Sequence Complementarity

The cornerstone of the interaction between miRNA and mRNA is the complementarity between the 3′ UTR and the seed region. While plant miRNAs typically bind to their targets with perfect complementarity, animal miRNAs exhibit partial complementarity, complicating the determination of potential miRNA targets with high specificity [19, 20]. The seed sequence represents the first 2–7 nucleotides in the 5′ region of miRNA. Most target prediction algorithms require Watson–Crick pairing between the miRNA and the targeted sites [21], which involves interactions between Adenine (A)–Uracil (U) and Guanine (G)–Cytosine (C). Although not always sufficient, a strong seed match is essential for effective targeting. Several types of seed matches can be considered [22]:

1. 8mer: A perfect Watson–Crick pairing between nucleotides 2 and 8 of the miRNA seed and an A aligned with the first miRNA nucleotide.
2. 7mer-A1: A perfect Watson–Crick match from nucleotides 2 to 7 of the miRNA seed, paired with an A opposite the first miRNA nucleotide.
3. 7mer-m8: A perfect Watson–Crick match from nucleotides 2 to 8 of the miRNA seed.
4. 6mer: A perfect six-nucleotide Watson–Crick match between the miRNA seed and the mRNA.

2.2 Conservation of the Target Sequence Site

The term "conserved sequence" refers to nucleic acid or protein sequences that are similar or identical across different species. In the case of miRNAs, prediction algorithms mainly focus on the 3′ UTR, determining orthologous sequences and checking whether the miRNA binding site is conserved among related species. Besides the sequence itself, the level of conservation is also considered to filter miRNA targets from a genomic sequence. Conservation is

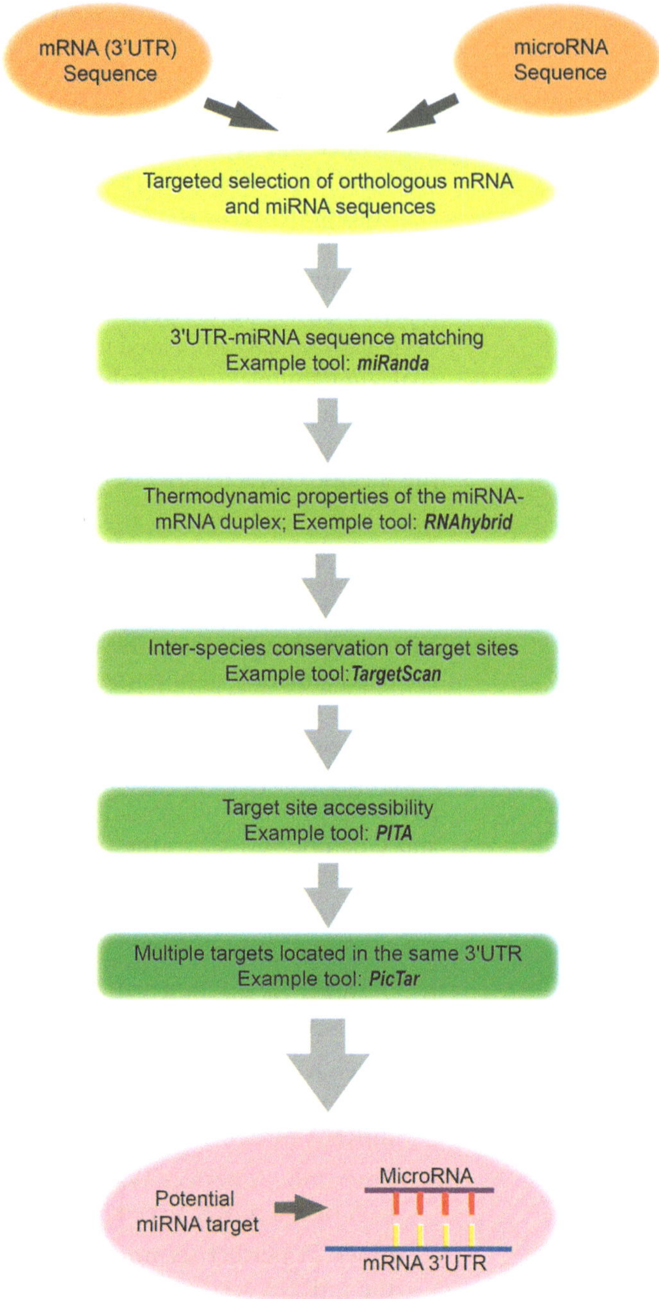

Fig. 1 Workflow for potential miRNA target prediction analysis

more pronounced in the miRNA seed region compared to other regions [23]. However, mismatches can sometimes occur in this region, and conserved pairing may happen at the 3′ terminal portion of the miRNA, defining what is known as a “compensatory site”

[24]. Sequence conservation can be assessed using software such as BLASTn [25] or AVID [26, 27] (https://genome.lbl.gov/cgi-bin/GenomeVista), which aligns 3′ UTR sequences from different species. Another helpful resource is the UCSC genome database, now updated to its 2024 version [28], which provides large-scale genomic alignments for a broad range of species and enables conservation analysis. Although conservation is one of the most potent measures for target prediction, it can sometimes be problematic, as some miRNAs have evolved to act on non-conserved 3′ UTR sequences [29].

2.3 Multiple Predicted Binding Sites in the Same 3′ UTR

Several studies have shown that miRNA families redundantly regulate targets with a matching site in the seed region. This highlights how multiple miRNAs can possess similar sequences in their seed regions, allowing them to bind to corresponding matching sites on the same target [30]. Multitargeting occurs more frequently than expected. Predictions based on multitargeting significantly reduce the number of predicted targets. Since this feature increases the likelihood of identifying true target genes, it is advantageous for experimental validation studies of miRNA targets [31]. The level of target regulation by miRNAs varies depending on the number of miRNAs binding to a target gene [32].

2.4 Target Site Accessibility

An additional parameter that target prediction software considers is the structural accessibility of binding sites, which includes both the 3′ UTR region of the mRNA and the miRNA seed region. The accessibility of the 3′ UTR region is primarily assessed as miRNAs are assembled into the RISC in their active state, ensuring that the seed region remains accessible. Some approaches use a cumulative score to estimate the accessibility of binding sites and rank predictions, considering the free energy required to make the site accessible and the hybridization energy [33]. Other methods, instead, use specific scores [34] or consider only the free energy [35].

2.5 Thermodynamic Properties

The thermodynamic properties of the duplex formed upon miRNA–mRNA binding are also features to consider in target prediction algorithms. Specifically, the stability of hybridization is indicated by the binding free energy. A low binding free energy implies a stable bond (thus considered accurate), which requires more energy to be disrupted [36]. Among the most widely used tools for thermodynamic analysis is the ViennaRNA package [37], which allows calculation of RNA secondary structures using energy models, prediction of the minimum free energy (MFE) structure, and calculation of partition functions. It includes a C library and standalone programs that can be used from the command line and integrated into other applications. The latest software update was on September 25, 2023, with version 2.6.4, improving stability, compiling in C++, and optimizing the code (https://www.tbi.univie.ac.at/RNA/#performance). Determining appropriate thresholds for free energy is challenging because datasets containing known miRNA–mRNA duplexes are limited. Moreover, a low free energy value does not necessarily guarantee accurate predictions of miRNA target genes. For this reason, it is essential also to consider other features, such as sequence conservation, to effectively predict miRNA targets [38].

3 Commonly Used Tools for miRNA Target Prediction

Several methods are currently available for the computational prediction of miRNA targets. These prediction platforms function based on different assumptions and implementations. Below, we summarize the most popular tools in this field (see Table 1 for their general characteristics).

1) **miRanda** is one of the most widely used target prediction tools. It is accessible via a web interface that allows users to predict and evaluate candidate targets using the mirSVR machine learning method, which is based on their ability to repress target genes [39]. miRanda is implemented in C, and the source code is available for download. As a result, it provides straightforward compilation and a wide range of adjustable parameters, enabling in-depth and customizable analysis. However, it requires command-line operation, which can complicate use for less experienced users. miRanda identifies target sites by using features such as sequence complementarity via a position-weighted local alignment algorithm, free energies of miRNA–mRNA duplexes computed with the Vienna RNAfold package [37], and target conservation across related genomes as a final filtering step. The algorithm accounts for the conservation of both the binding site and its position. Although miRanda was initially developed to identify targets in Drosophila, this algorithm has since been adapted to predict targets in vertebrates.
2) **TargetScan** [40, 41] is a web-based tool that searches for conserved 8mer, 7mer, and 6mer sites aligning with the seed region of each miRNA. After potential targets are identified, they are ranked based on the context scores of the sites [42]. The scoring system considers features such as 3′ compensatory pairing, local AU content, position contribution, target site abundance, seed-pairing stability, and the probability of conserved targeting. Furthermore, TargetScan predicts the RNA's secondary structure to calculate the free energy of the predicted duplexes. It also provides the option to include non-conserved targets or poorly conserved sites in its output. With the latest version (v8.0), TargetScan has introduced a new biochemical model for classifying predictions in mammals. This method integrates

Table 1
Bioinformatic tools for miRNA target prediction analysis.

						Site Coverage on miRNA						
Tool Name	Organism(s)	Type	Version or date of last update	Status	URL	P	5'UTR	CDS	3'UTR	mirBase version	Relative Citation Ratio (RCR)	References
Miranda	h, m, r, f, w	Database/ software[A]	2010[N]	Inaccessible	N/A				V	42318	29.89	[39,67]
TargetScan	h, m, r, d, cn, c, rh, cw, o, fr, z, f, w	Web Server[A]	8.0[y]	Active	http://www.targetscan.org/				V	10.1, 17,21	194.23	[40–43]
RNAhybrid	f,w	Web Server / software[A]	2006[N]	Active	http://bibiserv.techfak.uni-bielefeld.de/rnahybrid/				V	N/A	35.66	[45,46]
PicTar	h, m, f, w	Web server[NA]	2007[N]	Active	http://pictar.mdc-berlin.de/				V	6	5.93	[32,44]
rna22	h, m, f, w	Web server[NA]	2015[N]	Active	https://cm.jefferson.edu/rna22		V	V	V	16,18, 19,21	30.99	[48,49]
PITA	h, m, f, w	Web Server / software[A]	2008[N]	Inaccessible	N/A				V	11	39.24	[47]
miRDB	h, m, r, d, cn	Database / web server[NA]	6.0[y]	Active	http://mirdb.org/		V	V	V	9.1,10.0, 13.0, 18,21,22	11.9	[50–52]
DIANA	h, m, f, w	Web Server / software[A]	2023[y]	Active	http://www.microrna.gr/webServer			V	V	22.1	27.42	[55–59]

Table 1
(Continued)

						Site Coverage on miRNA						
Tool Name	**Organism(s)**	**Type**	**Version or date of last update**	**Status**	**URL**	**P**	**5'UTR**	**CDS**	**3'UTR**	**mirBase version**	**Relative Citation Ratio (RCR)**	**References**
STarMir	h, m, w	Web server[NA]	2014[N]	Active	https://sfold.wadsworth.org/cgi-bin/starmirWeb.pl		V	V	V	20	2.38	[53,54]
miRNAMap	h, m, r, d, cn, o, fr, z, f, w, p, mq	Database[NA]	2008[N]	Active	http://mirnamap.mbc.nctu.edu.tw/				V	6,9.2	3.14	[68,69]
MiRror Suite	h, m, r, f, w, z	Database[NA]	2014[N]	Active	http://www.proto.cs.huji.ac.il/mirror/index.php		V	V	V	N/A	1.05	[70,71]
miRTarBase	h	Web Server[NA]	8.0[y]	Active	https://mirtarbase.cuhk.edu.cn/~miRTarBase/miRTarBase_2022/php/index.php		V	V	V	15,21,22	2.61	[72,73]
miRWalk	h, m, r	Database[NA]	3.0 (in development)[y]	Active	http://mirwalk.umm.uni-heidelberg.de	V	V	V	V	22.1	37.41	[74]
mirDIP	N/A	Web Server[NA]	Tool 5.3.0.2 / Database 5.2.3.1[y]	Active	http://ophid.utoronto.ca/mirDIP/		V	V	V	N/A	4.29	[75,76]
ComiR	h, m, f, w	Web Server[A]	2014[N]	Active	http://www.benoslab.pitt.edu/comir/				V	N/A	4.42	[77]

Table 1
(Continued)

						Site Coverage on miRNA						
Tool Name	**Organism(s)**	**Type**	**Version or date of last update**	**Status**	**URL**	**P**	**5'UTR**	**CDS**	**3'UTR**	**mirBase version**	**Relative Citation Ratio (RCR)**	**References**
miRMap	h, m, r, cw, o, cn, c, z	Web Server / software[A]	2022_03[y]	Active	http://mirmap.ezlab.org/				V	18,19, 22.1	8.07	[78,79]
TarPmiR	Animals	Software-web[NA]	2016[N]	Active	https://hulab.ucf.edu/research/projects/miRNA/TarPmiR	V	V	V	V	N/A	*	[60]
P-TarPmiR	Plants	Software-web[A]	2023[y]	Active	https://ptarpmir.cu-bic.ca/analysis	V	V	V	V	N/A	*	[61]
miTAR	N/A	Code in python[A]	2021[y]	Active	https://github.com/tjgu/miTAR				V	N/A	*	[62]
DeepMirTar	H	Code in python[A]	2018[N]	Active	https://github.com/Bjoux2/DeepMirTar_SdA				V	N/A	*	[63]
TEC-mitarget	N/A	Code in python[A]	Version 2024[y]	Active	https://github.com/tingpeng17/TEC-miTarget?files=1		V	V	V	N/A	*	[64]

Description of the fields and conventions used in the table: "Type" indicates whether the sourcecode is available ("A") or not ("NA"); "Organisms" provides codes for the organisms analyzed, including h (human), m (mouse), r (rat), d (dog), cn (chicken), c (chimpanzee), rh (rhesus monkey), cw (cow), o (opossum), fr (frog), z (zebrafish), f (fly), w (worm), p (pufferfish), mq (mosquito); "Last updated" specifies whether the entry has been updated in the past 5 years ("Y") or not ("N"); and "RCR" uses "*" to denote new studies for which the Relative Citation Ratio has not been reported experimental measurements of miRNA–target mRNA affinity, accounting for relative affinity values, biological data integration, and a convolutional neural network [43].

3) **PicTar** [32, 44] is a computational method that predicts shared targets for miRNAs by leveraging comparative data across species. Moreover, it calculates the maximum likelihood that combinations of miRNAs bind to a specific RNA sequence (typically the 3′ UTR). Target genes are initially predicted based on a common criterion of optimal binding free energy and are then subjected to statistical testing using genome-wide alignments across eight vertebrate genomes to filter out genuine targets. This tool also identifies binding sites for an individual miRNA and multiple sites regulated by different miRNAs that act cooperatively.

4) **RNAhybrid** [45, 46] is available as a web tool and for download, offering users significant flexibility in their research. This program can be applied in various biological contexts, including cancer research, developmental biology, and studies of gene regulation in different tissues. It is designed to predict potential miRNA binding sites within large target RNAs, while accounting for the free energy of miRNA–mRNA duplexes. RNAhybrid efficiently identifies the optimal hybridization sites for a small RNA within a larger RNA molecule simultaneously. The ability to rapidly analyze multiple targets makes RNAhybrid an essential resource for researchers working with high-throughput sequencing data and large genomic datasets.

5) **PITA** [47] is a non-parametric tool for predicting miRNA–target interactions, focusing on target-site accessibility and acknowledging that mRNA secondary structures can significantly hinder miRNA binding. The tool calculates both the hybridization energy and the energy required to open up RNA structures, providing a comprehensive view of potential interactions. However, the accuracy of this method depends significantly on the ability to predict stable secondary structures, which becomes less reliable for long-distance interactions or highly complex RNA structures.

6) **RNA22** [48, 49] is based on a two-step analysis: first, it scans an RNA sequence to identify potential miRNA binding sites, and subsequently, it identifies the possible miRNAs that may bind to these sites. Utilizing these two distinct models, it can identify potential heteroduplexes without considering cross-species conservation. This allows RNA22 to predict miRNA-binding interactions even when the target miRNA is not yet known, broadening the scope for uncovering novel miRNA–target relationships.

7) **MiRDB** [50, 51] is an online database for predicting miRNA targets and providing functional annotations. It comprises two main components: (1) a wiki-style editing interface that offers an open platform for researchers to contribute manual annotations, and (2) a web interface for accessing computationally predicted miRNA targets generated using a Support Vector Machine (SVM)-based learning algorithm. MiRDB strongly emphasizes functional annotations, focusing specifically on mature miRNAs, the active agents that regulate miRNA-mediated gene expression. With the latest update, MiRDB showcases an improved user interface (UI), personalized predictions based on user-provided sequences, and enhanced transcriptome-level target predictions [52].

8) **StarMir** [53] predicts miRNA binding sites by evaluating a wide range of features, including sequence characteristics, thermodynamic properties, and the structural attributes of target sites. The scores associated with each binding site—located in the 3′ UTR, coding sequence (CDS), or 5′ UTR—are calculated using logistic regression models trained on miRNA binding data from CLIP studies [54]. This approach assigns a score that measures the accuracy of the prediction.

9) **DIANA-microT-CDS (now DIANA-micro-T-2023)** [55–57] is a web server that provides an intuitive interface, particularly beneficial for studies with less advanced computational infrastructures. This tool enables comprehensive analysis of microarray and next-generation sequencing (NGS) data and integrates the KEGG database to support pathway-specific searches based on KEGG annotations [58]. It is equipped with sophisticated, preconfigured pipelines designed to facilitate research on predicted or experimentally validated miRNA–gene interactions and their roles in metabolic and signaling pathways. One of the distinctive features of DIANA-microT-CDS is its ability to predict miRNA targets in both coding sequences (CDS) and the more commonly analyzed 3′ UTR regions. The algorithm considers various factors for target prediction, including binding category, proximity to the nearest region end (CDS or 3′ UTR), and adjacent binding sites.

Additionally, it considers the predicted free energy of miRNA–mRNA duplex formation, sequence conservation, and AU content. The DIANA-Taverna Plug-in is available for more advanced users, allowing seamless integration of the microT-CDS target prediction server into Taverna's graphical interface and facilitating the incorporation of advanced miRNA analysis functionalities into custom-designed pipelines. With the 2023 update, DIANA-microT has enhanced its capabilities by incorporating an updated dataset containing 83 million interactions across various species. It now integrates ENSEMBL, miRBASE, and MIRGENEDB annotations, significantly increasing prediction accuracy. Furthermore, the platform has introduced a section dedicated to predicting interactions for miRNAs encoded by viruses (https://dianalab.e-ce.uth.gr/microt_webserver/#/) [60].

4 Machine Learning and Deep Learning Approaches in miRNA Target Prediction

Machine learning- and deep learning-based methods for miRNA target prediction have significantly increased in recent years (Fig. 2). Below, we highlight some of the key methods that exemplify this progress:

1) **TarPmiR** identifies miRNA binding sites using two criteria: seed match and folding energy, the latter calculated using the Vienna RNA package to select sites with the lowest folding energy. For each candidate site, it computes 13 features that influence the likelihood of being a true target, combining both conventional and novel factors. A random forest model is then

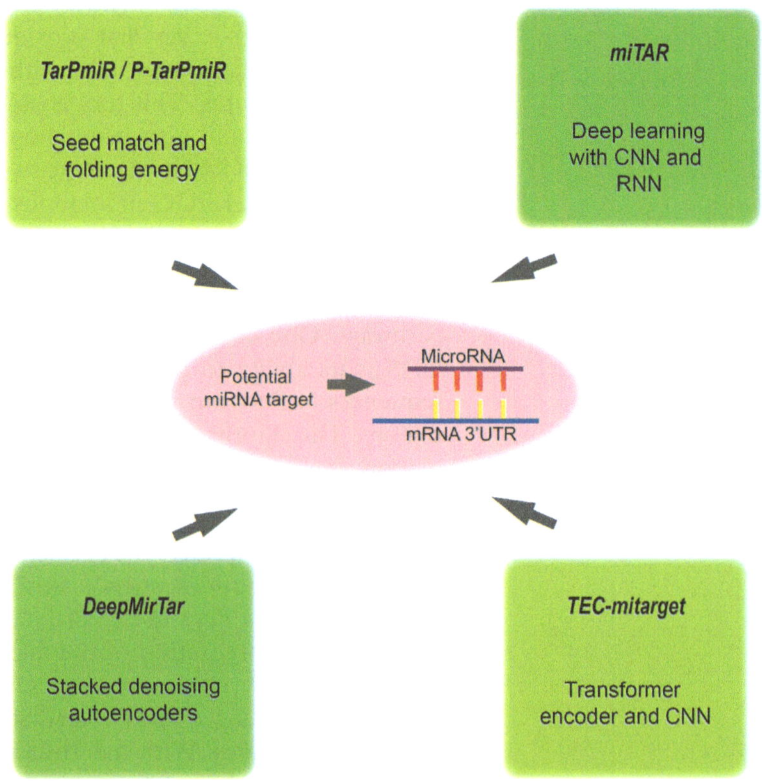

Fig. 2 miRNA target prediction tools employing machine learning and deep learning approaches

employed to predict the probability that a candidate site is a real target, testing different cut-offs and focusing on recall over precision. TarPmiR has been compared with tools such as miRanda and TargetScan, demonstrating superior recall and precision, particularly on independent datasets (https://hulab.ucf.edu/research/projects/miRNA/TarPmiR/) [73].

2) **P-TarPmiR** is a modified and retrained version of TarPmiR, specifically designed for use with plant species. It uses the TarDB database, which contains high-confidence plant miRNA–mRNA interactions. This version excludes animal data during training, significantly improving plant prediction accuracy. It has been validated on various plant species, demonstrating a significant improvement in predictive performance (https://ptarpmir.cu-bic.ca/analysis) [74].

3) **miTAR** was developed using a deep learning approach that combines convolutional neural networks (CNNs) to extract spatial features from input data and recurrent neural networks (RNNs) to process temporal sequences, allowing the model to capture complex relationships between miRNAs and their target genes. The model was trained on miRNA and gene datasets, with a random split into training and test sets repeated 30 times for robustness. Validation was performed by calculating average accuracy, sensitivity, and specificity, with miTAR achieving an average accuracy of 96.5% on the miRAW dataset and 97.9% on its model. These results demonstrate the model's superior performance compared to previous methods. miTAR can be found at the link: https://github.com/tjgu/miTAR [75].

4) **DeepMirTar** is a sophisticated tool for predicting microRNA (miRNA) targets using deep learning. It aims to identify potential miRNA binding sites on messenger RNAs (mRNAs). DeepMirTar employs a deep learning architecture based on stacked denoising autoencoders, allowing the model to learn complex patterns from input data and improving the accuracy of target predictions. The tool extracts approximately 750 features representing the characteristics of miRNA–target interactions, including high-level expert-designed features derived from prior knowledge, low-level expert-designed features based on basic sequence characteristics, and raw-data-level features directly obtained from the sequence data without prior processing. The primary input data for DeepMirTar consist of raw miRNA sequences and their potential target mRNAs, enabling flexible analysis without extensive preprocessing. DeepMirTar is freely accessible to researchers and available on platforms such as GitHub [76] (https://github.com/Bjoux2/DeepMirTar_SdA).

5) **TEC-miTarget** is emerging as one of the most innovative and promising approaches. This tool works by representing RNA sequences with a transformer encoder and combining the representations of a miRNA and its candidate target site sequences into a three-dimensional contact map. This map is then analyzed using a convolutional neural network (CNN) to extract features and predict interactions. The positional encoder in TEC-miTarget integrates positional information about RNA bases into its representations, enabling the model to understand the order of bases within RNA sequences. In contrast, the transformer encoder captures dependencies among RNA bases, generating rich, contextually enriched representations that enhance understanding of interactions between miRNAs and candidate target sites. Both encoders are crucial to improving the model's overall performance. The accuracy of the TEC-miTarget algorithm is promising, with a high level of accuracy and precision. Notably, the code for TEC-miTarget is open-source (https://github.com/tingpeng17/TEC-miTarget) and entirely written in Python, making it accessible for further research and development [77].

5 Discussion

The tools examined in this chapter each have distinct strengths and inherent limitations. While combining the outputs of multiple tools may enhance the quality of target predictions, there is no substantial evidence to suggest that this approach provides any advantages [78]. It is important to note that although evolutionary sequence conservation is a significant factor, relying solely on it may not be sufficient for conducting an accurate analysis, given the considerable number of functional but non-conserved 3′ UTRs [79]. Unfortunately, most miRNA target prediction tools lack consistent updates. Regular updates are essential due to frequent changes in miRNA nomenclature and the annual addition of new miRNAs to miRBase. For instance, miRanda, a widely used downloadable algorithm, has been awaiting updates for over 10 years. Similarly, the PicTar web servers have remained outdated for several years, and tools such as PITA are no longer usable. RNAhybrid is currently accessible via an outdated web server that has not received recent updates. This tool requires users to configure complex settings from their own input, which complicates its use for non-expert users.

In contrast, tools such as TargetScan, DIANA-microT-CDS, and miRDB are relatively well-maintained and frequently updated.

Notably, TargetScan and miRDB are among the most actively updated tools, leveraging the high-confidence status of a miRNA for functional identification. It is well established that miRNA binding sites within coding sequences (CDS) play a crucial role in regulating gene expression. Given the importance of interactions outside the 3′ UTR, it is essential to consider tools that verify interactions at CDS and 5′ UTR sites, such as DIANA-microT-2023, RNA22, miRDB, and STarMir. Finally, the diversity of tools in terms of functionality and implementation can pose a challenge for researchers seeking the most appropriate option. For this reason, users are encouraged to consider not only the intrinsic characteristics of the algorithms but also other factors, such as the consistent maintenance of algorithms and databases, available versions, and their respective strengths and weaknesses.

In addition, machine learning and deep learning tools are among the most promising approaches in miRNA target prediction, delivering exceptional results. However, given that these technologies are relatively new and rapidly evolving, monitoring them closely and understanding their mechanisms is crucial. Furthermore, staying up to date with the latest advancements is essential, as methodologies continue to improve. For example, TEC-miTarget has achieved remarkable performance, attaining 96.47% accuracy on the miRAW dataset and 97.42% on the DeepMirTar dataset. These results demonstrate its strong performance, surpassing the best outcomes of other models by 4.21% in accuracy and 3.95% in sensitivity, highlighting how models tend to improve over time.

Acknowledgments

This work was supported by the European Union's NextGenerationEU initiative under the Italian Ministry of University and Research as part of the PNRR – M4C2.-I1.3 PE00000019 "HEAL ITALIA," CUP B73C22001250006, awarded to S.D.B., M.T., and G.S. F.O. was supported by the European Union under the Italian Ministry of University and Research as part of the project: Health Extended Alliance for Innovative Therapies, Advanced Lab research, and Integrated Approaches of Precision Medicine (HEAL ITALIA) Spoke 3 – Task 1.2 CUP B73C22001250006. G.N. was supported in part by National Cancer Institute (National Institute of Health) grant (R21CA287180).

Conflict of Interest The authors have no conflicts of interest to declare that are relevant to the content of this chapter.

References

1. Lee RC, Feinbaum RL, Ambros V (1993) The C. elegans heterochronic gene lin-4 encodes small RNAs with antisense complementarity to lin-14. Cell 75(5):843–854. https://doi.org/10.1016/0092-8674(93)90529-y
2. Qadeer A, Wajid A, Rafey HA, Nawaz S, Khan S, Rahman SU, Alzahrani KJ, Khan MZ, Alsabi MNS, Ullah H, Safi SZ, Xia Z, Zahoor M (2024) Exploring extracellular vesicles in zoonotic helminth biology: implications for diagnosis, therapeutic and delivery. Front Cell Infect Microbiol 14:1424838. https://doi.org/10.3389/fcimb.2024.1424838
3. Mengistu AA, Tenkegna TA (2021) The role of miRNA in plant-virus interaction: a review. Mol Biol Rep 48(3):2853–2861. https://doi.org/10.1007/s11033-021-06290-4
4. Panda M, Kalita E, Singh S, Kumar K, Rao A, Prajapati VK (2022) MiRNA-SARS-CoV-2 dialogue and prospective anti-COVID-19 therapies. Life Sci 305:120761. https://doi.org/10.1016/j.lfs.2022.120761
5. Kozomara A, Birgaoanu M, Griffiths-Jones S (2019) miRBase: from microRNA sequences to function. Nucleic Acids Res 47(D1):D155–D162. https://doi.org/10.1093/nar/gky1141
6. Kronstein-Wiedemann R, Künzel SR, Thiel J, Tonn T (2024) Role of MiRNA in the regulation of blood group expression. Transfus Med Hemother 51(4):237–251. https://doi.org/10.1159/000538866
7. Chipman LB, Pasquinelli AE (2019) miRNA targeting: growing beyond the seed. Trends Genet 35(3):215–222. https://doi.org/10.1016/j.tig.2018.12.005
8. Bartel DP (2004) MicroRNAs: genomics, biogenesis, mechanism, and function. Cell 116(2):281–297. https://doi.org/10.1016/s0092-8674(04)00045-5
9. Thomson DW, Bracken CP, Goodall GJ (2011) Experimental strategies for microRNA target identification. Nucleic Acids Res 39(16):6845–6853. https://doi.org/10.1093/nar/gkr330
10. Chen S-L, Wu C-C, Li N, Weng T-H (2024) Post-transcriptional regulation of myogenic transcription factors during muscle development and pathogenesis. J Muscle Res Cell Motil 45(1):21–39. https://doi.org/10.1007/s10974-023-09663-3
11. Naaman H, Rall G, Matullo C, Veksler-Lublinsky I, Shemer-Avni Y, Gopas J (2017) MiRNA-124 is a link between measles virus persistent infection and cell division of human neuroblastoma cells. PlOS One 12(10):e0187077. https://doi.org/10.1371/journal.pone.0187077
12. Giordo R, Ahmadi FAM, Husaini NA, Al-Nuaimi NRAM, Ahmad SMS, Pintus G, Zayed H (2024) microRNA 21 and long non-coding RNAs interplays underlie cancer pathophysiology: a narrative review. Noncoding RNA Res 9(3):831–852. https://doi.org/10.1016/j.ncrna.2024.03.013
13. Alanazi IO, Ebrahimie E (2016) Computational systems biology approach predicts regulators and targets of microRNAs and their genomic hotspots in apoptosis process. Mol Biotechnol 58(7):460–479. https://doi.org/10.1007/s12033-016-9938-x
14. Agbu P, Carthew RW (2021) MicroRNA-mediated regulation of glucose and lipid metabolism. Nat Rev Mol Cell Biol 22(6):425–438. https://doi.org/10.1038/s41580-021-00354-w
15. Shvedova M, Kobayashi T (2020) MicroRNAs in cartilage development and dysplasia. Bone 140:115564. https://doi.org/10.1016/j.bone.2020.115564
16. Zhang J, Luo C, Long H, Zhang B, Shan H, Hou B (2024) Circulating exosomal miRNA-451 as an effective diagnostic biomarker and prognostic indicator for multiple myeloma. Int J Biol Markers 39(4):301–309. https://doi.org/10.1177/03936155241283747
17. Diener C, Keller A, Meese E (2022) Emerging concepts of miRNA therapeutics: from cells to clinic. Trends Genet 38(6):613–626. https://doi.org/10.1016/j.tig.2022.02.006
18. Monga I, Kumar M (2019) Computational resources for prediction and analysis of functional miRNA and their targetome. Methods Mol Biol 1912:215–250. https://doi.org/10.1007/978-1-4939-8982-9_9
19. Zhang Y (2005) miRU: an automated plant miRNA target prediction server. Nucleic Acids Res 33(Web Server issue):W701–W704. https://doi.org/10.1093/nar/gki383
20. Ambros V (2004) The functions of animal microRNAs. Nature 431(7006):350–355. https://doi.org/10.1038/nature02871
21. Bartel DP (2009) MicroRNAs: target recognition and regulatory functions. Cell 136(2):215–233. https://doi.org/10.1016/j.cell.2009.01.002
22. Peterson SM, Thompson JA, Ufkin ML, Sathyanarayana P, Liaw L, Congdon CB (2014) Common features of microRNA target prediction tools. Front Genet 5:23. https://doi.org/10.3389/fgene.2014.00023
23. Lewis BP, Shih IH, Jones-Rhoades MW, Bartel DP, Burge CB (2003) Prediction of mammalian microRNA targets. Cell

115(7):787–798. https://doi.org/10.1016/s0092-8674(03)01018-3
24. Xiong P, Schneider RF, Hulsey CD, Meyer A, Franchini P (2019) Conservation and novelty in the microRNA genomic landscape of hyperdiverse cichlid fishes. Sci Rep 9(1):13848. https://doi.org/10.1038/s41598-019-50124-0
25. Waterman MS, Eggert M (1987) A new algorithm for best subsequence alignments with application to tRNA-rRNA comparisons. J Mol Biol 197(4):723–728. https://doi.org/10.1016/0022-2836(87)90478-5
26. Bray N, Dubchak I, Pachter L (2003) AVID: a global alignment program. Genome Res 13(1):97–102. https://doi.org/10.1101/gr.789803
27. Couronne O, Poliakov A, Bray N, Ishkhanov T, Ryaboy D, Rubin E, Pachter L, Dubchak I (2003) Strategies and tools for whole-genome alignments. Genome Res 13(1):73–80. https://doi.org/10.1101/gr.762503
28. Raney BJ, Barber GP, Benet-Pagès A, Casper J, Clawson H, Cline MS, Diekhans M, Fischer C, Navarro Gonzalez J, Hickey G, Hinrichs AS, Kuhn RM, Lee BT, Lee CM, Le Mercier P, Miga KH, Nassar LR, Nejad P, Paten B, Perez G, Schmelter D, Speir ML, Wick BD, Zweig AS, Haussler D, Kent WJ, Haeussler M (2024) The UCSC Genome Browser database: 2024 update. Nucleic Acids Res 52(D1):D1082–D1088. https://doi.org/10.1093/nar/gkad987
29. Quillet A, Anouar Y, Lecroq T, Dubessy C (2021) Prediction methods for microRNA targets in bilaterian animals: toward a better understanding by biologists. Comput Struct Biotechnol J 19:5811–5825. https://doi.org/10.1016/j.csbj.2021.10.025
30. Hiers NM, Li T, Traugot CM, Xie M (2024) Target-directed microRNA degradation: mechanisms, significance, and functional implications. Wiley Interdiscip Rev RNA 15(2):e1832. https://doi.org/10.1002/wrna.1832
31. Ritchie W, Rasko JEJ, Flamant S (2013) MicroRNA target prediction and validation. Adv Exp Med Biol 774:39–53. https://doi.org/10.1007/978-94-007-5590-1_3
32. Krek A, Grün D, Poy MN, Wolf R, Rosenberg L, Epstein EJ, MacMenamin P, da Piedade I, Gunsalus KC, Stoffel M, Rajewsky N (2005) Combinatorial microRNA target predictions. Nat Genet 37(5):495–500. https://doi.org/10.1038/ng1536
33. Marín RM, Vanícek J (2011) Efficient use of accessibility in microRNA target prediction. Nucleic Acids Res 39(1):19–29. https://doi.org/10.1093/nar/gkq768
34. Robins H, Li Y, Padgett RW (2005) Incorporating structure to predict microRNA targets. Proc Natl Acad Sci USA 102(11):4006–4009. https://doi.org/10.1073/pnas.0500775102
35. Long D, Lee R, Williams P, Chan CY, Ambros V, Ding Y (2007) Potent effect of target structure on microRNA function. Nat Struct Mol Biol 14(4):287–294. https://doi.org/10.1038/nsmb1226
36. Homberg N, Galvão Ferrarini M, Gaspin C, Sagot M-F (2023) MicroRNA target identification: revisiting accessibility and seed anchoring. Genes 14(3):664. https://doi.org/10.3390/genes14030664
37. Tieng FYF, Abdullah-Zawawi M-R, Md Shahri NAA, Mohamed-Hussein Z-A, Lee L-H, Mutalib N-SA (2023) A Hitchhiker's guide to RNA-RNA structure and interaction prediction tools. Brief Bioinform 25(1):bbad421. https://doi.org/10.1093/bib/bbad421
38. Wuchty S, Fontana W, Hofacker IL, Schuster P (1999) Complete suboptimal folding of RNA and the stability of secondary structures. Biopolymers 49(2):145–165. 10.1002/(SICI)1097-0282(199902)49:2<145::AID-BIP4>3.0.CO;2-G
39. Betel D, Koppal A, Agius P, Sander C, Leslie C (2010) Comprehensive modeling of microRNA targets predicts functional non-conserved and non-canonical sites. Genome Biol 11(8):R90. https://doi.org/10.1186/gb-2010-11-8-r90
40. Lewis BP, Burge CB, Bartel DP (2005) Conserved seed pairing, often flanked by adenosines, indicates that thousands of human genes are microRNA targets. Cell 120(1):15–20. https://doi.org/10.1016/j.cell.2004.12.035
41. Friedman RC, Farh KK-H, Burge CB, Bartel DP (2009) Most mammalian mRNAs are conserved targets of microRNAs. Genome Res 19(1):92–105. https://doi.org/10.1101/gr.082701.108
42. Agarwal V, Bell GW, Nam J-W, Bartel DP (2015) Predicting effective microRNA target sites in mammalian mRNAs. eLife 4:e05005. https://doi.org/10.7554/eLife.05005
43. McGeary SE, Lin KS, Shi CY, Pham TM, Bisaria N, Kelley GM, Bartel DP (2019) The biochemical basis of microRNA targeting efficacy. Science(New York, NY) 366(6472):eaav1741. https://doi.org/10.1126/science.aav1741
44. Lall S, Grün D, Krek A, Chen K, Wang Y-L, Dewey CN, Sood P, Colombo T, Bray N, Macmenamin P, Kao H-L, Gunsalus KC, Pachter L, Piano F, Rajewsky N (2006) A genome-wide map of conserved microRNA targets in C. elegans. Curr Biol 16(5):460–471. https://doi.org/10.1016/j.cub.2006.01.050

45. Rehmsmeier M, Steffen P, Hochsmann M, Giegerich R (2004) Fast and effective prediction of microRNA/target duplexes. RNA (New York, NY) 10(10):1507–1517. https://doi.org/10.1261/rna.5248604
46. Krüger J, Rehmsmeier M (2006) RNAhybrid: microRNA target prediction easy, fast and flexible. Nucleic Acids Res 34(Web Server issue):W451–W454. https://doi.org/10.1093/nar/gkl243
47. Kertesz M, Iovino N, Unnerstall U, Gaul U, Segal E (2007) The role of site accessibility in microRNA target recognition. Nat Genet 39(10):1278–1284. https://doi.org/10.1038/ng2135
48. Miranda KC, Huynh T, Tay Y, Ang Y-S, Tam W-L, Thomson AM, Lim B, Rigoutsos I (2006) A pattern-based method for the identification of MicroRNA binding sites and their corresponding heteroduplexes. Cell 126(6):1203–1217. https://doi.org/10.1016/j.cell.2006.07.031
49. Loher P, Rigoutsos I (2012) Interactive exploration of RNA22 microRNA target predictions. Bioinformatics (Oxford, England) 28(24):3322–3323. https://doi.org/10.1093/bioinformatics/bts615
50. Wang X (2008) miRDB: a microRNA target prediction and functional annotation database with a wiki interface. RNA (New York, NY) 14(6):1012–1017. https://doi.org/10.1261/rna.965408
51. Wong N, Wang X (2015) miRDB: an online resource for microRNA target prediction and functional annotations. Nucleic Acids Res 43(Database issue):D146–D152. https://doi.org/10.1093/nar/gku1104
52. Chen Y, Wang X (2020) miRDB: an online database for prediction of functional microRNA targets. Nucleic Acids Res 48(D1):D127–D131. https://doi.org/10.1093/nar/gkz757
53. Rennie W, Liu C, Carmack CS, Wolenc A, Kanoria S, Lu J, Long D, Ding Y (2014) STarMir: a web server for prediction of microRNA binding sites. Nucleic Acids Res 42(Web Server issue):W114–W118. https://doi.org/10.1093/nar/gku376
54. Liu C, Mallick B, Long D, Rennie WA, Wolenc A, Carmack CS, Ding Y (2013) CLIP-based prediction of mammalian microRNA binding sites. Nucleic Acids Res 41(14):e138. https://doi.org/10.1093/nar/gkt435
55. Paraskevopoulou MD, Georgakilas G, Kostoulas N, Vlachos IS, Vergoulis T, Reczko M, Filippidis C, Dalamagas T, Hatzigeorgiou AG (2013) DIANA-microT web server v5.0: service integration into miRNA functional analysis workflows. Nucleic Acids Res 41 (Web Server issue):W169–W173. https://doi.org/10.1093/nar/gkt393
56. Maragkakis M, Reczko M, Simossis VA, Alexiou P, Papadopoulos GL, Dalamagas T, Giannopoulos G, Goumas G, Koukis E, Kourtis K, Vergoulis T, Koziris N, Sellis T, Tsanakas P, Hatzigeorgiou AG (2009) DIANA-microT web server: elucidating microRNA functions through target prediction. Nucleic Acids Res 37 (Web Server issue):W273–W276. https://doi.org/10.1093/nar/gkp292
57. Maragkakis M, Vergoulis T, Alexiou P, Reczko M, Plomaritou K, Gousis M, Kourtis K, Koziris N, Dalamagas T, Hatzigeorgiou AG (2011) DIANA-microT Web server upgrade supports Fly and Worm miRNA target prediction and bibliographic miRNA to disease association. Nucleic Acids Res 39(Web Server issue):W145–W148. https://doi.org/10.1093/nar/gkr294
58. Kanehisa M, Goto S, Sato Y, Furumichi M, Tanabe M (2012) KEGG for integration and interpretation of large-scale molecular data sets. Nucleic Acids Res 40(Database issue):D109–D114. https://doi.org/10.1093/nar/gkr988
59. Betel D, Wilson M, Gabow A, Marks DS, Sander C (2007) The microRNA.org resource: targets and expression. Nucleic Acids Res 36(Database):D149–D153. https://doi.org/10.1093/nar/gkm995
60. Tastsoglou S, Alexiou A, Karagkouni D, Skoufos G, Zacharopoulou E, Hatzigeorgiou AG (2023) DIANA-microT 2023: including predicted targets of virally encoded miRNAs. Nucleic Acids Res 51(W1):W148–W153. https://doi.org/10.1093/nar/gkad283
61. Hsu PWC (2006) miRNAMap: genomic maps of microRNA genes and their target genes in mammalian genomes. Nucleic Acids Res 34(90001):D135–D139. https://doi.org/10.1093/nar/gkj135
62. Hsu SD, Chu CH, Tsou AP, Chen SJ, Chen HC, Hsu PWC, Wong YH, Chen YH, Chen GH, Huang HD (2007) miRNAMap 2.0: genomic maps of microRNAs in metazoan genomes. Nucleic Acids Res 36(Database):D165–D169. https://doi.org/10.1093/nar/gkm1012
63. Friedman Y, Naamati G, Linial M (2010) MiRror: a combinatorial analysis web tool for ensembles of microRNAs and their targets. Bioinformatics 26(15):1920–1921. https://doi.org/10.1093/bioinformatics/btq298
64. Friedman Y, Karsenty S, Linial M (2014) miRror-Suite: decoding coordinated regulation by microRNAs. Database 2014(0):bau043–bau043. https://doi.org/10.1093/database/bau043
65. Hsu JB-K, Chiu C-M, Hsu S-D, Huang W-Y, Chien C-H, Lee T-Y, Huang H-D (2011) miRTar: an integrated system for identifying miRNA-target interactions in human.

BMC Bioinformatics 12(1):300. https://doi.org/10.1186/1471-2105-12-300
66. Huang H-Y, Lin Y-C-D, Li J, Huang K-Y, Shrestha S, Hong H-C, Tang Y, Chen Y-G, Jin C-N, Yu Y, Xu J-T, Li Y-M, Cai X-X, Zhou Z-Y, Chen X-H, Pei Y-Y, Hu L, Su J-J, Cui S-D, Wang F, Xie Y-Y, Ding S-Y, Luo M-F, Chou C-H, Chang N-W, Chen K-W, Cheng Y-H, Wan X-H, Hsu W-L, Lee T-Y, Wei F-X, Huang H-D (2019) miRTarBase 2020: updates to the experimentally validated microRNA–target interaction database. Nucleic Acids Res 48:D148–D154. https://doi.org/10.1093/nar/gkz896
67. Dweep H, Sticht C, Pandey P, Gretz N (2011) miRWalk–database: prediction of possible miRNA binding sites by "walking" the genes of three genomes. J Biomed Inform 44(5):839–847. https://doi.org/10.1016/j.jbi.2011.05.002
68. Shirdel EA, Xie W, Mak TW, Jurisica I (2011) NAViGaTing the micronome–using multiple microRNA prediction databases to identify signalling pathway-associated microRNAs. PlOS One 6(2):e17429. https://doi.org/10.1371/journal.pone.0017429
69. Tokar T, Pastrello C, Rossos AEM, Abovsky M, Hauschild A-C, Tsay M, Lu R, Jurisica I (2018) mirDIP 4.1—integrative database of human microRNA target predictions. Nucleic Acids Res 46(D1):D360–D370. https://doi.org/10.1093/nar/gkx1144
70. Coronnello C, Benos PV (2013) ComiR: combinatorial microRNA target prediction tool. Nucleic Acids Res 41(W1):W159–W164. https://doi.org/10.1093/nar/gkt379
71. Vejnar CE, Zdobnov EM (2012) miRmap: comprehensive prediction of microRNA target repression strength. Nucleic Acids Res 40(22):11673–11683. https://doi.org/10.1093/nar/gks901
72. Vejnar CE, Blum M, Zdobnov EM (2013) miRmap web: comprehensive microRNA target prediction online. Nucleic Acids Res 41(W1):W165–W168. https://doi.org/10.1093/nar/gkt430
73. Ding J, Li X, Hu H (2016) TarPmiR: a new approach for microRNA target site prediction. Bioinformatics 32(18):2768–2775. https://doi.org/10.1093/bioinformatics/btw318
74. Ajila V, Colley L, Ste-Croix DT, Nissan N, Golshani A, Cober ER, Mimee B, Samanfar B, Green JR (2023) P-TarPmiR accurately predicts plant-specific miRNA targets. Sci Rep 13(1):332. https://doi.org/10.1038/s41598-022-27283-8
75. Gu T, Zhao X, Barbazuk WB, Lee J-H (2021) miTAR: a hybrid deep learning-based approach for predicting miRNA targets. BMC Bioinformatics 22(1):96. https://doi.org/10.1186/s12859-021-04026-6
76. Wen M, Cong P, Zhang Z, Lu H, Li T (2018) DeepMirTar: a deep-learning approach for predicting human miRNA targets. Bioinformatics 34(22):3781–3787. https://doi.org/10.1093/bioinformatics/bty424
77. Yang T, Wang Y, He Y (2024) TEC-miTarget: enhancing microRNA target prediction based on deep learning of ribonucleic acid sequences. BMC Bioinformatics 25(1):159. https://doi.org/10.1186/s12859-024-05780-z
78. Ritchie W, Flamant S, Rasko JEJ (2009) Predicting microRNA targets and functions: traps for the unwary. Nat Methods 6(6):397–398. https://doi.org/10.1038/nmeth0609-397
79. Farh KK-H, Grimson A, Jan C, Lewis BP, Johnston WK, Lim LP, Burge CB, Bartel DP (2005) The widespread impact of mammalian microRNAs on mRNA repression and evolution. Science (New York, NY) 310(5755):1817–1821. https://doi.org/10.1126/science.1121158

Check for updates

Chapter 2

Target-Directed miRNA Degradation: Mechanisms and Significance

Michael Attathikhun, Ancuta Jurj and George A. Calin

Abstract

MicroRNAs are small noncoding RNAs that regulate gene expression within cells through the translational repression or degradation of targeted mRNA. miRNAs undergo a multiphase synthesis that includes conformational changes mediated by Drosha, Dicer, and Argonaute to integrate mature miRNA into the RNA-induced silencing complex to target messenger (m)RNA. miRNAs have been characterized through their transcription and biogenesis; however, the mechanism regarding miRNA turnover is less comprehensible because of the stability of each miRNA. Target-directed miRNA degradation (TDMD) is dependent on the stability of miRNAs, which modulates the decay rate of miRNAs. This mechanism is initiated through target RNAs' engagement with miRNAs via the 3' end complements, which induces Argonaute rearrangement. These rearrangements of Argonaute can be induced through tailing and trimming, with one of the central initiators being the ZSWIM8 E3 ubiquitin ligase. Tailing and trimming are often associated with TDMD, and the mechanistic role is context-dependent. This mechanism provides a regulation process in which mRNA is actively repressed and silenced by the miRNA. The significance of TDMD lies in its role in post-transcriptional regulation of miRNA expression, its implications for therapeutic treatments, and its association with various diseases.

Keywords miRNA, TDMD, Cancer

1 Introduction

In the 31 years since their discovery, microRNAs (miRNAs) have emerged as pivotal regulators of gene expression, functioning post-transcriptionally to modulate mRNA stability and translation [1]. Often referred to as the "Sculptors of the Transcriptome," miRNAs primarily silence or fine-tune the expression of mRNAs transported to the cytoplasm by the nuclear transcription and RNA processing machinery. Extensive research has illuminated their mechanisms of action, evolutionary origins, and critical roles in development and cellular homeostasis [2]. While significant efforts have been devoted to understanding miRNA transcription and biogenesis, relatively little attention has been focused on miRNA turnover, partly due to

Alessandro Laganà and Giovanni Nigita (eds.), *MicroRNA Target Identification*, Methods in Molecular Biology, vol. 3056,
https://doi.org/10.1007/978-1-0716-5392-0_2,

their general stability, with half-lives often spanning several days [3]. Compared to miRNAs, RNA sequences such as polyadenylated and homopolymer sequences are inherently susceptible to degradation by cellular ribonucleases (RNases) once exposed. This process is tightly regulated by Argonaute (AGO) proteins, which shield the 3′ and 5′ termini of miRNAs from RNase activity [4, 5].

Interestingly, miRNA stability can vary significantly, with active degradation observed in specific contexts such as neuronal cells, cell cycle transitions, and during viral infections [6]. These instances underscore the importance of miRNA turnover in transient and spatially confined alterations in gene expression [7]. Nevertheless, the mechanisms governing selective miRNA degradation, particularly among closely related miRNA family members, remain poorly understood. Emerging evidence identifies target-directed miRNA degradation (TDMD) as a key process, wherein RNA targets interact with the miRNA-AGO complex through Watson–Crick base-pairing at the seed region (nucleotides 2–8), promoting specific miRNA degradation [8]. Recent findings implicate the ZSWIM8 ubiquitin ligase complex as a critical mediator of TDMD that exhibits robust phylogenetic conservation at both the sequence and functional levels [9]. By inducing AGO polyubiquitination and subsequent degradation, this pathway facilitates the rapid turnover of miRNAs, enabling dynamic regulation of gene expression [10]. This chapter delves into the mechanistic details of TDMD and its broader implications for maintaining miRNA homeostasis in mammalian cells.

2 Let's Talk About miRNA

miRNAs are pivotal regulators of cellular homeostasis, exerting their influence through precise control of gene expression at the post-transcriptional level [11]. These small, noncoding (nc)RNAs are themselves tightly regulated through a series of stepwise maturation processes. Over the past few decades, extensive research has uncovered their critical roles in various biological processes, including the hallmarks of cancer, functional implications, molecular targets, and mechanisms of action. Victor Ambros and Gary Ruvkun, distinguished scientists and pioneers in the field of RNA biology, were the esteemed recipients of the 2024 Nobel Prize in Physiology or Medicine for their foundational work on microRNAs [12]. Their groundbreaking discovery of miRNAs and their role in post-transcriptional gene regulation revealed a previously unknown layer of gene control, revolutionizing our understanding of molecular biology and its implications for health and disease.

The formation of miRNAs begins in the nucleus with the transcription of primary miRNAs (pri-miRNAs) by RNA polymerase

II. Pri-miRNAs form characteristic stem-loop secondary structures, which are recognized and processed by the microprocessor complex, consisting of Drosha (a ribonuclease III enzyme) and its cofactor DGCR8. This cleavage generates a precursor miRNA (pre-miRNA), a short hairpin RNA of approximately 60 nucleotides in length with a characteristic 3′ overhang. The pre-miRNA is subsequently transported from the nucleus to the cytoplasm by the Exportin-5 (XPO5) protein in a process facilitated by Ran-GTP. In the cytoplasm, the pre-miRNA undergoes further processing by the ribonuclease Dicer, which removes the terminal loop to produce a double-stranded miRNA duplex. The duplex then associates with Argonaute (AGO) proteins, where strand selection occurs. The mature miRNA, referred to as the "guide strand," is preferentially loaded into the AGO protein, forming the core of the RNA-induced silencing complex (RISC). The "passenger strand," on the other hand, is typically degraded by cellular ribonucleases. This maturation process results in the formation of a functional miRNA-RISC complex capable of regulating target mRNAs through sequence-specific binding [1].

The stability of miRNAs is largely conferred by their binding to AGO proteins. The half-lives of RNAs vary across different classes, with mRNAs typically exhibiting half-lives of a few hours, while miRNAs can persist for several days [13]. Features such as the 5′ 7-methylguanosine cap and the poly(A) tail enhance mRNA stability by preventing degradation and limiting susceptibility to RNases. Similarly, the stability of miRNAs is closely linked to AGO binding, which involves the MID domain, protecting the 5′ end, and the Piwi/AGO/Zwille (PAZ) domain, which shields the 3′ end [14].

The most conserved and functionally significant region of miRNAs is the seed sequence, located at positions 2–8 of the miRNA guide strand (g2–g8). The presence of a complementary seed-matching site in the 3′ untranslated region (3′ UTR) of the target mRNA is a key determinant for miRNA-mediated regulation [15]. In addition to seed-matching sites, other binding modalities enhance targeting efficiency. For instance, the "supplementary" region (g13–g16) can strengthen miRNA-mediated repression [1], while the secondary structure of the target RNA can be "unzippered" by the RISC, potentially limiting seed accessibility [16].

An important alternative binding modality involves the engagement of the miRNA 3′ end. While the PAZ domain typically shields this region, some target RNAs have been shown to expose the 3′ end to cellular factors, such as terminal nucleotidyltransferases (TENTs) [17]. Once exposed, the 3′ end can be modified through the addition of non-templated nucleotides, such as adenosine and uridine, a process known as target-directed tailing and trimming (TDTT). This process, regulated by RNases, can lead to shortening or "trimming" of the 3′ end [17, 18].

Given the diversity of miRNAs and their target RNAs, we anticipate further insights into these complex mechanisms in the years to come.

3 Target-Directed miRNA Degradation

Target RNAs can induce selective miRNA turnover in animal cells via TDMD, first characterized in *Drosophila* and mammalian systems [19]. TDMD involves active degradation of mature miRNAs bound to Argonaute (AGO) and does not affect transcription or processing of pri- or pre-miRNAs [20]. Unlike canonical miRNA-mediated silencing, TDMD relies on a distinct miRNA–target interaction, favoring miRNA destabilization over mRNA repression. Extensive base pairing, especially in the miRNA's 3′ region, is required for TDMD, contrasting with the minimal pairing needed for silencing [1]. While natural TDMD-inducing targets are rare, endogenous examples are increasingly identified, highlighting their conservation across animals (Table 1).

The principles of TDMD were first identified using synthetic mRNA targets, such as antagomiRs and miRNA sponges, which exhibit extensive complementarity and accelerate miRNA decay in mammalian cells [21]. miRNA sponges, transcribed from stably integrated genetic elements, effectively induced TDMD in various cell types, significantly reducing miR-16 and miR-20 levels in 293T and HeLa cells [22], and miR-223 in hematopoietic stem cells [23]. Similarly, viruses exploit a TDMD-like mechanism by producing viral RNAs that destabilize host miRNAs, undermining host defense systems [24]. In 2010, Steitz and Zamore independently highlighted the significance of TDMD using distinct model systems. They demonstrated that *Herpesvirussaimiri*-encoded *HSUR1* binds to the 3′ end of the host-encoded miR-27a, leading to the degradation of miR-27a without destabilizing *HSUR1* [25–27]. Similarly, it was observed in both *Drosophila* and human cells that significant complementarity between a target RNA and an Argonaute1-bound miRNA induces miRNA tailing and 3′-to-5′ trimming, resulting in the destabilization of the associated miRNAs [19]. Further investigations during the same decade highlighted additional mechanisms of TDMD. Studies on *murine cytomegalovirus* (MCMV) suggested the involvement of the MCMV-encoded transcript *m169* in the degradation of miR-27a [28, 29] and miR-27b [30]. Shortly thereafter, it was shown that the HCMV-encoded transcript *UL144-5* mediates the degradation of the miR-17 family, facilitating enhanced viral production (Fig. 1) [26, 27, 31].

Table 1
Examples of TDMD and their functional effects

Example miRNA	Species	Mechanism	Disease	Reference
miR-7-5p	Mouse, human	lncRNA CYRANO binds to miR-7-5p recruiting ZSWIM8 to degrade AGO2	Neurodevelopmental disorders	[52]
miR-29b-3p	Mouse, human	3' UTRs of NREP, Serpine1, and BCL2L11through extensive pairing and 3' trimming of miR-29-3p	Neuromotor disorders	[57]
miR-27a-3p miR-27b-3p	Human (infected with herpesvirus), mouse	Herpesvirus saimiri (HVS) HSUR1 or murine cytomegalovirus (MCMV) m169 bind to miR-27a/b-3p	Viral Infection	[55]
miR-999-3p	*Drosophila melanogaster*	3' UTR of AGO1 mRNA binds to miR-999-3p inducing AGO1 conformational change with ZSWIM8 recruitment	N/A	[52]

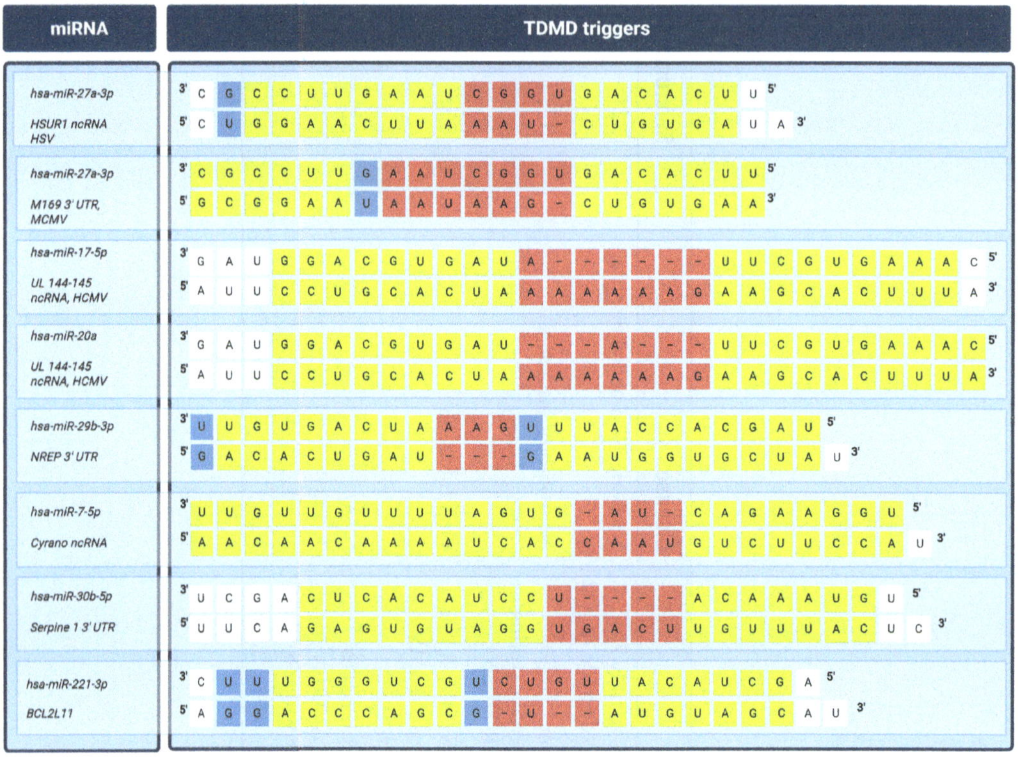

Fig. 1 Known and validated TDMD triggers among various model systems. U uracil, G guanine, A adenosine, C cytosine. Non-canonical base pairing is colored blue, complementary base pairing is colored yellow, central bulge is colored red, and unpaired nucleotides are uncolored

4 The Mechanism of TDMD

The first step in understanding TDMD is to explore *miRNA–target architecture*. Although there are differences in experimental systems, TDMD consistently involves significant complementarity between miRNAs and their targets, distinguishing it from canonical miRNA binding, which relies primarily on seed pairing. In flies and mammals, TDMD is triggered by target sites with broad pairing across the miRNA, particularly in the 3′ region, while sites with seed-only or partial pairing fail to induce degradation [32]. Structural features such as central bulges in the miRNA–target duplex, including a 4-nt bulge, further enhance TDMD efficacy, with the specific architecture playing a critical role in determining activity [33].

Thermodynamics, target abundance, and cell-type-specific factors also modulate TDMD efficacy. High target expression can compensate for lower complementarity in certain conditions. However, many mechanistic details, including the role of non-seed interactions and the precise contributions of RNA structure and cellular context, remain to be fully elucidated. This highlights the complexity and specificity of TDMD across systems and underscores the need for further investigation.

5 The Implication of ZSWIM8 E3 Ubiquitin Ligase in TDMD

The well-studied Cyrano and miR-7 TDMD interaction was further evaluated to understand other correlated factors. Cyrano functions as a long-noncoding RNA that interacts with the AGO complex to target miR-7. A pivotal discovery in 2020 identified ZSWIM8, the substrate-binding subunit of a Cullin-RING E3 ubiquitin ligase, as an essential mediator of TDMD through genome-wide CRISPR screens targeting CYRANO-directed miR-7 degradation [34]. ZSWIM8, via the ubiquitin-proteasome pathway, recognizes conformational changes in the AGO–miRNA complex induced by highly complementary targets (trigger RNAs), leading to AGO ubiquitination and proteasomal degradation (Fig. 2). This destabilizes the

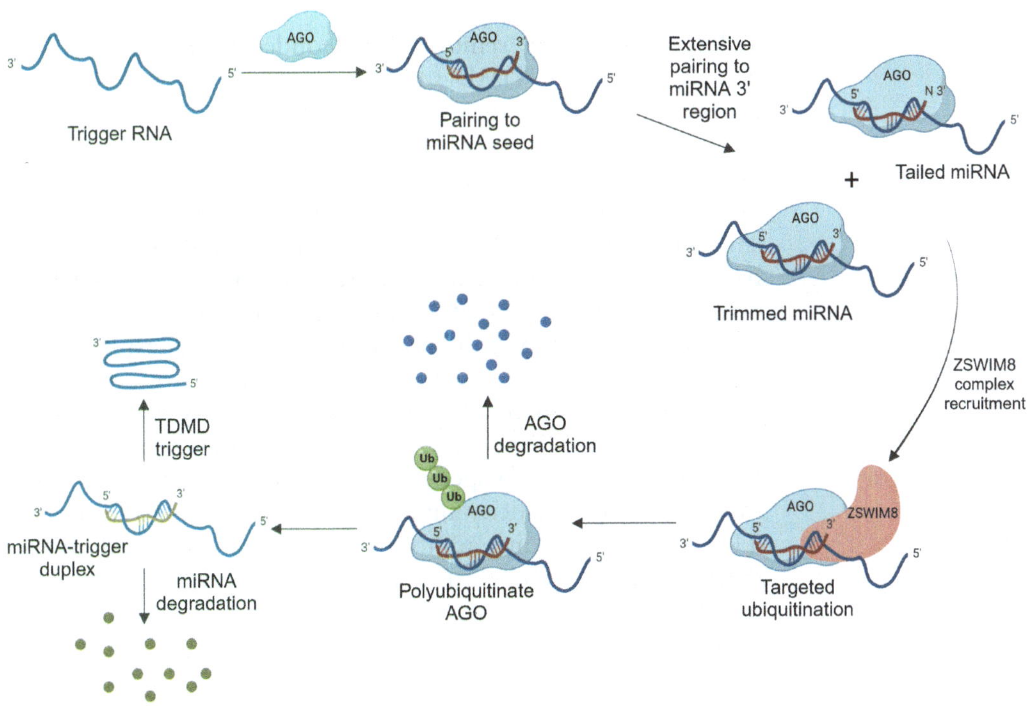

Fig. 2 The TDMD process is initiated when the miRNA binds to the complementary target or trigger RNA. The base pairings induce a structural change between the miRNAs and the AGO complex. Tailing by the terminal nucleotidyltransferases introduces non-templated nucleotides at the miRNA 3′ ends or by trimming. These modifications generate 3′-end diversity across various RNA types and have been implicated in regulating miRNA stability. The E3 ubiquitin ligase ZSWIM8 is recruited for targeted ubiquitination of the AGO-miRNA complex that degrades miRNAs through ubiquitin-proteasome complexes

miRNA, rendering it susceptible to exonucleases. Trigger RNAs can initiate multiple AGO–miRNA degradation cycles, acting with high efficiency. Notably, tailing and trimming are dispensable, as even 3′-O-methylated miRNAs remain vulnerable to TDMD [34]. ZSWIM8 interacts with AGO proteins to recognize TDMD interactions, resulting in degradation of the miRNA complex. The degradation is mediated by RNases that target cytoplasmic miRNAs for decay [34]. Global loss of function of ZSWIM8, coupled with small RNA sequencing, has uncovered various TDMD interactions among miRNAs in mammalian, *Drosophila*, and *C. elegans* tissues. Data collected from global loss-of-function ZSWIM8 studies in mice suggest perinatal lethality and impaired growth and organ development [35]. This is attributed to the upregulation of TDMD substrates such as miRNAs. These loss-of-function analyses of ZSWIM8 have confirmed the dependence of mammalian development and miRNA expression on the TDMD pathway [35].

6 Mechanisms of miRNA Turnover: Tailing, Trimming, and Exonucleolytic Degradation

Tailing and trimming (removal of nucleotides) of miRNA 3′ ends are defining features of TDMD, strongly correlating with miRNA degradation and suggesting their roles as intermediates in the degradation pathway mediated by 3′–5′ exonucleases. Tailing, driven by terminal nucleotidyltransferases (TNTases), introduces non-templated nucleotides—primarily adenosines or uridines—at the miRNA 3′ ends. This activity generates 3′-end diversity across various RNA types and has been implicated in regulating miRNA stability. Among TNTases, evidence has indicated that TUT1 co-purifies with TDMD-associated complexes, although knockdown studies suggest functional redundancy among TNTases [33]. Exonucleases play a critical role in miRNA degradation, with DIS3L2 being the only enzyme directly implicated in TDMD to date. DIS3L2, a member of the RNase II/R family conserved across species, exhibits a preference for uridylated miRNAs. This specificity aligns with findings in m169-mediated TDMD, where uridine additions are predominant, and miRNA degradation depends on DIS3L2 activity. Conversely, Cyrano-triggered TDMD, which predominantly induces adenine tailing, appears DIS3L2-independent, pointing to substrate-specific mechanisms for exonuclease recruitment [20]. It was reported that Cyrano, a highly conserved region within *OIP5-AS1*, can base-pair with and induce the degradation of miR-7, providing valuable insights into the decoupling of TDMD from miRNA tailing [20]. Furthermore, it was demonstrated that *Serpine1* mRNA may induce TDMD through the degradation of miR-30b/c [36]. Additionally, miR-221/222 were found to be degraded by *BCL2L11*, the mRNA encoding the tumor suppressor BIM, leading to increased cell death in human cells [29].

Evidence indicates that tailing and trimming occur while miRNAs are bound to AGO within the RISC complex, positioning the degradation machinery in close association with RISC components. For example, TDMD complexes include AGO proteins, exoribonucleases such as DIS3L2, and other RISC factors, facilitating miRNA turnover within this assembly. This parallels mechanisms in plants, where AGO-bound miRNAs undergo uridylation to initiate degradation. Furthermore, studies suggest that AGO-associated miRNAs are selectively tailed and trimmed, leading to their depletion from the RISC complex and subsequent turnover [32].

The interplay of TNTases and exonucleases in TDMD underscores the complexity of miRNA turnover. While highly complementary target RNAs induce tailing and trimming, the specific factors mediating these processes and their regulatory influence on TDMD efficiency across different systems remain incompletely understood. Additionally, the functional significance of auxiliary sequences and structural features within target RNAs, as well as potential cell-type-specific variations, warrants further investigation to fully elucidate the mechanisms underlying miRNA degradation in TDMD.

7 miRNA Degradation Occurs Independently of Tailing and Trimming

The correlation linking 3' tailing to miRNA stability control remains enigmatic. However, there remains uncertainty as to whether 3'-tailed and trimmed miRNA isoforms represent authentic intermediates in TDMD-driven degradation of miRNAs or if they are simply byproducts of separate, parallel processes. For instance, Cyrano-induced miRNA degradation appears to be associated with, yet independent of, 3′-end tailing and trimming [30]. Unlike most mammalian cases, where both adenylation and uridylation have been documented, Cyrano primarily triggers the attachment of two or more terminally added adenosines, with minimal impact on monoadenylation or tailing by other nucleotides [20]. This pattern mirrors the activity of GLD2, a key poly(A) polymerase functioning outside the nucleus [37]. Interestingly, although GLD2 may mediate Cyrano-induced A-tailing, its loss abolishes A-tailing without altering the fraction of miR-7 isoforms exhibiting 3' trimming or miR-7 levels. This suggests that adenylation and degradation are decoupled, and that tailing is unlikely to serve as an intermediate step in TDMD [20]. Alternatively, tailed miRNA species that accumulate after interaction with target RNAs exhibiting high sequence complementarity may reflect RISC complexes expressing enhanced stability or altered targeting specificity. Furthermore, evidence suggests target RNAs fulfill dual functions, activating through an independent pathway both tailing and miRNA separation from the AGO complex. Given the stability of AGO-miRNA complexes, highly complementary target RNAs featuring TDMD-permissive structures may facilitate efficient miRNA release from RISC, exposing them to nucleases for degradation [38]. Evidence supports this hypothesis. For instance, AGO2-bound guide RNAs can dissociate upon interaction with highly complementary target RNAs, enabling subsequent degradation by redundant nucleases. PARN, for example, degrades free miR-122 more effectively than miR-122 loaded onto AGO2, indicating that AGO protects miRNAs from exonucleolytic decay [39]. This protection aligns with findings in *C. elegans*, where XRN-1/XRN-2-mediated miRNA breakdown occurs outside RISC complexes [40].

Evaluating whether miRNA species with 3′ tails are bona fide intermediates formed during degradation pathways confronts several difficulties. First, TNTases mediate 3' nucleotide tailing in a miRNA- and stage-dependent fashion without necessarily affecting miRNA stability [41]. For example, ZCCHC6 (TUT7) and ZCCHC11 (TUT4) catalyze mono-uridylation of select miRNAs implicated in cell differentiation processes and regulation of Hox genes without altering miRNA abundance, even for preferred substrates [41]. Second, the effects of individual TNTases may vary depending on the tail length. While GLD2-mediated monoadenylation stabilizes miR-122 [42], oligoadenylation promotes PARN-mediated degradation, possibly involving adaptor proteins [39]. Additionally, redundancy among TNTases complicates analyses, as knockdown of one enzyme often upregulates others, masking individual contributions. Most studies to date have focused on steady-state levels of 3′-tailed miRNA species, leaving unanswered questions about how TNTases modulate nucleotide addition kinetics or influence degradation rates [41].

Overall, while tailing may modulate miRNA stability in certain contexts, its role as a necessary intermediate in TDMD remains uncertain. Further studies, particularly those addressing dynamic tailing and degradation processes, are required to clarify the interplay between tailing, RISC activity, and miRNA decay.

8 Contrasting miRNA and Target RNA Decay

The degradation of 5′ cleavage fragments derived from target RNAs frequently entails the addition of 3′ nucleotide tails downstream of the cleavage site. This observation implies that enzymatic activities responsible for 3′ tailing of both miRNAs and their corresponding target RNAs might be recruited during formation of the AGO protein-RNA complexes. Such coordinated tailing processes suggest a tightly regulated mechanism facilitating the turnover and processing of RNA molecules bound by AGO [43]. However, in mammals, miRNAs rarely exhibit sufficient pairing to trigger AGO2 slicing, and miRNA-mediated target decay generally follows distinct pathways [44]. Evidence argues against the co-degradation of miRNAs and corresponding targets in mediating TDMD. Conversely, TDMD mechanisms and associated target RNA destabilization are independent, competing processes influenced by the relative abundance of miRNAs and targets [32]. Furthermore, the 3'-end pairing of the targeted miRNA is necessary for TDMD interactions rather than the canonical base pairings that do not induce miRNA decay [32]. This was confirmed in neurons engineered to express mRNAs that induced a TDMD interaction with miRNAs. Results concluded that miRNA-binding sites lacking prolonged complementarity at the miRNA 3' base pairing end would not initiate decay [32]. In contrast to canonical miRNA-mediated mRNA decay, which requires cooperative engagement among several binding regions, TDMD enables individual target RNAs to independently degrade multiple miRNA molecules. This allows TDMD to modulate miRNAs independently without dependence on elevated levels of target RNA [45].

9 Role of AGO Proteins in the TDMD Process

AGO proteins protect the vulnerable unmodified 5′ and 3′ termini of miRNAs by binding them in specific pockets, preventing the accumulation of unbound mature cytoplasmic miRNAs [14]. Therefore, miRNA turnover predominantly occurs on AGO-loaded miRNAs. TDMD, which relies on extensive base-pair complementarity between miRNAs and their target RNAs, may trigger conformational changes in the AGO protein, thereby exposing the miRNA 3' terminus to nucleotidyltransferases from tailing and subsequent exonucleolytic degradation [46]. Alternatively, all miRNAs may undergo tailing and trimming upon target binding, but the slower dissociation of miRNAs with high-affinity sites could make them more susceptible to degradation. For instance, miRNAs with AU-rich seeds, such as miR-21, show slower dissociation rates upon extensive target pairing compared to GC-rich miRNAs, such as let-7 [47]. These differences, influenced by sequence composition and additional sequence-independent contacts, could explain why some miRNAs are more prone to TDMD than others.

In *Drosophila*, only unmodified miRNAs loaded onto AGO1 are susceptible to TDMD. In contrast, miRNAs and endogenous siRNAs (endo-siRNAs) bound to AGO2 are 2′-O-methylated at their 3′ ends, rendering them resistant to TDMD [48]. However, siRNAs become vulnerable in methylation-deficient flies (hen1 mutants), highlighting the protective role of 2′-O-methylation. Since animal miRNAs lack 2′-O-methylation, TDMD may have driven the evolution of partial complementarity in miRNA target sites [49]. In *Drosophila*, HEN1-mediated methylation likely contributed to the functional specialization of AGO proteins, enabling discrimination between self and non-self siRNA targets [50]. It is plausible that mammalian AGO proteins have also undergone functional specialization, potentially influencing TDMD susceptibility [51]. While current evidence does not strongly support this hypothesis, it remains an area for further investigation.

10 Functional Impact of TDMD Process

TDMD triggers play a fundamental role in the suppression of miRNAs because once the triggers are removed, miRNAs can accumulate and target mRNA. These triggers are specific RNA sequences that can contain specialized binding sites for miRNAs, facilitating degradation [52]. Without TDMD triggers, miRNA can be overexpressed, which can target vital genes. In *Drosophila*, TDMD triggers, such as AGO1, degrade miR-999 to prevent the accumulation of the targeted mRNA [52]. Key characteristics of these triggers include high conservation across different species, central mismatches within the miRNA sequences, and flanking sequences that promote degradation upstream and downstream of the trigger sequence [52]. miR-999 in *Drosophila* is not clearly understood, but it is classified as a miRNA that regulates gene expression, in which the central mismatches and the flanking sequences are required to promote degradation. The central mismatches are a defining characteristic of TDMD triggers that distinguish TDMD from canonical RISC mechanisms. The mismatches within the miRNA sequence prevent normal miRNA-mediated gene silencing by inhibiting canonical RISC activity. Therefore, the mismatches extend the duration of the miRNA's interaction with the specific trigger RNA, allowing for targeted degradation [52]. The flanking sequences on either end play a complementary role in stabilizing miRNA-TDMD trigger complex while recruiting degradation enzymes such as ribonucleases [53]. These characteristics ensure that miRNA levels are regulated through the prevention of overaccumulation and disruption of target genes.

11 TDMD and Diseases

11.1 Developmental Impact in Mice

TDMD has been linked to developmental abnormalities in early mouse embryos that lack ZSWIM8. Furthermore, extensive miRNA disruption was observed, especially with higher levels of miR-322 and miR-503. The increase in miR-322 and miR-503 directly causes abnormal growth patterns of organs such as the heart and lungs, growth restrictions, and perinatal lethality among the mouse population [35]. Consequently, when there is a complete genetic deletion of these two miRNAs, there is an increase in body size among various genotypes, emphasizing the dose-dependency of miR-322 and miR-503 in ZSWIM8-negative mice. These findings suggest a rescue of embryonic size development, which highlights the TDMD pathway as a regulator of embryonic growth and organogenesis [35].

The expression level of the lncRNA Cyrano in mouse embryos is elevated in early-stage central and peripheral nervous systems, but in adult mice, Cyrano is expressed five to ten times higher in brain regions [20]. Therefore, the prediction is that higher amounts of Cyrano contribute to more stable neuronal activity in mice. Cyrano-deficient mice with a >90% decrease in Cyrano RNA were used to examine miR-7 levels. In the absence of Cyrano, miR-7 levels increase, leading to decreased expression of various mRNA targets of miR-7, such as phosphoinositide 3-kinase (PI3K), protein kinase B (Akt), Raf (rapidly accelerated fibrosarcoma kinase), and extracellular signal-regulated kinase (ERK). Another outcome of Cyrano is its ability to prevent cytoplasmic destruction of the circular RNA Cerebellar degeneration-related protein 1 antisense (Cdr1as), which influences synaptic plasticity and neuronal activity [20, 54].

11.2 Viral Exploitation of TDMD

Contrary to its developmental roles in mice, TDMD is also exploited during lytic murine cytomegalovirus (MCMV) infection, in which the virus actively uses host cells to replicate, leading to host-cell destruction and release of viral particles. Furthermore, there is a rapid decrease of cellular miR-27a and miR-27b through a single binding site in the MCMV m169 transcript. miR-27a/b undergo further tailing and trimming, which rapidly modify and degrade the miRNAs within 48 h [30]. The significance lies in how viral exploitation hijacks the TDMD mechanism to degrade host miRNAs. Consequently, global ribosome inhibition increases in the presence of coding sequence (CDS) triggers, which can decrease miR-16 in human embryonic kidney (HEK) 293T cells. Therefore, mouse striatal cell models of Huntington disease (HD) were used to evaluate the efficacy of CDS triggers against miR-16. Northern blot data demonstrated that CDS triggers can induce miR-16 degradation in mice with HD-homozygous cells compared to control cells [55]. This ribosome stalling correlates with the presence of proteotoxic components such as misfolded proteins or cellular stress and with the progression of neurodegenerative conditions [55].

11.3 Broader Implications in Disease Biology

There have been few links between TDMD and specific human diseases, but developmental defects remain the most studied. Through its function in miRNA regulation, TDMD is an essential mediator of gene expression. As previously mentioned, TDMD is associated with neurological disorders, such as miR-7's restriction of Cdr1as accumulation for neuronal activity [8]. Defects in cardiovascular and pulmonary formation were observed in ZSWIM8 knockout (KO) embryos, as mice sustained lung sacculation defects and heart ventricular septal defects (VSDs). The miR-17~92 cluster was significantly increased in these ZSWMIM8-KO mice [35, 56].

Furthermore, TDMD triggers have been associated with the regulation of the cell cycle and apoptosis through TDMD interactions between Serpine1 and miR-30b/c. Serpine1 encodes a serine protease inhibitor protein that plays a role in early G0 and G1 as a transcriptional driver [36]. Interactions between Serpine1 and miR-30b-5p and miR-30c-5p involve the formation of a 5-nucleotide internal bulge 3' base pairing. This is likely to underlie Serpine1-mediated degradation of miR-30b/c [36]. The loss or degradation of miR-30b/c promotes cell cycle progression into S phase while increasing susceptibility to apoptosis following serum deprivation or doxorubicin treatment [8, 36]. Direct clinical implications and links to human diseases are still lacking, but the role of TDMD interactions in shaping miRNA dynamics is a developing field of study, supported by expanding evidence in growth abnormalities, developmental defects, and virology.

12 Conclusion

Discoveries of miRNAs have reshaped our understanding of gene expression at the post-transcriptional level. Consequently, miRNAs are stabilized by their interaction with AGO proteins; the stronger the binding, the higher the stability of the miRNA. The TDMD process is initiated through interactions of AGO-bound miRNAs with highly complementary target RNAs. When these interactions are engaged, they can cause upregulation of mRNA levels by depleting specific miRNAs. Characterization of miRNA-target complements at the 3' end of the miRNA defines the TDMD mechanism because of the non-canonical interactions. Recent findings identify the ZSWIM8 ligase as a key central mediator of AGO ubiquitination that leads to miRNA degradation through the ubiquitin-proteasome complex. There are nearly 80 miRNAs that are controlled by ZSWIM8 with fewer confirmed trigger RNAs. TDMD sites are challenging to discover because of the lengthy experimental validation that each trigger site requires [8]. Tailing and trimming of miRNAs at the 3' end is a crucial feature of TDMD. The correlation is with the involvement of TNTases as intermediates in miRNA degradation. Tailing and trimming are initiated when miRNAs are bound to AGO within the RISC complex. The association of AGO proteins, exoribonucleases, and other RISC factors, which facilitate miRNA turnover, makes up the TDMD complex.

The relevance of TDMD lies in its broad implications for gene expression control and development. For instance, both aspects are illustrated by ZSWIM8 knockout in mice, which can cause developmental complications such as perinatal lethality, cardiac defects, and restricted growth [35]. Despite the progress in understanding TDMD interactions, there remain gaps in knowledge of key molecular regulators of TDMD, specific cell type dependency of degradation, and the regulation of tailing and trimming. Furthermore, the current catalog of TDMD triggers, including viral RNAs such as MCMV m169, may underestimate the functional relevance of TDMD within an organism [36].

Acknowledgments

Michael Attathikhun is supported by the MD Anderson CATALYST program and was previously by the 1R25CA240137 UPWARDS Training Program and CPRIT Training Program RP210028. George A. Calin is the Charles B. Barker Chair. Work in George A. Calin's laboratory is supported by NCI grant 1R01CA222007-01A1, NIDCR grant R01DE032018, DoD CDRMP Idea Award BC200208P1 (W81XWH-21-1-0030), DoD CDRMP Idea Award BC200208P1 (W81XWH-21-1-0030), Team DOD grant in Gastric Cancer CA200990P2 (W81XWH-21-1-0715), DoD Idea Award PC230419 (HT9425-24-1-0052), the 2019 Faculty Achievement Award, CLL Global Research Foundation grants (2019, 2020, 2022, 2024), The G. Harold & Leila Y. Mathers Foundation, two grants from Torrey Coast Foundation, an Institutional Research Grant 2024, a Development Grant associated with the Brain SPORE 2P50CA127001, an Institutional Bridge Funding grant 2023, and the Ben and Catherine Ivy Foundation grant.

Conflict of Interest Statement This study was funded by Dr. George Calin, who is the Charles B. Barker Chair. Michael Attathikhun and Ancuta Juri have no relevant financial relationships to disclose. The authors have no conflicts of interest to declare that are relevant to the content of this chapter.

References

1. Bartel DP (2018) Metazoan MicroRNAs. Cell 173(1):20–51
2. Grimson A, Srivastava M, Fahey B, Woodcroft BJ, Chiang HR, King N et al (2008) Early origins and evolution of microRNAs and Piwi-interacting RNAs in animals. Nature 455(7217):1193–1197
3. Bail S, Swerdel M, Liu H, Jiao X, Goff LA, Hart RP, Kiledjian M (2010) Differential regulation of microRNA stability. RNA 16(5):1032–1039
4. Winter J, Diederichs S (2011) Argonaute proteins regulate microRNA stability: increased microRNA abundance by Argonaute proteins is due to microRNA stabilization. RNA Biol 8(6):1149–1157
5. Vincent HA, Deutscher MP (2009) Insights into how RNase R degrades structured RNA: analysis of the nuclease domain. J Mol Biol 387(3):570–583
6. Rissland OS, Hong SJ, Bartel DP (2011) MicroRNA destabilization enables dynamic regulation of the miR-16 family in response to cell-cycle changes. Mol Cell 43(6):993–1004
7. Hausser J, Syed AP, Selevsek N, van Nimwegen E, Jaskiewicz L, Aebersold R, Zavolan M (2013) Timescales and bottlenecks in miRNA-dependent gene regulation. Mol Syst Biol 9:711
8. Buhagiar AF, Kleaveland B (2024) To kill a microRNA: emerging concepts in target-

directed microRNA degradation. Nucleic Acids Res 52(4):1558–1574
9. Hassan IU, Rehman HM, Liu Z, Zhou L, Samma MK, Wang C et al (2023) Genome-wide identification and spatiotemporal expression profiling of zinc finger SWIM domain-containing protein family genes. Zool Res 44(3): 663–674
10. Han J, LaVigne CA, Jones BT, Zhang H, Gillett F, Mendell JT (2020) A ubiquitin ligase mediates target-directed microRNA decay independently of tailing and trimming. Science 370(6523):eabc9546
11. Carrington JC, Ambros V (2003) Role of microRNAs in plant and animal development. Science 301(5631):336–338
12. Drula R, Calin GA (2024) MicroRNAs: where brilliance, perseverance, and ambition converged. J Clin Invest 135(2):e189625. https://doi.org/10.1172/JCI189625. PMID: 39656547; PMCID: PMC11735086
13. Gantier MP, McCoy CE, Rusinova I, Saulep D, Wang D, Xu D et al (2011) Analysis of microRNA turnover in mammalian cells following Dicer1 ablation. Nucleic Acids Res 39(13):5692–5703
14. Schirle NT, Sheu-Gruttadauria J, MacRae IJ (2014) Structural basis for microRNA targeting. Science 346(6209):608–613
15. Agarwal V, Bell GW, Nam JW, Bartel DP (2015) Predicting effective microRNA target sites in mammalian mRNAs. Elife 4: e05005
16. Kosek DM, Banijamali E, Becker W, Petzold K, Andersson ER (2023) Efficient 3'-pairing renders microRNA targeting less sensitive to mRNA seed accessibility. Nucleic Acids Res 51(20):11162–11177
17. Han J, Mendell JT (2023) MicroRNA turnover: a tale of tailing, trimming, and targets. Trends Biochem Sci 48(1):26–39
18. Yu S, Kim VN (2020) A tale of non-canonical tails: gene regulation by post-transcriptional RNA tailing. Nat Rev Mol Cell Biol 21(9):542–556
19. Ameres SL, Horwich MD, Hung JH, Xu J, Ghildiyal M, Weng Z, Zamore PD (2010) Target RNA-directed trimming and tailing of small silencing RNAs. Science 328(5985):1534–1539
20. Kleaveland B, Shi CY, Stefano J, Bartel DP (2018) A network of noncoding regulatory RNAs acts in the mammalian brain. Cell 174(2):350–362 . e17
21. Baccarini A, Chauhan H, Gardner TJ, Jayaprakash AD, Sachidanandam R, Brown BD (2011) Kinetic analysis reveals the fate of a microRNA following target regulation in mammalian cells. Curr Biol 21(5):369–376
22. Ebert MS, Neilson JR, Sharp PA (2007) MicroRNA sponges: competitive inhibitors of small RNAs in mammalian cells. Nat Methods 4(9):721–726
23. Gentner B, Schira G, Giustacchini A, Amendola M, Brown BD, Ponzoni M, Naldini L (2009) Stable knockdown of microRNA in vivo by lentiviral vectors. Nat Methods 6(1):63–66
24. McCaskill J, Praihirunkit P, Sharp PM, Buck AH (2015) RNA-mediated degradation of microRNAs: a widespread viral strategy?. RNA Biol 12(6):579–585
25. Cazalla D, Yario T, Steitz JA (2010) Down-regulation of a host microRNA by a Herpesvirus saimiri noncoding RNA. Science 328(5985):1563–1566
26. Hiers NM, Li T, Traugot CM, Xie M (2024) Target-directed microRNA degradation: mechanisms, significance, and functional implications. Wiley Interdiscip Rev RNA 15(2):e1832
27. Pawlica P, Sheu-Gruttadauria J, MacRae IJ, Steitz JA (2019) How complementary targets expose the microRNA 3' end for tailing and trimming during target-directed microRNA degradation. Cold Spring Harb Symp Quant Biol 84:179–183
28. Libri V, Helwak A, Miesen P, Santhakumar D, Borger JG, Kudla G et al (2012) Murine cytomegalovirus encodes a miR-27 inhibitor disguised as a target. Proc Natl Acad Sci USA 109(1):279–284
29. Li L, Sheng P, Li T, Fields CJ, Hiers NM, Wang Y et al (2021) Widespread microRNA degradation elements in target mRNAs can assist the encoded proteins. Genes Dev 35(23–24):1595–1609
30. Marcinowski L, Tanguy M, Krmpotic A, Radle B, Lisnic VJ, Tuddenham L et al (2012) Degradation of cellular mir-27 by a novel, highly abundant viral transcript is important for efficient virus replication in vivo. PLOS Pathog 8(2):e1002510
31. Lee S, Song J, Kim S, Kim J, Hong Y, Kim Y et al (2013) Selective degradation of host MicroRNAs by an intergenic HCMV noncoding RNA accelerates virus production. Cell Host Microbe 13(6):678–690
32. de la Mata M, Gaidatzis D, Vitanescu M, Stadler MB, Wentzel C, Scheiffele P et al (2015) Potent degradation of neuronal miRNAs induced by highly complementary targets. EMBO Rep 16(4):500–511
33. Haas G, Cetin S, Messmer M, Chane-Woon-Ming B, Terenzi O, Chicher J et al (2016) Identification of factors involved in target RNA-directed microRNA degradation. Nucleic Acids Res 44(6):2873–2887
34 Shi CY, Kingston ER, Kleaveland B, Lin DH, Stubna MW, Bartel DP (2020) The

ZSWIM8 ubiquitin ligase mediates target-directed microRNA degradation. Science 370(6523):eabc9359
35. Jones BT, Han J, Zhang H, Hammer RE, Evers BM, Rakheja D et al (2023) Target-directed microRNA degradation regulates developmental microRNA expression and embryonic growth in mammals. Genes Dev 37(13-14):661–674
36. Ghini F, Rubolino C, Climent M, Simeone I, Marzi MJ, Nicassio F (2018) Endogenous transcripts control miRNA levels and activity in mammalian cells by target-directed miRNA degradation. Nat Commun 9(1):3119
37. Mansur F, Ivshina M, Gu W, Schaevitz L, Stackpole E, Gujja S et al (2016) Gld2-catalyzed 3' monoadenylation of miRNAs in the hippocampus has no detectable effect on their stability or on animal behavior. RNA 22(10):1492–1499
38. De N, Young L, Lau PW, Meisner NC, Morrissey DV, MacRae IJ (2013) Highly complementary target RNAs promote release of guide RNAs from human Argonaute2. Mol Cell 50(3):344–355
39. Katoh T, Hojo H, Suzuki T (2015) Destabilization of microRNAs in human cells by 3' deadenylation mediated by PARN and CUGBP1. Nucleic Acids Res 43(15):7521–7534
40. Bosse GD, Ruegger S, Ow MC, Vasquez-Rifo A, Rondeau EL, Ambros VR et al (2013) The decapping scavenger enzyme DCS-1 controls microRNA levels in Caenorhabditis elegans. Mol Cell 50(2):281–287
41. Thornton JE, Du P, Jing L, Sjekloca L, Lin S, Grossi E et al (2014) Selective microRNA uridylation by Zcchc6 (TUT7) and Zcchc11 (TUT4). Nucleic Acids Res 42(18):11777–11791
42. Burns DM, D'Ambrogio A, Nottrott S, Richter JD (2011) CPEB and two poly(A) polymerases control miR-122 stability and p53 mRNA translation. Nature 473(7345):105–108
43. Ibrahim F, Rymarquis LA, Kim EJ, Becker J, Balassa E, Green PJ, Cerutti H (2010) Uridylation of mature miRNAs and siRNAs by the MUT68 nucleotidyltransferase promotes their degradation in Chlamydomonas. Proc Natl Acad Sci USA 107(8):3906–3911
44. Jonas S, Izaurralde E (2015) Towards a molecular understanding of microRNA-mediated gene silencing. Nat Rev Genet 16(7):421–433
45. Broderick JA, Salomon WE, Ryder SP, Aronin N, Zamore PD (2011) Argonaute protein identity and pairing geometry determine cooperativity in mammalian RNA silencing. RNA 17(10):1858–1869
46. Wang Y, Sheng G, Juranek S, Tuschl T, Patel DJ (2008) Structure of the guide-strand-containing argonaute silencing complex. Nature 456(7219):209–213
47. Salomon WE, Jolly SM, Moore MJ, Zamore PD, Serebrov V (2015) Single-molecule imaging reveals that argonaute reshapes the binding properties of its nucleic acid guides. Cell 162(1):84–95
48. Pelisson A, Sarot E, Payen-Groschene G, Bucheton A (2007) A novel repeat-associated small interfering RNA-mediated silencing pathway downregulates complementary sense gypsy transcripts in somatic cells of the Drosophila ovary. J Virol 81(4):1951–1960
49. Yang R, Li J, Wu Y, Jiang X, Qu S, Wang Q et al (2022) A novel method to purify small RNAs from human tissues for methylation analysis by LC-MS/MS. Front Mol Biosci 9:949181
50. Forstemann K, Horwich MD, Wee L, Tomari Y, Zamore PD (2007) Drosophila microRNAs are sorted into functionally distinct argonaute complexes after production by dicer-1. Cell 130(2):287–297
51. Swarts DC, Makarova K, Wang Y, Nakanishi K, Ketting RF, Koonin EV et al (2014) The evolutionary journey of Argonaute proteins. Nat Struct Mol Biol 21(9):743–753
52. Sheng P, Li L, Li T, Wang Y, Hiers NM, Mejia JS et al (2023) Screening of Drosophila microRNA-degradation sequences reveals Argonaute1 mRNA's role in regulating miR-999. Nat Commun 14(1):2108
53. Ran FA, Hsu PD, Wright J, Agarwala V, Scott DA, Zhang F (2013) Genome engineering using the CRISPR-Cas9 system. Nat Protoc 8(11):2281–2308
54. Zhao J, Tao Y, Zhou Y, Qin N, Chen C, Tian D, Xu L (2015) MicroRNA-7: a promising new target in cancer therapy. Cancer Cell Int 15:103
55. Li T, Li L, Hiers NM, Sheng P, Wang Y, Traugot CM et al (2025) Translation suppresses exogenous target RNA-mediated microRNA decay. Nat Commun 16(1):5257
56. Shi CY, Elcavage LE, Chivukula RR, Stefano J, Kleaveland B, Bartel DP (2023) ZSWIM8 destabilizes many murine microRNAs and is required for proper embryonic growth and development. Genome Res 33(9):1482–1496
57. Bitetti A, Mallory AC, Golini E, Carrieri C, Carreno Gutierrez H, Perlas E et al (2018) MicroRNA degradation by a conserved target RNA regulates animal behavior. Nat Struct Mol Biol 25(3):244–251

Chapter 3

Sfold Tools for MicroRNA Target Prediction

William Rennie, Shaveta Kanoria, Jun Lu and Ye Ding

Abstract

Computational prediction of miRNA-binding sites on target mRNAs facilitates experimental investigation of miRNA functions. In this chapter, we describe the command-line use of STarMir, an application module of the Sfold RNA package. STarMir performs miRNA-binding site predictions for target RNAs. Output data include comprehensive sequence, thermodynamic, and target structure features, a logistic probability measuring confidence for each predicted site and a quantitative score. Based on a large 3′ UTR assay dataset, this score provides a quantitative measure of the overall regulatory effects of both seed and seedless sites on the target. Sfold is now on Wikipedia at https://en.wikipedia.org/wiki/Sfold. STarMir and the rest of the Sfold package can be downloaded from GitHub at https://github.com/Ding-RNA-Lab/Sfold.

Keywords miRNA, CLIP, Target mRNA, RNA secondary structure, miRNA-binding site, Efficacy, Quantitative score

1 Introduction

MicroRNAs (miRNAs) are a class of naturally occurring, small non-coding RNAs (ncRNAs) of ~22 nucleotides (nt) in length, found in plants, animals, and some viruses [1, 2] A mature miRNA recognizes a target by guiding the RNA-induced silencing complex (RISC) to hybridize with partially complementary sequences typically in the 3′ untranslated regions (3′ UTRs) of the target mRNAs. miRNA binding induces translational repression and/or mRNA degradation of the target mRNA [1, 2]. miRNA-mediated gene regulation is extensive, as one miRNA may regulate hundreds of targets, whereas an individual mRNA can be targeted by multiple miRNAs [3]. miRNAs play important roles in various biological processes such as development, differentiation, apoptosis, and proliferation [2, 4]. Moreover, dysregulation in miRNA activity has been associated with numerous human diseases [5, 6].

Most existing algorithms for miRNA target prediction are primarily based on the seed rule [7]. With the development of the

Alessandro Laganà and Giovanni Nigita (eds.), *MicroRNA Target Identification*, Methods in Molecular Biology, vol. 3056, https://doi.org/10.1007/978-1-0716-5392-0_3,

cross-linking immunoprecipitation (CLIP) technique [8], it has become possible to identify short Argonaute (AGO) cross-linked sequences containing miRNA-binding sites. CLIP involves UV irradiation of tissues, organisms, or cells to covalently cross-link miRNA targets to the AGO proteins (the catalytic component of the RISC complex). The cross-linked RNAs are shortened by partial digestion with RNase to ~50 nt and further amplified by RT-PCR. The shortened RNA fragments are then sequenced to identify AGO tags containing miRNA-binding sites on the target mRNAs.

Several CLIP studies have been published, including HITS-CLIP for the mouse brain [8], PAR-CLIP in human cell lines [9], variants of PAR-CLIP [10], and a study in *C. elegans* [11]. These CLIP studies provide a genome-wide map of miRNA target interactions by generating short target fragments containing miRNA-binding sites. The high-throughput data from the CLIP studies were successfully utilized in the development of logistic models for improved miRNA-binding site predictions [12]. These models are based on a comprehensive list of sequence, thermodynamic, and target-structure features found to be enriched for miRNA-binding sites identified by CLIP. They were validated through intra-dataset, inter-dataset, and cross-species validations [12].

The models have been implemented into the STarMir application module of the Sfold RNA package, which predicts miRNA-binding sites on a target mRNA [13]. The use of STarMir through the Sfold web server was described in an earlier chapter [14]. In this chapter, we describe the command-line use of STarMir after installing the Sfold package on a local computer.

2 Materials

Sfold is now on Wikipedia at https://en.wikipedia.org/wiki/Sfold. Figure 1 shows the core of the Sfold Wikipedia page. STarMir, along with the rest of Sfold package, can be downloaded from GitHub at https://github.com/Ding-RNA-Lab/Sfold. Figure 2 is a screenshot of the files and folders in the Sfold GitHub repository.

3 Methods

STarMir now provides a quantitative score to assess the combined regulatory effects of multiple seed and seedless sites. This score, developed based on a high-throughput luciferase reporter dataset that includes 461 miRNAs and 11 3′ UTRs for a total of 4994 miRNA:3′ UTR pairs, provides a quantitative measure of the overall regulatory effects of both seed and seedless sites of one miRNA on a given target 3′ UTR [15].

Sfold is a software program developed to predict probable RNA secondary structures through structure ensemble sampling and centroid predictions[1][2] with a focus on assessment of RNA target accessibility,[3] for major applications to the rational design of siRNAs[4] in the suppression of gene expressions, and to the identification of targets for regulatory RNAs particularly microRNAs.[5][6]

Original author(s)	Ye Ding and Charles E. Lawrence
Developer(s)	Dang Long and Chaochun Liu (application modeling); Clarence Chan, Adam Wolenc, William A. Rennie and Charles S. Carmack (software development)
Initial release	1 April 2003; 21 years ago
Repository	github.com/Ding-RNA-Lab/Sfold
Operating system	Linux
Website	www.healthresearch.org/sfold-software-for-sirna/

Development [edit]

The core RNA secondary structure prediction algorithm is based on rigorous statistical (stochastic) sampling of Boltzmann ensemble of RNA secondary structures, enabling statistical characterization of any local structural features of potential interest to experimental investigators. In a review on nucleic acid structure and prediction,[7] the potential of structure sampling described in a prototype algorithm[8] was highlighted. With the publication of the mature algorithms for Sfold,[1][2] the sampling approach became the focus of a review[9] Both the sampling approach and the centroid predictions were discussed in a comprehensive review.[10] As an application module of the Sfold package, the STarMir program[11] has been widely used for its capability in modeling target accessibility.[6] STarMir was described in an independent study on microRNA target prediction[12] STarMir predictions have been used in an attempt to derive improved predictions.[13] Predictions by Sfold have led to new biological insights.[14] The novel ideas of ensemble sampling and centroids have been adopted by others not only for RNA problems, but also for other fundamental problems in computational biology and genomics.[15][16][17][18][19]

An implementation of stochastic sampling has been included in two widely used RNA software packages, RNA Structure[20] and the ViennaRNA Package,[21] which are also based on the Turner RNA thermodynamic parameters.[22] Sfold was featured on a *Nucleic Acids Research* cover,[23] and was highlighted in *Science* NetWatch.[24] The underlying novel model for STarMir[11] was featured in the Cell Biology section of *Nature Research Highlights*.[25]

Distribution [edit]

Sfold runs on Linux, and is freely available to the scientific community for non-commercial applications, and is available under license for commercial applications. Both the source code and the executables are available at GitHub.

External links [edit]

- Sfold GitHub repository
- Sfold commercial licensing

Fig. 1 The core section of the Sfold Wikipedia page

3.1 Operating System and GitHub Sfold Package Download

For running on a user's local computer, STarMir is only available for the Linux operating system. It runs on the Ubuntu distribution, version 18.04 or newer. Ubuntu's tutorial for installation is available at https://ubuntu.com/tutorials/install-ubuntu-desktop#1-overview.

The zip archive for STarMir can be downloaded from https://github.com/Ding-RNA-Lab/Sfold, by selecting the green "Code" button. Experienced GitHub users can also "fork" the archive. The zip file should be saved in a directory where the user plans to install the archive. The zip file can be extracted using an unzip utility tool. After unzipping, a directory called "Sfold-main" is created, containing the components of the Sfold package, including the STarMir program.

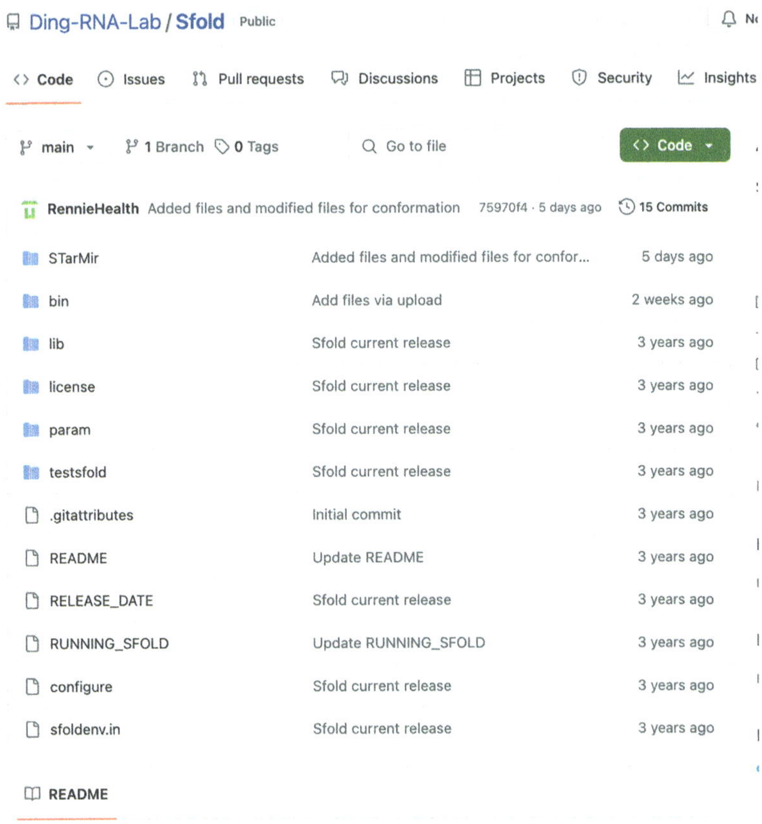

Fig. 2 The files and folders in Sfold GitHub repository

3.2 Installation of Other Required Software and Utilities

RNAhybrid

RNAhybrid [16] is used by STarMir to create a set of candidate binding sites. It can be downloaded from https://bibiserv.cebitec.uni-bielefeld.de/rnahybrid. Download and installation instructions are available on this site.

R

R is a statistical package (https://www.r-project.org) used by STarMir for executing the prediction models. R is usually available on most Linux systems and can be downloaded and installed using the Linux distribution's package manager. The installed executable must be globally accessible on the host computer.

Perl

The bulk of the STarMir code was written in Perl. Version 5 or newer is required. Perl is usually pre-installed on any Linux system. Two Perl modules used by STarMir, Bio::Seq and Bio::SeqIO, must be installed, typically using CPAN. The main Bioperl installation page provides recommendations and links to the main CPAN page. The URL is https://github.com/bioperl/bioperl-live/blob/master/

README.md. The CPAN package is included with the Perl installation.

Shell

STarMir can be run on the command line within a console window. It was developed using Bash and thus should run with any shell.

3.3 Configuration

Sfold

To configure Sfold, the user must enter the "`Sfold-main/bin`" directory. This directory contains the Sfold executable and a document called "`Running_Sfold`," which provides instructions for configuring and running Sfold. The document also explains how to use the testing utility to confirm that the Sfold package is installed correctly.

STarMir

STarMir does not have an automated configuration utility. The user must manually edit a few parameter files located in the "`Sfold-main/STarMir`" directory. The README file in that directory contains relevant information.

The file "`starmir_param.pl`" needs to be edited. The user needs to set the `$RNA_bindir` path to the location of the RNAhybrid executable. Typing "which RNAhybrid" on the command line will display the bin directory of the RNAhybrid executable. Only the directory (path) is needed, not the program name.

The user must also set the path to `disruptEn`, a program that calculates the free energy required to open a local structure. This program is part of the Sfold package, and the binary can be found in the `Sfold-main/bin` directory. To ensure proper functionality, use the full path, such as: "`/home/williamrennie/development/Sfold_main/bin/.`" The final slash is important and must be included. Additionally, the line "`$SFOLDBIN="/home/bill/Desktop/Sfold-main/bin/`" must be set to the same `Sfold-main/bin` directory specified above.

The steps above complete the installation and configuration of the programs required for STarMir.

3.4 Procedures for Executing STarMir

Executing Sfold

Sfold must be run first for predicting target structures. Running Sfold is straightforward, and the default parameters are sufficient for STarMir to use predicted structures. It has many configuration options, which can be viewed by running Sfold without any arguments. Sfold should always be executed using the full path to the executable, for example, "`/home/bill/Sfold-main/sfold.`"

- The input to Sfold is a file containing a single RNA sequence in FASTA format.

- Sfold produces a directory of output files, some of which are required by STarMir. The user shall select the directory that needs to be passed to STarMir.
- An example command to execute Sfold is "`/home/william-rennie/Sfold-main/bin/sfold -o myoutputdir myseq.fasta`"
- Running Sfold can take anywhere from a few minutes to a few hours, depending on both the computational power of the host computer and the length of the target sequence (*see* **Note 1**).

Executing STarMir

- STarMir is a system of Perl scripts and a helper application that predicts and ranks miRNA-binding sites on a target mRNA.
- The main command script is located in the distribution's `Sfold main/STarMir` directory. The user can run the program through that script.
- The script must be executed from the directory that contains it. Otherwise, it may not find all the necessary supporting scripts and may not run correctly.
- The main command script, "`starmir_research.pl`", requires nine arguments. Arguments that point to files must give the full path to the file (absolute path). The code has no facility for deducing the path to the file. The arguments, in the order they appear, are the following:

Required Arguments (in Order)

1. <working directory>: The directory where output files will be stored. By default, the intermediate output files will also be saved here. This argument must end with a backslash (\) to set an absolute path to the directory.
2. <miRNA file>: A file containing one or more miRNA sequences in FASTA format (absolute path).
3. <mRNA file>: A file containing a single mRNA sequence in FASTA format (absolute path).
4. <sfold output directory>: The directory containing Sfold results for the mRNA sequence can be found (absolute path, must end with a forward slash (/).
5. <target species>: The species name used by RNAhybrid. Must be one of fly, human, or worm (these are the only species supported by RNAhybrid).
6. <model species>: The species used by the STarMir prediction model. It must be one of human, mouse, or worm. STarMir predicts miRNA-binding sites using models built for human (*Homo sapiens*), mouse (*Mus musculus*), and worm (*Caenorhabditis elegans*). These models were trained on V-CLIP data for humans [9], HITS-CLIP data for mice [8], and ALG-1 CLIP data for worms [11]. The human and mouse

models were cross-validated and can be broadly used for other species [12].

7. <CDS start>: The start of the coding region.
8. <CDS end>: The end of the coding region.

Optional Argument

An optional ninth argument can be either 1 or 0. Setting this argument to 1 deletes all the intermediate files, whereas the default (0) preserves the intermediate files.

The command line, which must be run in the same directory as the scripts, is:

```
<path to script directory>/starmir_research.
pl <working directory> <miRNA file> <mRNA file>
<sfold output directory> <target species> <model
species> <CDS start> <CDS end> <optional 1 for
deleting intermediate files>
```

For best practice, use absolute path names for all input files and directories. Below is an example of a STarMir command-line run with intermediate files deleted:

```
./starmir_research.pl Data/myseq mirnas.fasta
myseq.fasta ~/runs/sfoldDataMyseq/ human human
234 874 1
```

3.5 Output

A successful local installation and execution of STarMir will generate separate final output files for miRNA-binding sites in the 5′ UTR, the coding region, and the 3′ UTR, and for both seed and seedless sites (Table 1). These output files are prefixed with "Final-." If no prediction is made for a specific site, e.g., in the case of the lack of a single seed, the corresponding file will not be generated.

Examples of seed and seedless output files are presented in Fig. 3. For the seed output file, several seed-specific features (e.g., Seed_Access) are provided (see Fig. 3A and Table 2). In each file, the binding sites are listed in descending order of their logistic probabilities.

Table 1
Final prediction files from STarMir

File name	File content description
Final-3pUTR_seed_sites.txt	Features and predictions for 3′ UTR seed sites
Final-3pUTR_seedless_sites.txt	Features and predictions for 3′ UTR seedless sites
Final-CDS_seed_sites.txt	Features and predictions for CDS seed sites
Final-CDS_seedless_sites.txt	Features and predictions for CDS seedless sites
Final-5pUTR_seed_sites.txt	Features and predictions for 5′ UTR seed sites
Final-5pUTR_seedless_sites.txt	Features and predictions for 5′ UTR seedless sites
conformation_all_sites.out	Predicted miRNA-target conformations for all sites

a

```
"SiteID"        "Target"        "miRNA" "Target_Len"    "Site_Position"
"Seed_Position" "Seed_Type*"    "3'_bp" "dG_hybrid"     "dG_nucl"
"dG_total"      "Site_Access"   "Seed_Access"   "Upstream_Access(5nt)"
"Dwstream_Access(5nt)"  "Upstream_AU(5nt)"      "Dwstream_AU(5nt)"
"Upstream_Access(10nt)" "Dwstream_Access(10nt)" "Upstream_AU(10nt)"
"Dwstream_AU(10nt)"     "Upstream_Access(15nt)" "Dwstream_Access(15nt)"
"Upstream_AU(15nt)"     "Dwstream_AU(15nt)"     "Upstream_Access(20nt)"
"Dwstream_Access(20nt)" "Upstream_AU(20nt)"     "Dwstream_AU(20nt)"
"Upstream_Access(25nt)" "Dwstream_Access(25nt)" "Upstream_AU(25nt)"
"Dwstream_AU(25nt)"     "Upstream_Access(30nt)" "Dwstream_Access(30nt)"
"Upstream_AU(30nt)"     "Dwstream_AU(30nt)"     "Site_Location"
"LogitProb"     "Upstream_Access(30nt)" "Dwstream_Access(30nt)"
"Upstream_AU(20nt)"     "Dwstream_AU(30nt)"     Score
8       NM_017589.4     hsa-miR-501-3p  1050    383-411 405-411 7mer-m8
1       -23.1   -8.413  -0.137  0.357   0.283   0.033   0.099   0.6
0.4     0.119   0.46    0.5     0.6     0.096   0.634   0.467   0.667
0.14    0.55    0.45    0.65    0.297   0.526   0.48    0.72    0.407
0.476   0.533   0.667   0.224702380952381       0.655441321236195
0.407   0.476   0.45    0.667   0.0407421894466
```

b

```
"SiteID"        "Target"        "miRNA" "Target_Len"    "Site_Position" "3'_bp"
"dG_hybrid"     "dG_nucl"       "dG_total"      "Site_Access"
"Upstream_Access(5nt)"  "Dwstream_Access(5nt)"  "Upstream_AU(5nt)"
"Dwstream_AU(5nt)"      "Upstream_Access(10nt)" "Dwstream_Access(10nt)"
"Upstream_AU(10nt)"     "Dwstream_AU(10nt)"     "Upstream_Access(15nt)"
"Dwstream_Access(15nt)" "Upstream_AU(15nt)"     "Dwstream_AU(15nt)"
"Upstream_Access(20nt)" "Dwstream_Access(20nt)" "Upstream_AU(20nt)"
"Dwstream_AU(20nt)"     "Upstream_Access(25nt)" "Dwstream_Access(25nt)"
"Upstream_AU(25nt)"     "Dwstream_AU(25nt)"     "Upstream_Access(30nt)"
"Dwstream_Access(30nt)" "Upstream_AU(30nt)"     "Dwstream_AU(30nt)"
"Site_Location" "LogitProb"     "Upstream_Access(10nt)" "Dwstream_Access(30nt)"
"Upstream_AU(30nt)"     "Dwstream_AU(30nt)"     Score
20      NM_017589.4     hsa-miR-501-3p  1050    906-928 0       -17.6   -0.224
-7.354  0.379   0.636   0.499   0.8     0.2     0.481   0.379   0.9     0.4
0.483   0.379   0.8     0.467   0.441   0.317   0.65    0.45    0.411   0.349
0.56    0.44    0.366   0.363   0.567   0.433   0.0136986301369863
0.528740278449575       0.481   0.363   0.567   0.433   0.0407421894466
22      NM_017589.4,    hsa-miR-501-3p  1050    996-1008        0       -17
-0.115  -2.25   0.266   0.416   0.222   0.6     0.8     0.51    0.115   0.6
0.6     0.452   0.182   0.6     0.533   0.352   0.227   0.55    0.6     0.329
0.34    0.52    0.68    0.36    0.444   0.5     0.7     0.63013698630137
0.568329262359457       0.51    0.444   0.5     0.7     0.0407421894466
23      NM_017589.4,    hsa-miR-501-3p  1050    1011-1022       0       -15.4
-0.371  5.848   0.091   0.237   0.377   0.6     0.8     0.199   0.541   0.5
0.9     0.302   0.678   0.6     0.867   0.33    0.749   0.6     0.85    0.385
0.788   0.6     0.8     0.377   0.819   0.6     0.8     0.732876712328767
0.569609983804448       0.199   0.819   0.6     0.8     0.0407421894466
21      NM_017589.4,    hsa-miR-501-3p  1050    949-981 0       -20.4   -0.167
3.465   0.344   0.129   0.461   0.4     0.4     0.255   0.492   0.5     0.5
0.256   0.512   0.533   0.6     0.317   0.466   0.45    0.65    0.317   0.403
0.48    0.56    0.29    0.376   0.533   0.6     0.308219178082192
0.434462062811596       0.255   0.376   0.533   0.6     0.0407421894466
```

Fig. 3 An example of the seed site output file that provide several seed-specific features (**A**) and an example of the output file for seedless sites (**B**)

In addition, a file containing miRNA:target hybrid conformations in text format is generated (Table 1 and Fig. 4).

Publication-quality hybrid diagrams in PDF format can also be produced using the "`create_PDF.pl`" script, located in the STarMir subdirectory of the Sfold GitHub repository. The input file for this script is "`Total-En-Hyb-Fil-mRNA id.out`" file (where "mRNA id" corresponds to the target name). This file will be located in the output directory specified during the execution of STarMir. The resulting PDF file will be named "`SiteX.pdf`," where X refers to the specific site number provided for the `create_PDF.pl` script. For example, the command "`./create_pdf.pl../../`

Table 2
Sequence, structural features, and predictions for target sites from STarMir

Site ID	Predicted sites are sequentially numbered along the target sequence
Target	Accession number of the target mRNA
miRNA	Name of the microRNA (miRNA)
Target_Len	Length of the target
Site_Position	Start and end position of the target region (site) predicted to be bound by miRNA
Seed_Position	Start and end position of the target sub-region complementary to the miRNA seed (i.e., positions 2–7/8 of the miRNA)
Seed_Type	6mer, offset 6mer, 7mer-A1, 7mer-m8, and 8mer seed sites [17]
Site_Access	A measure of structural accessibility as computed by the average probability of a nucleotide being single-stranded (i.e., unpaired) for the nucleotides in the predicted binding site
Seed_Access	A measure of structural accessibility as computed by the average of single-stranded probabilities of the nucleotides in the target sub-region complementary to the miRNA seed
Upstream_Access (# nt)	A measure of structural accessibility as computed by the average of single-stranded probabilities for the block of nucleotides upstream of the predicted binding site (#is the block size)
Dwstream_Access (# nt)	A measure of structural accessibility as computed by the average of single-stranded probabilities for the block of nucleotides downstream of the predicted binding site (#is the block size)
Upstream_AU (#nt)	Percentage of AU for the block of nucleotides upstream of the binding site (# is the block size)
Dwstream_AU (#nt)	Percentage of AU for the block of nucleotides downstream of the binding site (#is the block size)
Site_Location	Relative starting location of the predicted binding site along the length of the sequence (e.g., for 3′ UTR, 0 indicates the 5′ end of the UTR, and 1 corresponds to the 3′ end)
3′_bp	Presence of contiguous Watson Crick base pairing for miRNA nucleotide positions 12-17 (sites with 3′_bp are also called 3′ compensatory/supplementary sites) [17]
dG_hybrid = ΔG_{hybrid}	A measure of stability for miRNA:target hybrid as computed by RNAhybrid [16]
dG_nucl = ΔG_{nucl}	A measure of the potential of nucleation for miRNA:target hybridization [18]
dG_total = ΔG_{total}	A measure of the total energy change of the hybridization [18]
LogitProb	Probability of the site being an miRNA-binding site as predicted by nonlinear logistic model [19]
Score	A quantitative measure of the overall regulatory effects of both seed and seedless sites on the target [15]

```
Target name   = NM_017589.4

Site 1 -- hsa-miR-501-3p

    5'->3'        A   GG  GAGAGAAA  G   U      G
    Target     3   GGA  CU        GC CUG GG GCG    29
                   |||  ||        || ||| || |||
    miRNA     22   UCU  GG        CG GGC CC CGU    1
    3'->5'            UA  AA                A   AA


Site 2 -- hsa-miR-501-3p

    5'->3'        U   CG     CUGAAGG    CCAA   A
    Target    25   GGG  GUCC       GCCU    GGG        51
                   |||  ||||       ||||    |||
    miRNA     22   UCU  UAGG       CGGG    CCC        1
    3'->5'                  AA                ACGUAA


Site 3 -- hsa-miR-501-3p

    5'->3'          N    C  AA    C
    Target    30    GUCC UG   GGG        41
                    |||| ||   |||
    miRNA     22    UAGG AC   CCC        1
    3'->5'        UCU    A  GGG   ACGUAA


Site 4 -- hsa-miR-501-3p

    5'->3'        N   CG    A AAGA   A    N
    Target    52   AGA  CCU G    UGG G GC      72
                   |||  ||| |    ||| | ||
    miRNA     22   UCU  GGA C    GCC C CG      1
    3'->5'            UA    A GG       A  UAA


Site 5 -- hsa-miR-501-3p

    5'->3'        N   CG    AGAAGAUGGAG    A
    Target    52   AGA  CCU           GCCC           74
                   |||  |||           ||||
    miRNA     22   UCU  GGA           CGGG           1
    3'->5'            UA    A             CCCACGUAA
```

Fig. 4 Hybridization diagrams in text format for the first five binding sites predicted for hsa-miR-501-3p on target NM_017589.4 (gene BTG4)

`OutputDir2/TotalEn-Hyb-Fil-NM_017589.4.out 3`" will generate the PDF diagram for site 3, saved as "`Site3.pdf`" in the same directory in which the command was executed. Examples of diagrams for both a seed site and a seedless site are shown in Fig. 5, with the miRNA seed highlighted.

For each binding site, STarMir provides a comprehensive list of site features. Several are unique to STarMir: structure-based free-energy measures, a logistic probability, and a score for the miRNA:target pair (Table 2). The logistic probability is a measure of confidence for a predicted site [19]. The score is a measure of predicted regulatory

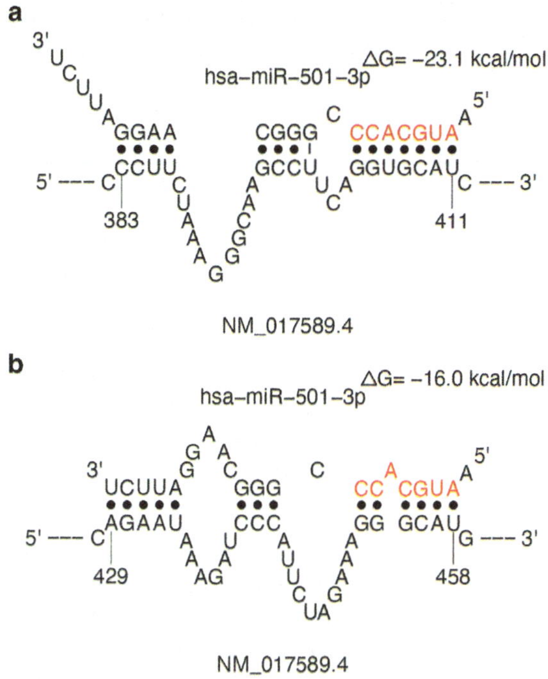

Fig. 5 Examples of high-quality hybridization diagrams: one for an hsa-miR-501-3p seed site on NM_017589.4 (**A**) and one for a seedless site (**B**). Note that the seed region of the miRNA is highlighted in red

Table 3
Examples of quantitative scores with the number of different types of binding sites

Transcript name	microRNA name	Number of 8mer sites	Number of 7mer sites	Number of 6mer sites	Number of seedless sites	Score
BTG4	hsa-miR-501-3p	0	1	0	22	0.040742190
hKLF4	hsa-mir-153-2	0	0	2	13	0.037326758
hDICER	hsa-let-7c	0	2	3	319	0.470307676
mBMI1	hsa-mir-154	1	0	4	77	0.144362275
hOCT4	hsa-mir-335	1	0	0	19	0.021329511
hLIN28	hsa-mir-9-2	0	1	6	196	0.282444367

efficacy of the miRNA on the target, based on a linear combination of contributions from both seed and seedless binding sites [15]. Examples of these scores are given in Table 3. Notably, although 8-mer sites are often considered the most effective among all types of sites, they do not necessarily ensure high scores that are predictive of effective regulation. On the other hand, large numbers of seedless sites can influence the combined score, potentially leading to a strong regulatory impact.

4 Notes

1. The Sfold/STarMir calculations are both CPU and memory-intensive. For example, the Sfold web server utilizes a ThinkStation P700 Tower equipped with two Xeon E5-2650 v3 processors (2.3 GHz), 128 GB of memory, and an 8 TB SATA disk. Typical calculation times are as follows: 3 min for 500 nts, 5 min for 1000 nts, 30 min for 2000 nts, 2 h for 3000 nts, 5 h for 4000 nts, and 9 h for 5000 nts.
2. The logistic probability is a measure of confidence in a predicted binding site, calculated using models developed from cross-linking immunoprecipitation (CLIP) studies. While it is challenging to directly assign a specific level of contribution to the overall functional efficacy based on the logistic probability of a given binding site, we recommend considering a logistic probability of 0.75 or higher for further experimental investigation.
3. The quantitative score was developed using a large reporter expression dataset. This dataset indicated that when the score is 0.09 or higher (≥ 0.09), 54% of the targets were effectively downregulated.

Acknowledgments

This work is supported in part by the National Institutes of Health (grant GM138856 to J. L. and Y.D.).

Conflict of Interest The authors have no conflicts of interest to declare that are relevant to the content of this chapter.

References

1. Ambros V (2004) The functions of animal microRNAs. Nature 431(7006):350–355
2. Bartel DP (2004) MicroRNAs: genomics, biogenesis, mechanism, and function. Cell 116(2):281–297
3. Friedman RC, Farh KK, Burge CB, Bartel DP (2009) Most mammalian mRNAs are conserved targets of microRNAs. Genome Res 19(1):92–105. https://doi.org/10.1101/gr.082701.108
4. Harfe BD (2005) MicroRNAs in vertebrate development. Curr Opin Genet Dev 15(4):410–415. https://doi.org/10.1016/j.gde.2005.06.012
5. Esau CC, Monia BP (2007) Therapeutic potential for microRNAs. Adv Drug Deliv

Rev 59(2–3):101–114. https://doi.org/10.1016/j.addr.2007.03.007
6. Erson AE, Petty EM (2008) MicroRNAs in development and disease. Clin Genet 74(4):296–306. https://doi.org/10.1111/j.1399-0004.2008.01076.x
7. Lewis BP, Burge CB, Bartel DP (2005) Conserved seed pairing, often flanked by adenosines, indicates that thousands of human genes are microRNA targets. Cell 120(1):15–20
8. Chi SW, Zang JB, Mele A, Darnell RB (2009) Argonaute HITS-CLIP decodes microRNA-mRNA interaction maps. Nature 460(7254):479–486. https://doi.org/10.1038/nature08170
9. Hafner M, Landthaler M, Burger L, Khorshid M, Hausser J, Berninger P, Rothballer A, Ascano Jr.M, Jungkamp AC, Munschauer M, Ulrich A, Wardle GS, Dewell S, Zavolan M, Tuschl T (2010) Transcriptome-wide identification of RNA-binding protein and microRNA target sites by PAR-CLIP. Cell 141(1):129–141. https://doi.org/10.1016/j.cell.2010.03.009
10. Kishore S, Jaskiewicz L, Burger L, Hausser J, Khorshid M, Zavolan M (2011) A quantitative analysis of CLIP methods for identifying binding sites of RNA-binding proteins. Nat Methods 8(7):559–564. https://doi.org/10.1038/nmeth.1608
11. Zisoulis DG, Lovci MT, Wilbert ML, Hutt KR, Liang TY, Pasquinelli AE, Yeo GW (2010) Comprehensive discovery of endogenous Argonaute binding sites in Caenorhabditis elegans. Nat Struct Mol Biol 17(2):173–179. https://doi.org/10.1038/nsmb.1745
12. Liu C, Mallick B, Long D, Rennie WA, Wolenc A, Carmack CS, Ding Y (2013) CLIP-based prediction of mammalian microRNA binding sites. Nucleic Acids Res 41(14):e138. https://doi.org/10.1093/nar/gkt435
13. Rennie W, Liu C, Carmack CS, Wolenc A, Kanoria S, Lu J, Long D, Ding Y (2014) STarMir: a web server for prediction of microRNA binding sites. Nucleic Acids Res 42(Web Server issue):W114–118. https://doi.org/10.1093/nar/gku376
14. Rennie W, Kanoria S, Liu C, Carmack CS, Lu J, Ding Y (2019) Sfold tools for MicroRNA target prediction. Methods Mol Biol 1970:31–42. https://doi.org/10.1007/978-1-4939-9207-2_3
15. Mastriano S, Kanoria S, Rennie W, Liu C, Li D, Cheng J, Ding Y, Lu J (2024) High-throughput quantification of miRNA-3'-untranslated-region regulatory effects. bioRxiv. https://doi.org/10.1101/2024.12.05.626985
16. Rehmsmeier M, Steffen P, Hochsmann M, Giegerich R (2004) Fast and effective prediction of microRNA/target duplexes. RNA 10(10):1507–1517. https://doi.org/10.1261/rna.5248604
17. Bartel DP (2009) MicroRNAs: target recognition and regulatory functions. Cell 136(2):215–233. https://doi.org/10.1016/j.cell.2009.01.002
18. Long D, Lee R, Williams P, Chan CY, Ambros V, Ding Y (2007) Potent effect of target structure on microRNA function. Nat Struct Mol Biol 14(4):287–294. https://doi.org/10.1038/nsmb1226
19. Liu C, Mallick B, Long D, Rennie WA, Wolenc A, Carmack CS, Ding Y (2013) CLIP-based prediction of mammalian microRNA binding sites. Nucleic Acids Res 41(14):e138. https://doi.org/10.1093/nar/gkt435

Chapter 4

Predicting miRNA Targets at the isomiR Level

Xiaoman Li and Haiyan Hu

Abstract

Understanding how microRNAs (miRNAs) regulate gene expression through target binding is critical for comprehending gene regulation mechanisms. Despite many experimental and computational studies on miRNAs and the availability of various tools for identifying their targets, accurately pinpointing these targets remains a significant challenge. The presence of various miRNA isoforms, known as isomiRs, further complicates this task due to their differing abundances in tissues. To address this challenge, we introduce DMISO, the first and currently only computational tool specifically designed for miRNA and isomiR target identification. DMISO employs deep learning to predict target sites for both isomiRs and miRNAs using a model trained on CLASH (cross-linking, ligation, and sequencing of hybrids) data. In this chapter, we start with a background introduction to isomiR and miRNA target prediction. We then outline the data available for identifying isomiR and miRNA targets and provide a detailed overview of DMISO, including its key features, performance compared to existing tools, and instructions for its installation and use. Finally, we discuss potential applications of the DMISO tool. This chapter aims to provide a foundational understanding for further research on isomiRs and their roles in gene regulation.

Keywords miRNA, isomiR, isomiR target prediction, DMISO, Deep learning

1 Introduction

MicroRNAs (miRNAs) are approximately 22-nucleotide-long non-coding RNAs that bind to target mRNAs. This binding can lead to mRNA degradation or inhibit the translation of the mRNAs [1]. Because of such an important role in post-transcriptional regulation, it is imperative to identify and study target mRNAs of miRNAs.

A large number of computational tools have been developed to identify miRNA targets [2–15]. Most of these tools rely on traditional features such as seed matching, conservation, and miRNA–mRNA duplex binding energy [1, 2, 14, 15]. However, a few newer tools incorporate advanced features derived from recent cross-linking immunoprecipitation (CLIP) data, allowing them to detect novel target sites beyond those identified by traditional methods [16–20]. Despite these advancements, the performance of existing tools

Alessandro Laganà and Giovanni Nigita (eds.), *MicroRNA Target Identification*, Methods in Molecular Biology, vol. 3056,
https://doi.org/10.1007/978-1-0716-5392-0_4,

remains suboptimal, particularly in accurately identifying target sites within mRNAs [2, 14].

One significant factor that can greatly impact the performance of these computational tools is the presence of isomiRs [21–23]. IsomiRs are isoforms of miRNAs that differ from their canonical counterparts at specific locations, such as the 5' end [24]. During miRNA biogenesis, miRNA genes are transcribed into primary miRNA [25–29]. These long primary miRNAs are then processed by the Drosha–DGCR8 complex into approximately 70-nucleotide precursor miRNAs [1]. The precursor miRNAs are exported from the nucleus to the cytoplasm, where they are further cleaved by the enzyme Dicer to produce miRNA duplexes. From these duplexes, one or both strands become mature miRNAs or isomiRs [24]. The resulting isoforms may exhibit variations in their 5′, 3′, and/or middle regions, potentially differing in length from the canonical miRNAs [24]. These isoforms are widely distributed and display tissue-specific expression patterns. The coexistence of isomiRs and canonical miRNAs in tissues and organs means that observed mRNA expression reflects the combined effects of both isoforms, rather than solely the wild-type miRNAs. Therefore, it is crucial to identify the target mRNAs of both isomiRs and miRNAs.

The CLASH (cross-linking ligation and sequencing of hybrids) data and the CLEAR-CLIP (covalent ligation of endogenous Argonaute-bound RNAs with CLIP) data provide the first set of experimental data to study how isomiRs interact with their target mRNAs [17, 18]. Both types of data contain chimeric reads, which ligate isomiRs/miRNA with their corresponding target mRNA sequences into a single read. In essence, each chimeric read consists of a segment from the isomiR/miRNA sequence and a segment from the target mRNA sequence to which the isomiR/miRNA binds under the given experimental conditions. Traditionally, the sequences of isomiRs in chimeric reads were often overlooked and treated as if they were canonical miRNA sequences. To our knowledge, DMISO is the first tool to specifically utilize isomiR sequence information from chimeric reads to identify target mRNAs for isomiRs [18, 24].

In this chapter, we introduce the DMISO tool. We begin with an overview of CLASH and other data used to train and test DMISO. Next, we detail the DMISO methodology, its performance on different datasets, and its comparison with other methods and tools. Finally, we provide instructions for installing and using DMISO for mRNA target identification, along with additional notes for optimizing its application and interpreting results. Our goal is to establish a foundation for advancing research in the identification of isomiR target mRNAs.

2 Material and Methods

2.1 The Training Dataset

DMISO was trained with the CLASH dataset from Helwak et al., specifically dataset GSE50452, which includes six samples of single reads and chimeric reads. Only the chimeric reads were utilized because they contain segments from both miRNA variants and their corresponding target mRNAs. In other words, each chimeric read represents a subsequence formed by concatenating an isomiR/miRNA sequence with its target mRNA sequence. By comparing these chimeric reads to miRNA sequences from miRBase [30] and the mRNA reference sequences from GENCODE [31], DMISO obtained isomiR sequences and their corresponding target mRNA sequences as pairs of interacting sequences. Here, the isomiR sequences in these chimeric reads corresponded to miRNA sequences in miRBase, but with some mismatches. To reduce the impact of sequencing errors and false isomiRs, DMISO required that each isomiR be present in at least ten distinct chimeric reads across the six samples. Since full-length target sites were often not fully represented in the chimeric reads, the matching mRNA sequences were extended equally from both ends to ensure each sequence was at least 60 nucleotides long. This approach resulted in the extraction of 12,170 miRNA–mRNA pairs and 58,043 isomiR–mRNA pairs from the CLASH dataset.

The above 70,213 isomiR/miRNA–mRNA pairs were positive interacting pairs. For each positive pair, a corresponding negative pair was created, where the isomiR/miRNA sequence remained the same, but the mRNA was replaced with a carefully selected 60-nucleotide subsequence from the same mRNA. Consequently, DMISO had a balanced dataset of 70,213 positive and 70,213 negative pairs. To train the DMISO model, 80% of these pairs were randomly selected, while the remaining 20% were reserved for testing.

2.2 The Testing Dataset

In addition to the 20% of the CLASH dataset aside for testing, DMISO was evaluated on two independent datasets. The first was the CLEAR-CLIP dataset from the HuH-7.5 cell line containing chimeric reads (GSE73059), which was processed similarly to the CLASH data [18]. From the CLEAR-CLIP data, 14,684 positive isomiR/miRNA–mRNA pairs were identified, with corresponding negative pairs generated in the same manner as for the CLASH data. The second independent dataset was sourced from miRTarBase 8.0 [32], which provided 14,144 miRNA–mRNA sequence pairs.

2.3 The DMISO Method and Performance

DMISO is a deep learning-based tool designed to predict miRNA and isomiR interactions with target mRNAs [24]. Its neural network architecture begins with two convolutional neural network (CNN) layers: one processes the isomiR/miRNA sequences, and the other handles the mRNA target sequences. These CNN layers extract relevant features from the paired isomiR/miRNA and mRNA sequences. The extracted features are then integrated by a long short-term memory (LSTM) layer, which evaluates the "connection" or "interaction" between the features from the two CNNs. Finally, the integrated features are passed through a fully connected layer with 100 neurons, followed by a sigmoid function, to compute the probability of interaction between input sequence pairs.

DMISO was trained using 80% of the 70,213 positive and negative isomiR/miRNA–mRNA pairs from the CLASH data. Each sequence was encoded as a matrix with four rows and *n* columns, where *n* represents the sequence length and the four rows correspond to the four nucleotide types [24]. For testing, each sequence pair was encoded in the same manner and input into the DMISO model. The model then generated a prediction score between 0 and 1, indicating the likelihood that the input isomiR/miRNA interacts with the input mRNA sequence.

DMISO was evaluated using tenfold cross-validation and the three independent testing datasets previously mentioned. In all cases, DMISO demonstrated high recall and precision, with values at or above 0.90 (Table 1). When compared to five other tools—miRanda, TargetScan, RNA22, miRAW, and miTAR [14, 15, 33–35], DMISO consistently outperformed these methods in terms of either precision or recall.

2.4 Tool Installation and Running

Installation of DMISO

The source code of DMISO is freely available at https://github.com/amlantalukder/DMISO. It is written in Python, which can be installed and run on Linux and Windows. To run DMISO, Python 3.7 or higher needs to be installed.

Table 1
Performance comparison of six tools

	20% CLASH		CLEAR-CLIP		miRTarBase
Tools	**Precision**	**Recall**	**Precision**	**Recall**	**Recall**
DMISO	0.9286	0.9895	0.9844	0.8990	0.9164
miRanda	0.9851	0.1792	0.9790	0.0603	0.7045
TargetScan	0.9563	0.1231	0.9901	0.2252	0.7632
RNA22	0.7273	0.0011	0.6000	0.0002	0.0199
miRAW	0.7305	0.5940	0.9367	0.5626	0.6734
miTAR	0.7639	0.4057	0.9793	0.4972	0.0090

Note: Since there is no negative data from miRTarBase, precision is not measured on this dataset

Users also need to have the make utility installed on their computer. On Linux and macOS systems, make is typically available by default. To install DMISO, users should download the software, open a terminal, and navigate to the DMISO directory. Then, they can run the command make all to install all required dependencies.

For Windows, users must install make through Chocolatey. First, open Windows PowerShell as an Administrator (search for PowerShell, right-click on it, and select "Run as Administrator"). Then, paste and execute the following commands in PowerShell to install Chocolatey and subsequently install make:

1. Set-ExecutionPolicy Bypass -Scope Process -Force; [System.Net.ServicePointManager]::SecurityProtocol = [System.Net.ServicePointManager]::SecurityProtocol -bor 3072; iex ((New-Object System.Net.WebClient).DownloadString('https://community.chocolatey.org/install.ps1'))
2. choco install make -y

Once "make" is installed on Windows, users need to navigate to the DMISO directory in PowerShell and run the command "make all" to install all dependencies.

Running DMISO

Users can run DMISO in two ways. The first method involves inputting a single file in TSV format. Each row of this file must contain the following tab-separated columns: miRNA ID, mRNA ID, miRNA sequence, and mRNA sequence. An example input file is included in the DMISO package. To process this file, users should execute the following command: python3 dmiso.py -p <the single file>, where the path to the input single file needs to be specified together with the single file name. DMISO will then determine the isomiR/miRNA target subsequences within the corresponding mRNA sequences listed in the file.

The second method for running DMISO involves using two separate files: one for isomiRs/miRNAs and another for mRNAs. Both files should be in FASTA format, where each sequence is represented by two lines: the first line specifies the sequence name, and the second line contains the sequence itself. To execute DMISO with two separate files, use the following command: python3 dmiso.py -m <isomiR/miRNA file> -t <mRNA file>. In this setup, DMISO evaluates every possible pair of isomiR/miRNA and mRNA sequences from the provided files to predict which isomiRs/miRNAs bind to which mRNAs. For both input methods, DMISO may report multiple binding instances for a single mRNA sequence.

For both types of input sequences, DMISO will output a file in tsv format with six columns:

1. isomiR/miRNA ID: ID of the isomiR/miRNA sequence
2. Target ID: ID of the mRNA sequence

3. isomiR/miRNA Sequence: The entire isomiR/miRNA sequence
4. Target Sequence: The section of the mRNA sequence interacting with the miRNA
5. Prediction Score: Confidence score for the miRNA to bind the target sequence
6. Prediction: 1 if the interaction is predicted to be true, 0 otherwise

If the input is a single file, the output file will include predictions for every isomiR/miRNA–mRNA pair listed in the input file. If the input consists of two separate files, the output file will contain the predicted binding pairs, showing which isomiRs/miRNAs bind to which mRNAs.

3 Notes

1. DMISO is recommended for use with Python versions 3.7–3.11. While errors may be reported with Python versions higher than 3.11, DMISO typically operates normally and produces the expected predictions.
2. In principle, DMISO can handle any number of sequences and any sequence length. However, as the number and length of sequences increase—particularly when using two separate files—the memory requirements can become substantial. If memory limitations arise, it is advisable to split the sequences into multiple files and process each file separately to manage memory usage more effectively.
3. The DMISO study demonstrated that the tool has a high recall rate, indicating that it effectively identifies many of the true positive interactions. To reduce potential false positives, future improvements could involve incorporating more training data and considering the abundance of isomiRs/miRNAs, which are not yet accounted for in the current version of DMISO.
4. Despite being trained solely on CLASH data, which includes approximately 400 miRNAs, DMISO can reliably be applied to other isomiRs and miRNAs. It consistently outperforms other tools that focus exclusively on miRNAs, demonstrating its effectiveness in handling both miRNAs and isomiRs.
5. The isomiR/miRNA–mRNA pairs predicted by DMISO should be regarded as "static," meaning the predicted target sites are not specific to any particular condition. To enhance the relevance of predictions, users can input active isomiRs/miRNAs and mRNAs specific to their experimental conditions, which may help reduce non-condition-specific binding predictions.

Conflict of Interest Statements

The authors have no conflicts of interest to declare that are relevant to the content of this chapter.

References

1. Bartel DP (2004) MicroRNAs: genomics, biogenesis, mechanism, and function. cell 116(2):281–297
2. Ding J, Li X, Hu H (2016) TarPmiR: a newapproach for microRNA target site prediction. Bioinformatics 32(18):2768–2775
3. Lewis BP, Burge CB, Bartel DP (2005) Conserved seed pairing, often flanked by adenosines, indicates that thousands of human genes are microRNA targets. cell 120(1):15–20
4. Lewis BP et al. (2003) Prediction of mammalian microRNA targets. Cell 115(7):787–798
5. Ule J et al (2003) CLIP identifies Nova-regulated RNA networks in the brain. Science 302(5648):1212–1215
6. Vejnar CE, Zdobnov EM (2012) MiRmap: comprehensive prediction of microRNA target repression strength. Nucleic Acids Res 40(22):11673–11683
7. Enright AJ et al (2004) MicroRNA targets in Drosophila. Genome Biol 5(1):R1
8. Paraskevopoulou MD et al (2013) DIANA-microT web server v5. 0: serviceintegration into miRNA functional analysis workflows. Nucleic Acids Res 41:W169–W173
9. Loher P, Rigoutsos I (2012) Interactive exploration of RNA22 microRNA target predictions. Bioinformatics 28(24):3322–3
10. Friedman RC et al (2009) Most mammalian mRNAs are conserved targets of microRNAs. Genome Res 19(1):92–105
11. Wang X (2010) Computational prediction of microRNA targets. Methods Mol Biol 667:283–95
12. Talukder A, Li XM, Hu HY (2020) Position-wise binding preference is important for miRNA target site prediction. Bioinformatics 36(12):3680–3686
13. Li X, Hu H (2019) Improving miRNA Target Prediction Using CLASH Data. In: Laganà A (Ed.) MicroRNA Target Identification. Methods in Molecular Biology. Humana Press, New York, NY, vol 1970 https://doi.org/10.1007/978-1-4939-9207-2_6
14. Agarwal V et al (2015) Predicting effective microRNA target sites in mammalianmRNAs. Elife 4:e05005
15. Miranda KC et al (2006) A pattern-based method for the identification of microRNA binding sites and their corresponding heteroduplexes. Cell 126(6):1203–1217
16. Hafner M et al (2010) Transcriptome-wide identification of RNA-binding protein and microRNA target sites by PAR-CLIP. Cell 141(1):129–141
17. Helwak A et al (2013) Mapping the human miRNA interactome by CLASH reveals frequent noncanonical binding. Cell 153(3):654–665
18. Moore MJ et al (2015) miRNA-target chimeras reveal miRNA 3'-end pairing as a major determinant of argonaute target specificity. Nat Commun 6: 8864
19. Chi SW et al (2009) Argonaute HITS-CLIP decodes microRNA-mRNA interaction maps. Nature 460(7254):479–86
20. Ding J, Li X, Hu H (2017) CCmiR: A computational approach forcompetitive and cooperative microRNA binding prediction. Bioinformatics 34(2):198–206
21. Burroughs AM et al (2010) A comprehensive survey of 3′ animal miRNA modification events and a possible role for 3′ adenylation in modulating miRNA targeting effectiveness. Genome Res 20(10):1398–1410
22. Loher P, Londin ER, Rigoutsos I (2014) IsomiR expression profiles in human lymphoblastoid cell lines exhibit population and gender dependencies. Oncotarget 5(18):8790–8802
23. Wyman SK et al (2011) Post-transcriptional generation of miRNA variants by multiple nucleotidyl transferases contributes to miRNA transcriptome complexity. Genome Res 21(9):1450–1461
24. Talukder A et al (2022) A deep learning method for miRNA/isomiR target detection. Sci Rep 12(1), 10618.
25. Schanen BC, Li X (2011) Transcriptional regulation of mammalian miRNA genes. Genomics 97(1):1–6
26. Wang S et al (2021) Computational annotation of miRNA transcription start sites. Brief Bioinform 22(1):380–392
27. Barham C et al (2019) Application of Deep Learning Models to MicroRNA Transcription Start Site Identification. In: Proceedings of 2019 IEEE 7th international conference on bioinformatics and computational biology (ICBCB 2019), pp 22–28

28. Cha MY, et al (2021) A two-stream convolutional neural network formicroRNA transcription start site feature integration and identification. Sci Rep 11(1):5625
29. Zheng HS et al (2024) A computational modeling of pri-miRNA expression. PLOS One 19(1):e0290768
30. Kozomara A, Birgaoanu M, Griffiths-Jones S (2019) miRBase: from microRNA sequences to function. Nucleic Acids Research 47(D1):D155–D162
31. Frankish A et al (2023) GENCODE: reference annotation for the human and mouse genomes in 2023. Nucleic Acids Research 51(D1):D942–D949
32. Huang HY et al (2020) miRTarBase 2020: updates to the experimentally validated microRNA-target interaction database. Nucleic Acids Res 48(D1):D148–D154
33. Gu TJ et al (2021) miTAR: a hybrid deep learning-based approach for predicting miRNA targets. BMC Bioinform 22(1):96
34. Pla A, Zhong XF, Rayner S (2018) miRAW: a deep learning-based approach to predict microRNA targets by analyzing whole microRNA transcripts. PLOS Comput Biol 14(7):e1006185
35. Betel D et al (2010) Comprehensive modeling of microRNA targets predicts functional non-conserved and non-canonical sites. Genome Biol 2010. 11(8):R90

Chapter 5

isoTar: miRNA-Target Consensus Prediction with Enrichment Analysis for Canonical and Modified miRNAs

Sebastiano Di Bella, Francesco Orilio, Shaopeng Gu, Rosario Distefano, Junhao Liu, Miriam Gaggianesi, Rosario Nicola Brancaccio, Shaohong Feng, Alessandro Laganà, Mario Acunzo, Qin Ma and Giovanni Nigita

Abstract

Over the past few decades, microRNAs (miRNAs) have gained significant attention for their role in regulating gene expression and their association with various diseases, including cancer. These small noncoding RNAs (sncRNAs) have been identified as crucial regulators of different biological functions. However, recent advancements in high-throughput sequencing technologies have revealed a more complex miRNAome, driven by phenomena such as RNA editing and isomiRs, which modify miRNA sequences. This increasing complexity has necessitated the development of new tools to analyze the miRNAome and better understand their biological roles. To address this challenge, isoTar was developed—a high-performance, web-based containerized application designed for miRNA consensus-targeting prediction and functional enrichment analyses. This chapter provides a comprehensive overview of isoTar (https://ncrnaome.osumc.edu/isotar/), including benchmarks of its performance and a guide to its usage.

Keywords MicroRNA target prediction, isomiRs, A-to-I RNA editing, Consensus prediction, Functional enrichment analysis

1 Introduction

MicroRNAs (miRNAs) are small, single-stranded, noncoding RNA molecules, approximately 21 nucleotides in length, found in eukaryotes [1]. These molecules play a crucial role in regulating gene expression [1]. The process of miRNA biogenesis is a multistep mechanism that starts with the transcription of the primary miRNA (pri-miRNA) and culminates in the production of the mature miRNA [2]. This intricate process begins in the nucleus, where RNA polymerase II transcribes the pri-miRNA. The pri-miRNA

These authors equally contributed to this work.

Alessandro Laganà and Giovanni Nigita (eds.), *MicroRNA Target Identification*, Methods in Molecular Biology, vol. 3056, https://doi.org/10.1007/978-1-0716-5392-0_5,

is then recognized by DGCR8 and cleaved by Drosha, an RNase III-type enzyme [1], generating a ~70-nucleotide hairpin-like double-stranded structure known as pre-miRNA. The pre-miRNA is subsequently transported to the cytoplasm by Exportin 5, where it is further processed by Dicer, another RNase III-type enzyme, resulting in a 20–25 nucleotide-long mature miRNA [3, 4]. The duplex generated is then securely anchored to an Argonaute (AGO) protein, where the passenger strand (miRNA*) is discarded and eventually degraded. The retained miRNA strand is incorporated into the RNA-induced silencing complex (RISC), a key regulator of gene expression. RISC targets messenger RNA (mRNA) molecules by binding to their complementary sequences, typically in the 3′ untranslated region (UTR) [5]. A critical component of this recognition is the miRNA seed region, located at the 5′ end of the miRNA and spanning nucleotides 2–7, which plays an essential role in target recognition and gene expression regulation. Modifications occurring within the seed region of miRNAs, as well as in other critical areas, can significantly alter their targetome, giving rise to novel miRNA–gene interactions [6, 7]. This phenomenon extends beyond sequence variations in miRNAs to include DNA-level modifications, such as single-nucleotide polymorphisms (SNPs), and post-transcriptional modifications, such as RNA editing [8, 9]. Adenosine-to-Inosine (A-to-I) RNA editing, mediated by ADAR (adenosine deaminases acting on RNA) enzymes, is one of the most common forms of RNA editing in mammals and can profoundly impact miRNA function [8, 9]. Sequence variations in miRNAs, such as isomiRs [7, 10, 11], characterized by changes in sequence length, especially at the 5′ end, can shift their targetome [12]. These variations can lead to different biological functions compared to the "canonical" form defined in miRBase [13], thereby altering their effect on miRNA-mediated gene regulation. RNA editing events in mature miRNAs, particularly within the seed region, play a critical role in determining their targetome, modifying their interactions and the associated gene regulatory effects [14–16].

To address the increasing complexity of modified miRNAs and to effectively study these novel targetomes, isoTar has been developed as a web-based tool for miRNA consensus targeting prediction and functional enrichment analysis. isoTar is a containerized [17], user-friendly application that requires neither advanced technical skills nor complex installation processes. Results are efficiently presented through intuitive graphical user interfaces, making the tool accessible to researchers without bioinformatics expertise. In this chapter, we introduce isoTar, outlining its functionality and providing benchmarks through a specific case study.

2 Materials

The results presented in this study are based on the analysis of the mature miRNA (mature sequence: hsa-miR-411-5p UAGUAGACCGUAUAGCGUACG and pre-miR hsa-mir-411) with an A-to-I editing modification at position 5. Although not used directly as an example of tool application, isoTar also allows the definition of isomiRs during the pre-analysis phase via a dedicated section in the dropdown menu. This feature will be further illustrated in Fig. 4. The case study can be summarized as (FASTA-like format):

```
> hsa-miR-411-5p hsa-mir-411; 5A|C
AAACAAACAUGGUGCACUUCUU
```

Developed to simplify and enhance miRNA analysis, isoTar is available as a freely downloadable containerized application at https://ncrnaome.osumc.edu/isotar. Its integration with Docker [18] enables installation across multiple operating systems, including Windows 10 Pro, macOS 10.12+, and Ubuntu Linux 16.04+. The setup requires no advanced technical skills, libraries, or additional software, making isoTar accessible to a wide range of users. A key strength of isoTar lies in its ability to integrate biological tools, prediction resources, software packages, and user-friendly interfaces into a single platform. This unified approach significantly simplifies miRNA consensus-targeting prediction and functional enrichment analysis, reducing operational complexity while enhancing workflow efficiency. Designed with user experience in mind, isoTar maximizes hardware resources by utilizing all available CPU cores. Its multiprocessing architecture ensures rapid execution of computationally intensive tasks, establishing it as a high-performance solution. The application features an intuitive web interface that allows users to fully customize both input data and analysis results. These results are securely stored locally, ensuring data safety and eliminating the need to transfer information to external platforms. Although isoTar operates primarily as a standalone application, its reliance on web-based resources is minimal, limited to a single step in the enrichment analysis process, thereby maintaining local data control. With these capabilities, isoTar emerges as a cutting-edge tool for researchers aiming to analyze miRNAs efficiently and effectively, facilitating the discovery of novel biological targets and advancing the understanding of their functions.

2.1 System Requirements for isoTar

To efficiently run isoTar, a computing system must have at least 4 GB of RAM, approximately 10 GB of available disk space for the Docker image, the latest version of Docker installed, a stable internet connection, and an updated web browser such as Google Chrome, Mozilla Firefox, Microsoft Edge, or Apple Safari. These requirements ensure optimal performance and a seamless user experience.

3 Methods

3.1 Architecture and Functionalities of isoTar

The isoTar application is a bioinformatics platform meticulously designed to integrate open-source tools, libraries, biological datasets, and custom-developed scripts. Its software foundation is built on Python and leverages the Flask web framework (http://flask.pocoo.org), ensuring high-performance, secure delivery of web-based applications. To achieve a seamless, interactive user experience, isoTar employs advanced web technologies such as HTML5, Twitter Bootstrap [19], and jQuery [20]. For enhanced data visualization, the application incorporates JavaScript libraries, including Highcharts.js (http://www.highcharts.com) and jvenn.js [21]. These libraries enable the generation of dynamic and interactive charts that users can export in multiple formats, such as PNG, SVG, and PDF, ensuring flexibility in data dissemination and presentation. isoTar functionalities can be accessed through interactive web-based interfaces using any modern web browser. Additionally, the platform supports installation, input customization (for both prediction and functional analyses), task execution, data download, and reloading results from previous analyses, offering a comprehensive suite for biological data exploration.

3.2 Core Functions of isoTar

isoTar is designed to perform two key functions that complement each other in biological data analysis. The first function, Consensus miRNA Target Prediction Analysis, identifies potential miRNA targets using advanced prediction tools. These tools employ a variety of approaches, including thermodynamic stability-based methods (RNAhybrid [22], miRanda [23], PITA [24], and miRmap [25]), sequence-based methods (TargetScan [26], miRmap [25], and miRanda [24]), and evolutionary conservation-based methods (TargetScan [26]). isoTar supports the latest versions of these tools, including miRmap (v1.1) [25], TargetScan (v7.0) [26], PITA (v6) [24], RNAhybrid (v2.1.2) [23], and miRanda (v3.3a) [23]. To ensure high precision, predictions are restricted to seed regions of 7–8 nucleotides in length (7mer-(A1 and m8), 8mer) with no mismatches or G wobble base pairs. Furthermore, miRanda, RNAhybrid, and PITA parameters are fine-tuned to improve their accuracy [27]. Specifically, PITA predictions are constrained to a $\Delta\Delta G$ threshold of ≤ 10 kcal/mol; miRanda employs an energy cutoff of ≤ 20 kcal/mol, with a gap opening penalty of 9 and a gap extension penalty of 4; and RNAhybrid considers only predictions with a $\Delta G \leq 20$ kcal/mol. The input customization process for this task draws from two primary resources. The first is miRBase (v21), a curated list of canonical mature miRNAs and their pre-miRNAs [13]. The second includes two datasets detailing DNA and RNA modifications in mature miRNAs. One dataset, derived from dbSNP (v150) [28], reports 89 single-nucleotide variants (SNVs) within the seed region and 189 outside it, while the other dataset documents known A-to-I RNA editing sites, including 69 within the seed region and 69 outside it.

The second function, Functional Enrichment Analysis, extends the biological insights from the predicted targets by associating them with their biological functions and the processes in which they participate. This task integrates data from the NCBI Gene Ontology [29] (release date December 20, 2017), as well as pathway datasets from Reactome [30–32] and KEGG [33], both updated in 2016. The implementation of this functionality is powered by the Python library GOATOOLS (v0.7.11) [34] for Gene Ontology analyses and the R package RamiGO (v1.22.0) [35], available in Bioconductor (v3.5) [36], for visual exploration of functional data.

3.3 Optimizations and Accessibility

isoTar stands out for its emphasis on precision and usability. The integration of optimized prediction parameters ensures reliable results, while the customization of input datasets enables flexible analyses tailored to specific research needs. Additionally, its visualization features allow users to easily interpret and share results through interactive charts that can be downloaded in widely supported formats, including PNG, SVG, and PDF. isoTar thus combines robust analytical capabilities with a user-centric design, making it a valuable tool for researchers seeking insights into miRNA target prediction and functional enrichment analysis.

3.4 isoTar Installation and Execution

isoTar employs a containerized architecture powered by Docker, which must be installed on the user's system to ensure seamless deployment and operation. The Docker installation process varies depending on the operating system. Users can access detailed instructions in the official Docker documentation at https://docs.docker.com/install. Once Docker is appropriately configured, users can download the latest isoTar image from the designated repository at https://ncrnaome.osumc.edu/isotar. The installation workflow is straightforward, with only a few key steps. For Linux and macOS users, the Terminal console serves as the primary interface, while Windows users can run commands in PowerShell. By executing the command:

```
docker load -i /PATH_TO_ISOTAR_IMAGES/isotar.tar
```

The system will begin installing isoTar. Upon completion, the latest isoTar image will be installed, and users will receive an output log resembling the following example.

This approach ensures that isoTar is easily deployable across diverse computing environments, offering both simplicity and reliability. Furthermore, the containerized model enhances reproducibility, making it an ideal choice for bioinformatics applications that demand consistent results regardless of the underlying system configuration. See installation section for details.

```
34394faab172: Loading layer [====================
=============================>] 17.92kB/17.92kB
fbf57c7a838c: Loading layer [====================
=============================>] 11.78kB/11.78kB
2d63176664ff: Loading layer [====================
=============================>] 5.632kB/5.632kB
5ab785ed50da: Loading layer [====================
=============================>] 3.072kB/3.072kB
097adadfb283: Loading layer [====================
=============================>] 7.429GB/7.429GB
86c87be6face: Loading layer [====================
=============================>] 108.6MB/108.6MB
2dcd42fb11cf: Loading layer [====================
=============================>] 720.3MB/720.3MB
a64e27523e49: Loading layer [====================
=============================>] 103.6MB/103.6MB
1975cd33bc84: Loading layer [====================
=============================>] 1.536kB/1.536kB
e518cf78e956: Loading layer [====================
=============================>] 77.04MB/77.04MB
890769017077: Loading layer [====================
=============================>] 817.8MB/817.8MB
8085b1d466a4: Loading layer [====================
=============================>] 84.26MB/84.26MB
42d84f02a1ed: Loading layer [====================
=============================>] 1.778MB/1.778MB
c01029da65f8: Loading layer [====================
=============================>] 26.83MB/26.83MB
e6aaf08a8d15: Loading layer [====================
=============================>] 238.7MB/238.7MB
7ec2c40ad70d: Loading layer [====================
=============================>] 3.402MB/3.402MB
2890d18c1855: Loading layer [====================
=============================>] 784.9kB/784.9kB
fff38d1ba0f3: Loading layer [====================
=============================>] 6.212MB/6.212MB
e854dde956f4: Loading layer [====================
=============================>] 4.008MB/4.008MB
0dc1195a50cd: Loading layer [====================
=============================>] 156.6MB/156.6MB
17f94a00c15c: Loading layer [====================
=============================>] 337.4MB/337.4MB
f1699f58bd58: Loading layer [====================
=============================>] 2.048kB/2.048kB
dcc74bbe2463: Loading layer [====================
=============================>] 3.072kB/3.072kB
9eb14f476dd0: Loading layer [====================
=============================>] 3.584kB/3.584kB
```

```
30ff6d340466: Loading layer [====================
================================>] 3.072kB/3.072kB
aa61ab025cdc: Loading layer [====================
================================>] 4.096kB/4.096kB
8494dd294d5f: Loading layer [====================
================================>] 4.096kB/4.096kB
0dc19cc08570: Loading layer [====================
================================>] 3.584kB/3.584kB
90acf0e9a3a3: Loading layer [===================
==================================>] 2.56kB/2.56kB
416dc1488da0: Loading layer [===================
==================================>] 2.56kB/2.56kB
ec4f7b5e60cb: Loading layer [===================
==================================>] 2.56kB/2.56kB
21adbf10d522: Loading layer [===================
==================================>] 2.56kB/2.56kB
46acdfccfc0a: Loading layer [===================
================================>] 4.096kB/4.096kB
Loaded image: isotar:v1.1
```

To confirm that the isoTar installation was completed successfully, users can execute the command:

```
docker images
```

This command lists all Docker images currently installed on the system. The output will display information such as repository name, tag, image ID, creation date, and size. For example:

```
REPOSITORY TAG IMAGE ID CREATEDSIZE
IsoTar v1.1 15e0f7fa60fe 3minutes ago 10.2GB
```

It is important to note that the IMAGE ID shown is specific to the user's environment and may differ from the example provided. The final step involves launching the isoTar application. This is achieved by running the command:

```
docker run -d -p 80:8080 isotar:v1.1
```

This command initiates a temporary instance of isoTar using the installed Docker image (isotar:v1.1). The instance remains active until the host machine is powered down, meaning the command must be re-executed each time the application is needed.

Breaking down the command: the -d option ensures the instance runs in detached (background) mode, while the -p 80:8080 flag binds port 80 on the host system to port 8080 within the instance. Users can verify that the isoTar instance has been successfully created by executing the command:

```
docker ps
```

This will return a list of running Docker containers, with an output similar to the following:

```
CONTAINER ID IMAGE COMMAND CREATED STATUS PORTS
NAMES
```

```
019f68c97531 isotar:v1.1 "/usr/bin/supervisord"
5minutes ago Up 5 minutes 0.0.0.0:80->8080/
tcphopeful_brown
```

Once the isoTar container is running, the application can be accessed by entering localhost in the address bar of a web browser (see Fig. 1). A detailed guide to these steps is available at https://ncrnaome.osumc.edu/isotar. This streamlined process ensures that isoTar is operational with minimal setup, allowing users to focus on its powerful analytical features for miRNA target prediction and functional enrichment.

3.5 Consensus Prediction: Input Setup

The *Consensus Prediction Analysis* module in isoTar provides a versatile way for users to input miRNA information and their variations, offering three distinct methods for loading these data into the system (see Fig. 2). The first option allows users to utilize the "Search and customize a miRNA" form. This method provides extensive customization, enabling users to specify detailed information about the miRNA and its variations, making it the most comprehensive choice. Alternatively, users can choose the "Known Variations Databases" button. This option grants access to two predefined datasets: one containing known single-nucleotide polymorphisms (SNPs) and another containing A-to-I RNA editing modifications, both specific to mature miRNAs. These datasets facilitate a quick and straightforward approach to loading miRNA variations. Lastly, users can enter miRNA data directly into the text area at the bottom of the module, following the isoTar input format outlined in the previous section. For further details on these methods, users are encouraged to consult the isoTar documentation at https://ncrnaome.osumc.edu/isotar. Focusing on the most advanced technique, we will now

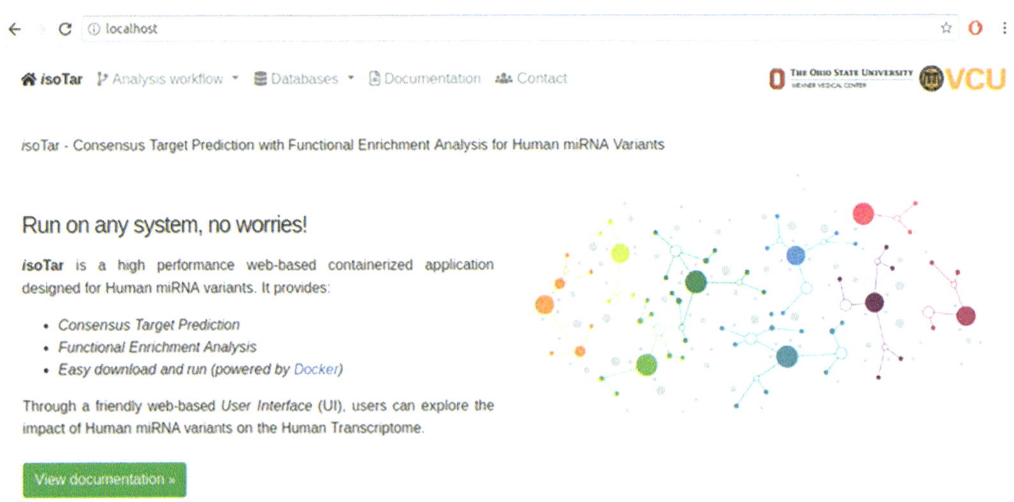

Fig. 1 isoTar application interface

explain how to use the "Search and customize a miRNA" form for preparing miRNA data for prediction analysis. In this process, the first step is to search for a specific miRNA–pre-miRNA pair (see Fig. 3). For illustration, we will demonstrate how to search for the mature miRNA *hsa-miR-411-5p*, which is associated with the pre-miRNA *hsa-mir-411* and carries an A-to-I editing modification at position 5.

1) Upon clicking the "Search" button, the system will display a temporary customization form on the right-hand side, as illustrated in Fig. 4.
2) In this form, users can specify variations grouped by type and color. isomiRs are displayed in cyan, while single-nucleotide modifications are shown in black. For example, an A-to-I RNA editing modification (5: A|C) can be specified by selecting the corresponding option in the "Specify a novel single nucleotide variation" dropdown menu. For illustrative purposes, the figure also shows isomiRs +2 and +1, which are added by selecting the appropriate option in the "Specify a novel isomiR" dropdown menu. Once the user is satisfied with their customizations, they can add the variations using the "+ Add to the list" button. After adding the variations, they will appear at the bottom of the form, just above the "Done" and "Add the customized miRNA" buttons (see Fig. 4). Once all variations have been specified, users can add the customized information to the textarea (as shown in Fig. 5) by clicking on the "Add the customized miRNA" button.
3) In the second step, users will need to choose which prediction tools to use and set the minimum consensus threshold (the minimum number of tools that must identify the same target).

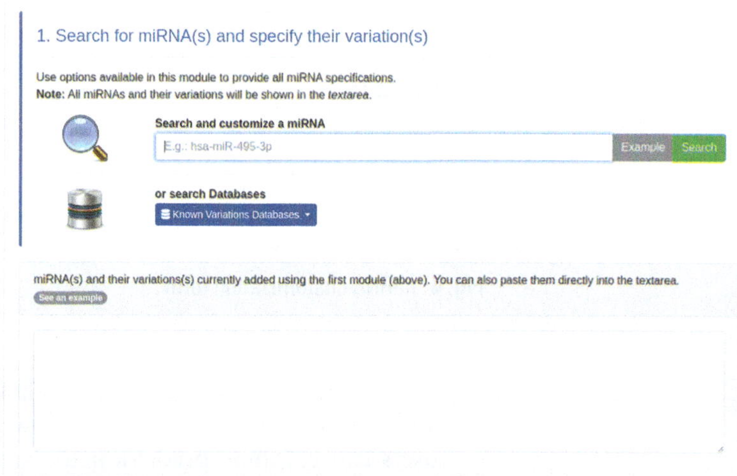

Fig. 2 Consensus prediction module of isoTar

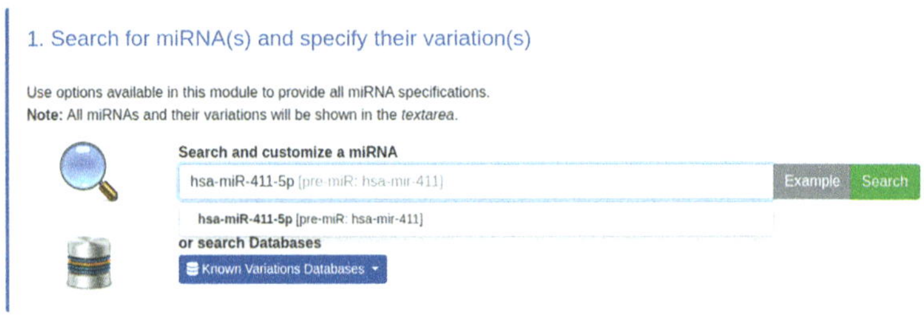

Fig. 3 miRNA search and personalization form. Demonstration of miRNA search

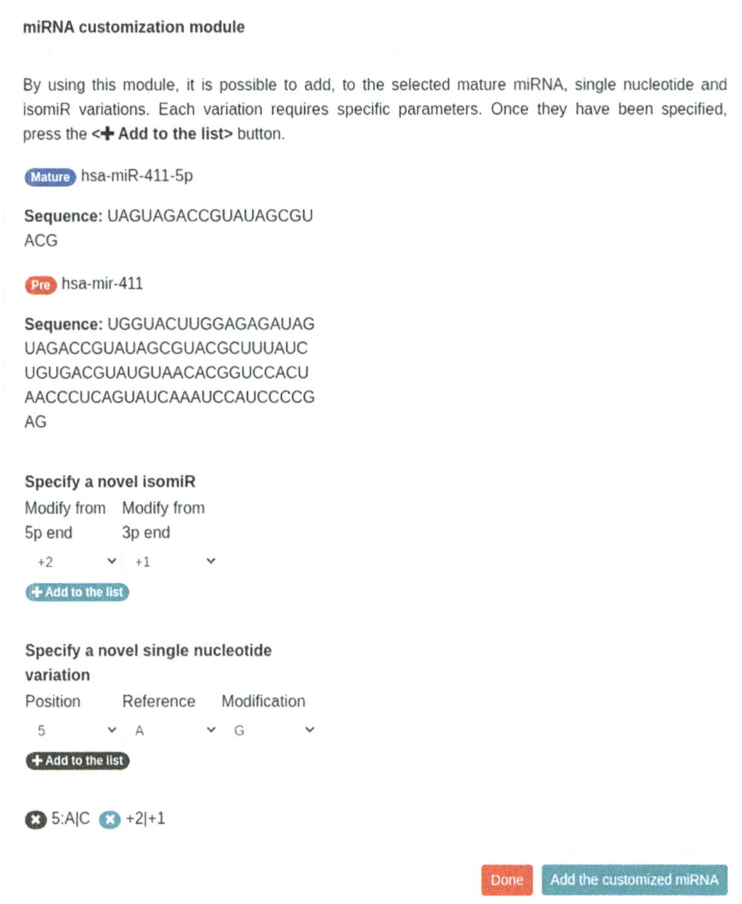

Fig. 4 miRNA customization form

In this example, we select all available prediction tools with a minimum consensus of 4, as shown in Fig. 6. Additionally, users can combine prediction tools by specifying the AND/OR option, which determines whether one or more tools must or can identify the same predicted targets.

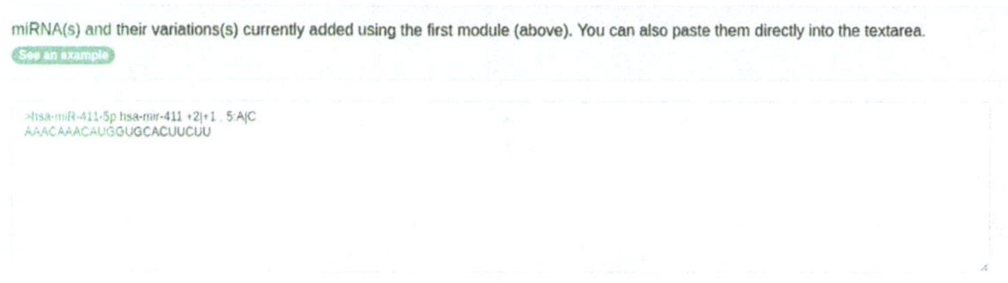

Fig. 5 The miRNA and its variations generated using the customization form

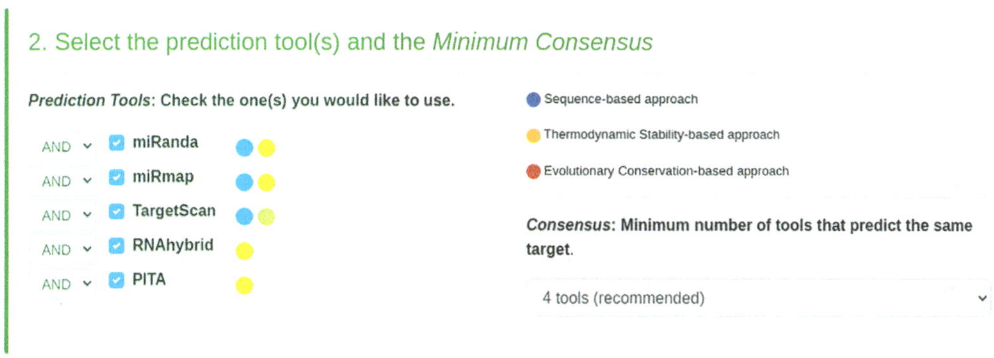

Fig. 6 Form for selecting prediction tools and minimum consensus

4) Users can then specify the number of CPU cores to be employed for the analysis (see Fig. 7).

5) Finally, users can confirm their input data by clicking the "Submit" button. This action prompts isoTar to begin the input verification process. During this phase, the system checks the data format to ensure it is correct before proceeding with the prediction analysis. If any errors are identified, they will be clearly displayed. Upon completion of the input verification, isoTar will provide a summary (see Fig. 8) of the specified miRNAs and their variations. The variations will be organized in a table, categorized by type (e.g., isomiRs and single-nucleotide modifications).

3.6 Function Analysis: Input Setup

The Functional Analysis module (see Fig. 9) enables users to enrich the predicted targets obtained through Prediction Analysis by integrating additional information from datasets such as GO terms (Gene Ontology) and pathways (KEGG and Reactome). For further details, users can refer to the isoTar documentation available at https://ncrnaome.osumc.edu/isotar. The following is a step-by-step guide to enhance predicted targets using all integrated datasets, including the steps for performing functional analysis.

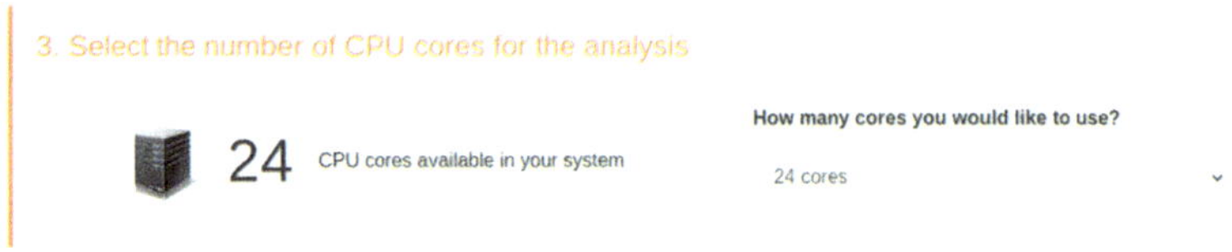

Fig. 7 Form for selecting CPU cores

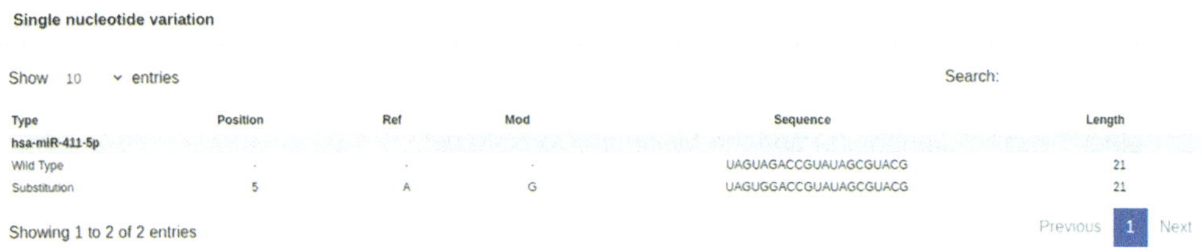

Fig. 8 Input summary

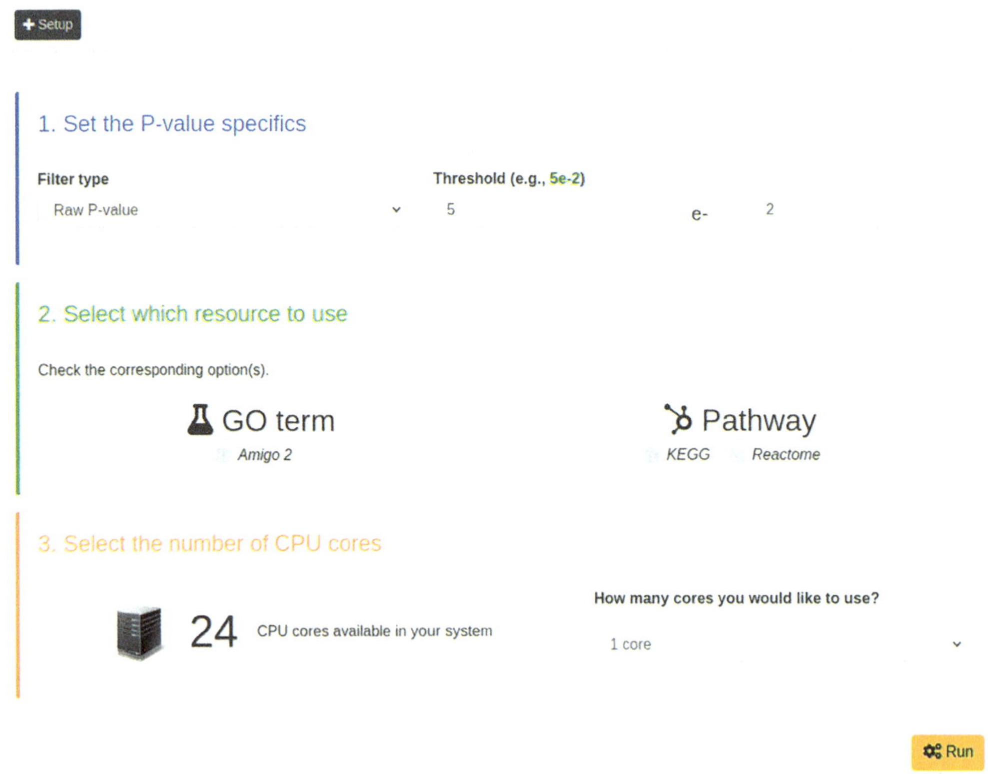

Fig. 9 Functional analysis module: *P*-value settings, dataset selection, and form for selecting CPU cores

1) Initially, users need to specify the *P*-value type (default: Raw *P*-value) and set the *P*-value threshold (default: <0.05) for filtering purposes (Fig. 9). In this example, we will choose the "Raw *P*-value" and set the threshold to <0.01.
2) Next, users can select the datasets to be used for the functional analysis (Fig. 9).
3) In the third step, users can specify the number of CPU cores to use for the analysis (Fig. 9).
4) Finally, users can execute the analysis by clicking the "Run" button (see Fig. 10).

3.7 Prediction Analysis: Results

Once the analysis is initiated, execution time will depend on factors such as the number of miRNAs and variations included in the input, the number of selected prediction tools, and the number of CPU cores allocated to the analysis. The results will be displayed in a new tab labeled "Results" (see Fig. 11). The prediction results are organized by variation type and prediction tool. For each variation, isoTar provides details including the type of variation (e.g., substitution or isomiR) and its position (for substitutions), the number of predicted targets identified based on the minimum consensus (shown in the "Targets" column), the number of predicted targets for each selected prediction tool, and a Venn diagram illustrating the interaction of predicted targets across all chosen tools. Users can click the number of targets to view additional information in a modal box (see Fig. 12). This information includes the Gene Symbol and Gene ID and provides links to online resources such as NCBI Gene (https://www.ncbi.nlm.nih.gov/gene/) and GeneCards (https://www.genecards.org/).

Additionally, by clicking on the chart icons in the "Statistics" column (located at the far right), users can view the Venn diagram (see Fig. 13), illustrating the intersection of predicted targets across the selected prediction tools. The diagram is accompanied by a histogram

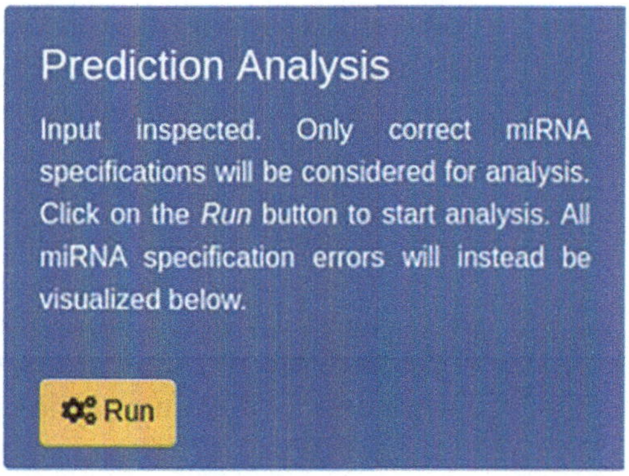

Fig. 10 Prediction analysis execution button

Fig. 11 Prediction analysis results grouped per variation type

showing the distribution of targets among these tools. Users can also click on any number displayed within the Venn diagram areas to view the corresponding targets on the right side of the diagram.

3.8 Functional Analysis: Results

The Functional Analysis module in isoTar enables the enrichment of predicted targets with information derived from Gene Ontology (GO), KEGG, and Reactome datasets. The analysis duration depends on the number of CPU cores used and the load on the AmiGO Web Service, which generates GO term data for the Gene Ontology categories: Biological Process (BP), Cellular Component (CC), and Molecular Function (MF). Results are displayed through tab panels (Fig. 14), each dedicated to a specific dataset or resource, such as GO terms, KEGG pathways, or Reactome pathways.

The analysis is performed for each Wild Type–Variation miRNA pair provided as input during the Prediction Analysis. For GO terms, the results are organized into three Gene Ontology categories: Biological Process, Cellular Component, and Molecular Function. The outputs are further divided into two key perspectives. The first focuses on Wild Type or Variation, providing separate GO term lists for each Gene Ontology category based on the consensus-predicted targets associated with either the wild-type miRNA or its variants. The second comparison, Wild Type vs. Variation, generates GO term lists highlighting differences between the two. These lists emphasize GO terms exclusive to the wild type, exclusive to the variation, or shared between the wild type and the variation, with visual differentiation using cyan, red, and green, respectively. isoTar provides detailed information for each variation, including its type and positional data. It also identifies the number of GO terms filtered by the selected *P*-value type and threshold and categorizes them by Biological Process, Cellular Component, and Molecular Function. Additionally, the number of GO terms specific to the wild type, the variation, or their intersection is reported, allowing users to explore the functional differences among them.

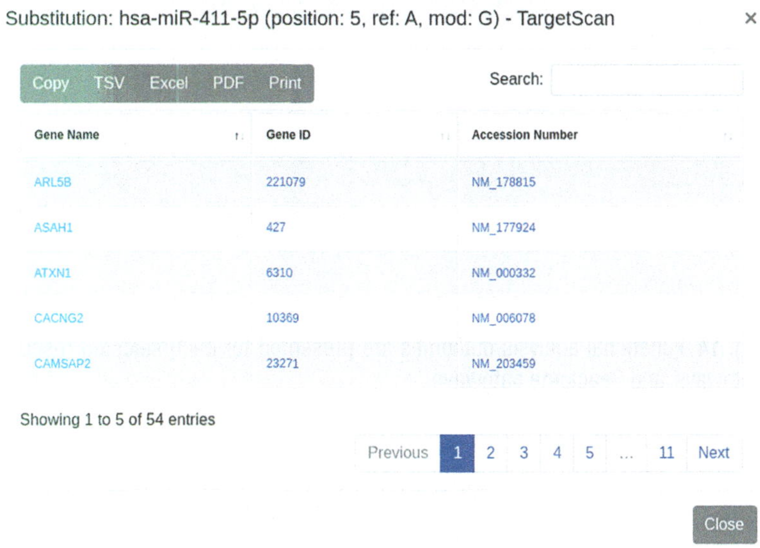

Fig. 12 The modal box reports additional information on predicted targets identified by TargetScan. The predicted targets list can be downloaded in various formats, including PDF and Excel

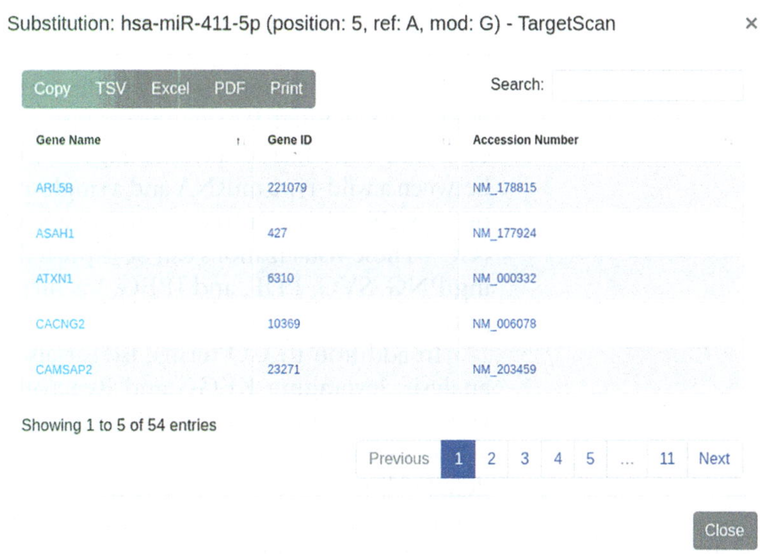

Fig. 13 The modal box reports the Venn diagram of predicted targets identified by different tools for the substitution. The diagram and the list of targets can be downloaded in SVG, PNG, and CSV formats

The results for each resource, including GO terms, KEGG pathways, and Reactome pathways, are grouped based on the type of variation. This enables a detailed exploration of functional implications

Fig. 14 Functional analysis outcomes are presented for each selected resource, including GO terms, KEGG pathways, and Reactome pathways

for each miRNA variation. For further guidance, users are advised to refer to the isoTar main page, which offers comprehensive documentation on navigating and interpreting these results. The Functional Analysis module provides a comprehensive view of the pathways and associated variations for each miRNA specified during the analysis. Users can access detailed insights into GO terms by clicking on them, which opens a modal box (see Fig. 15). This box displays relevant information, such as the GO term and its corresponding GO name, and links to external resources, such as the AmiGO Gene Ontology database.

Histograms summarizing significant GO terms, filtered according to user-defined *P*-value thresholds, are also provided. As in this case study, the histogram (see Fig. 16) illustrates the comparison between a wild-type miRNA and a single-nucleotide variant, focusing on a specific Gene Ontology category, such as Cellular Component (CC). These visualizations can be exported in various formats, including PNG, SVG, PDF, and JPEG, for further analysis and reporting purposes.

In addition to GO terms, isoTar also supports pathway-based analysis, leveraging KEGG and Reactome databases. These analyses are performed for each Wild Type–Variation pair, following a structure similar to that used for GO terms. The results include detailed lists of pathways associated with either the wild-type or variation-specific miRNAs, as well as pathways common to both. This classification is color-coded: cyan for wild-type-specific pathways, red for variation-specific pathways, and green for their intersection. For each variation, isoTar provides additional details, including the type of variation (e.g., substitution or isomiR) and its position. It also presents the number of identified pathways filtered by the user-specified *P*-value type and threshold. These pathways are categorized according to their association with the wild-type miRNA, the variation, or both. Visual summaries, such as histograms, accompany these findings and illustrate the distribution of pathways across the

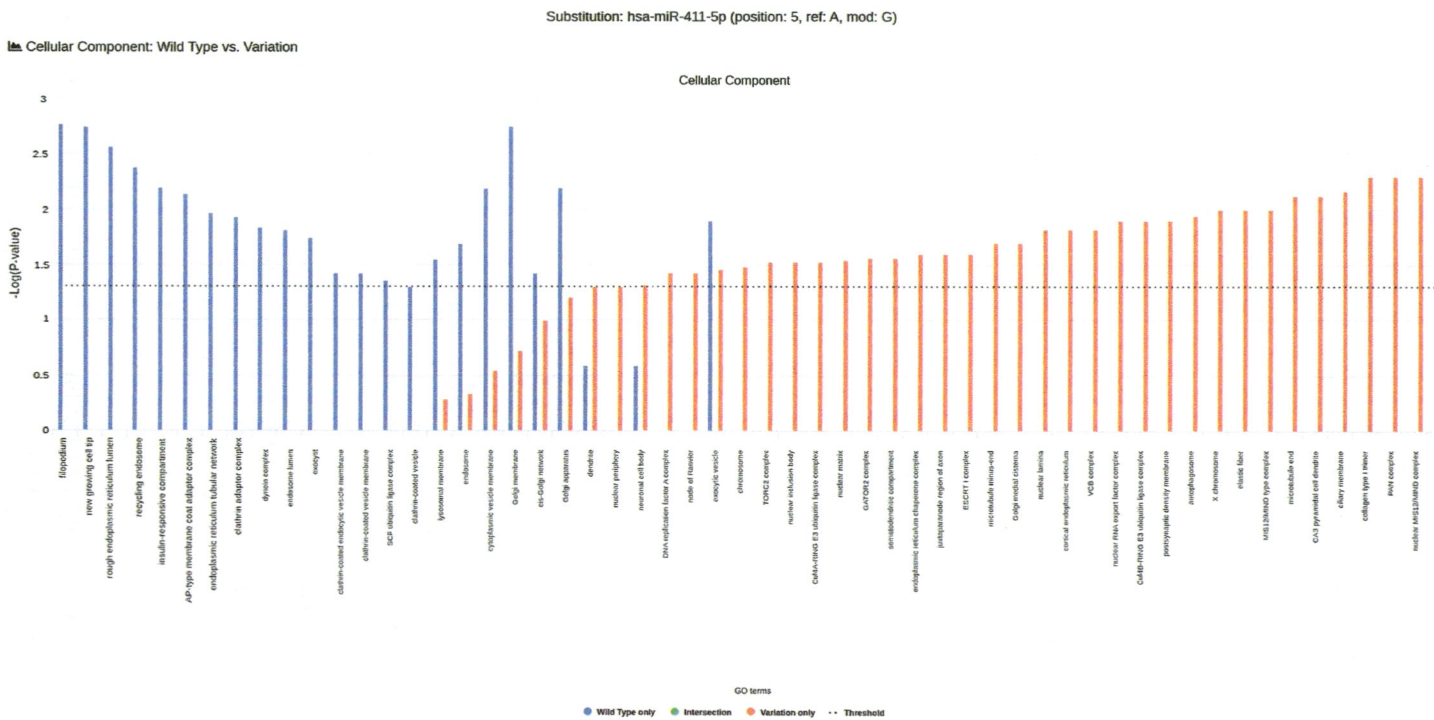

Fig. 15 The modal box provides detailed information on GO terms explicitly associated with the wild type (represented in green) and the substitution within the Gene Ontology category "Biological Process" (BP). Additionally, the GO terms list can be exported in various formats, including Excel, PDF, and TSV, allowing users to analyze the data further or integrate it into other workflows

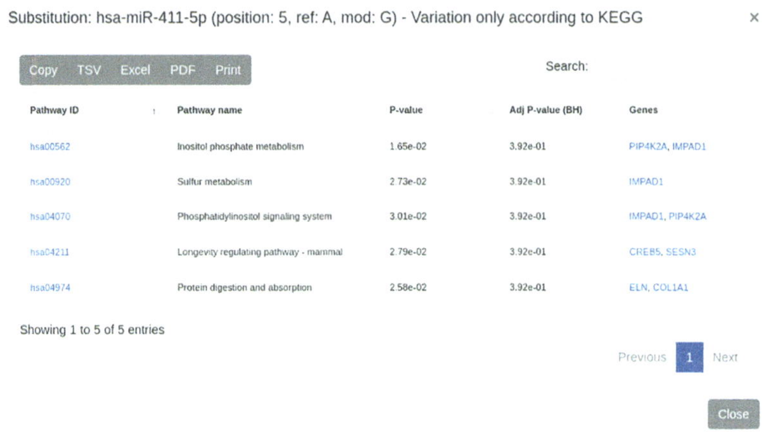

Fig. 16 Wild type vs. substitution histogram according to the Gene Ontology category CC

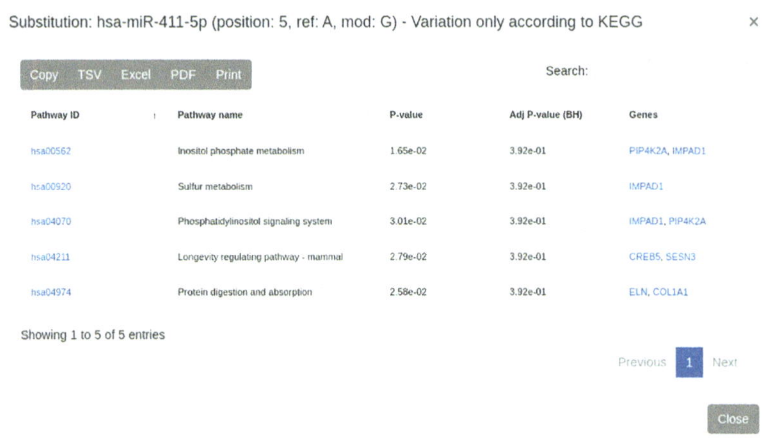

Fig. 17 Modal box reporting additional information on pathways according to the variation miRNA only and the KEGG resource. The list of pathways can be downloaded in various formats, including Excel, PDF, and TSV

various categories. These comprehensive results enable researchers to systematically and efficiently explore the functional implications of miRNA variations.

Pathway-specific details can be accessed interactively in isoTar, allowing users to explore additional information about single pathways. By clicking on a pathway name, a modal box is displayed (see Fig. 17), providing details such as the Pathway ID and name. These details are linked to online resources, such as KEGG or Reactome, depending on the dataset in use.

For instance, Fig. 18 shows a histogram that compares a wild-type miRNA with the substitution in association with Reactome; the same plot can be created according to KEGG (Fig. 19). Similarly,

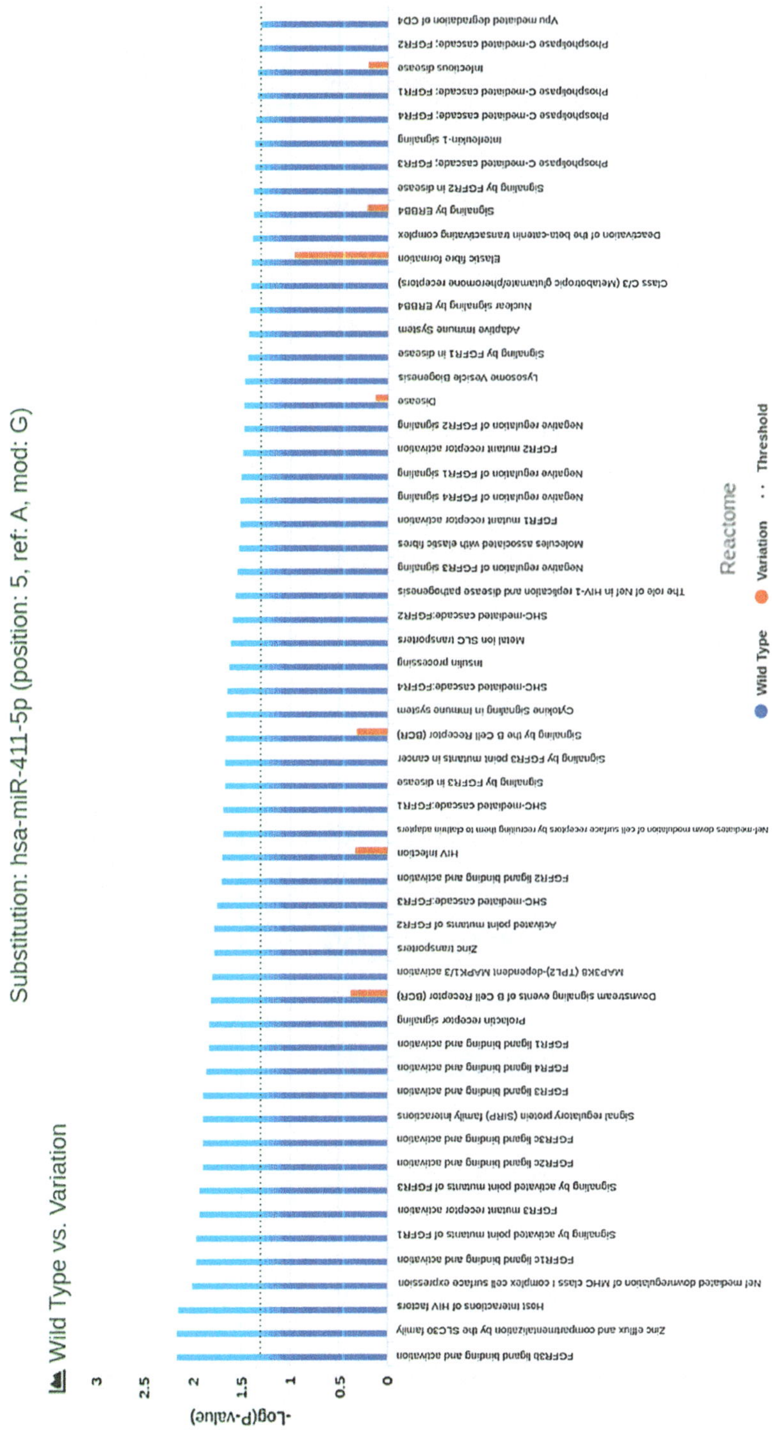

Fig. 18 Wild type vs. variation histogram in association with Reactome

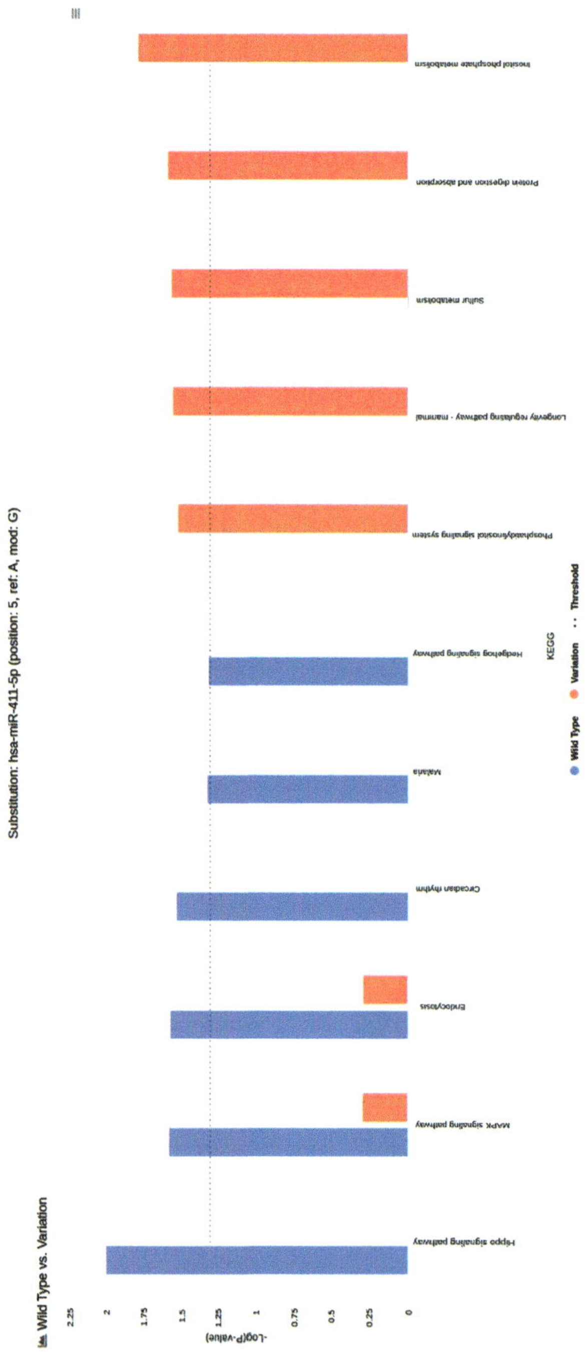

Fig. 19 Wild type vs. variation histogram in association with KEGG

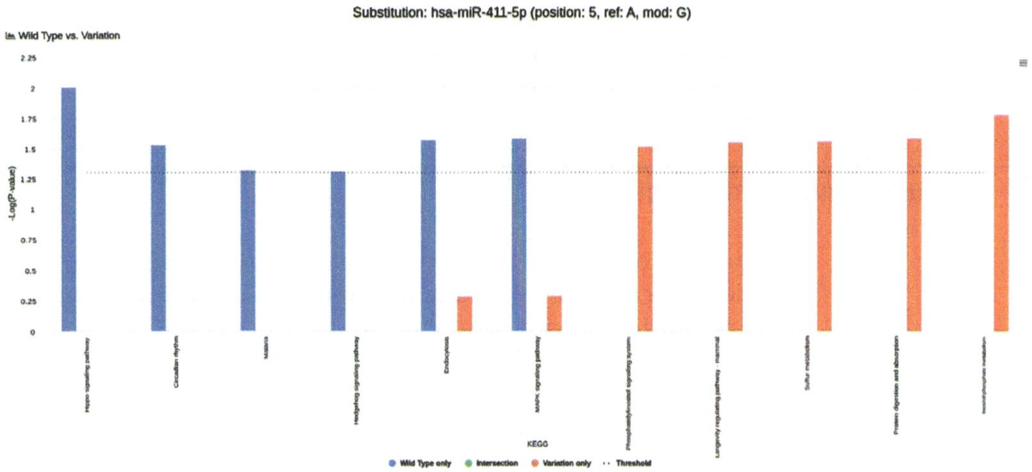

Fig. 20 Histogram comparing pathway subsets, with distinct bars representing pathways unique to the wild type, unique to the variation, and shared between the two (absent in the example taken from the case study). The bars are color-coded: cyan for wild type, red for variation, and green for the intersection. The histogram highlights the distribution of pathways across these categories, emphasizing differences and overlaps

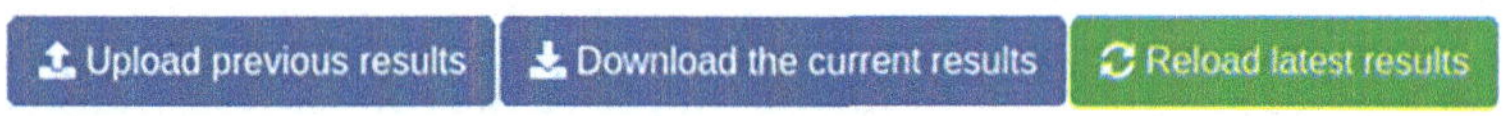

Fig. 21 Upload previous results, Download the current results, and Reload latest results buttons

Fig. 20 illustrates a histogram that compares pathway subsets, distinguishing between those unique to the wild type, the variation, and their intersection. These visualizations, filtered by the user-defined *P*-value type and threshold, are available for download in PNG, SVG, PDF, and JPEG formats.

To facilitate data management, isoTar provides tools for downloading and storing results. Users can save all current analysis results by clicking the "Download the current results" button (see Fig. 21), which compiles the outputs into a compressed ZIP file named output.zip. It is important to note that subsequent analyses overwrite existing results within the same instance. For example, initiating a new prediction analysis will replace both the prediction and functional results, whereas a new functional analysis will replace only the previously generated functional data.

To address potential interruptions, isoTar allows users to reload past analyses. By uploading a previously downloaded output.zip file through the "Upload previous results" button (see Fig. 21), users can restore prior results seamlessly. Additionally, if the isoTar web page is closed accidentally or intentionally during lengthy analyses, users can recover their latest results within the current session by clicking the "Reload latest results" button (see Fig. 21). This feature proves particularly useful for avoiding the need to keep the isoTar interface open continuously, offering convenience and flexibility during extended computational workflows.

4 Conclusion and Future Directions

Building on its current implementation as a containerized, user-friendly framework for consensus miRNA–target prediction and downstream functional enrichment, isoTar [17] is well positioned to evolve into a broader platform for studying the increasingly complex "miRNAome," including sequence-edited and isoform-shifted molecules and their context-dependent regulatory effects. Future development will first focus on expanding the range of supported organisms and annotation sources, enabling analyses across additional species commonly used in biomedical and basic research (e.g., mouse and other model systems), while maintaining consistent identifiers and orthology-aware reporting to facilitate cross-species hypothesis generation. In parallel, we plan to integrate additional prediction resources and updated versions of the currently implemented algorithms, alongside harmonized parameterization and benchmarking, to further improve the robustness, transparency, and reproducibility of consensus calling. Beyond target prediction for single canonical or singly modified miRNAs, an important future direction is the systematic support of "mixed" miRNA variants—molecules harboring multiple point modifications (e.g., combinations of editing events and/or SNVs) together with 5′ end shifts that redefine the seed sequence. These compound variants are increasingly observed in deep-sequencing datasets and can generate targetomes that are not well approximated by analyzing each modification in isolation; therefore, isoTar will be extended to accept composite miRNA definitions, enumerate plausible combinations when appropriate, and report both shared and unique target subsets across variant sets to help users prioritize experimentally testable interactions.

Another key enhancement will be the integration of competing endogenous RNA (ceRNA)-oriented analyses by adding dedicated modules to predict and explore long noncoding RNAs (lncRNAs) as potential miRNA "sponges." Concretely, this will involve incorporating lncRNA sequence/annotation databases and implementing sponge-aware prediction strategies (e.g., identifying high-confidence miRNA response elements on lncRNAs, evaluating site multiplicity and seed-match classes, and enabling joint visualization of mRNA and lncRNA target landscapes). This extension will allow users to move from a "miRNA→mRNA" view toward a network perspective in which lncRNAs, mRNAs, and potentially other RNA species contribute to miRNA availability and regulatory buffering, thereby supporting exploratory reconstruction of candidate ceRNA circuits.

Finally, because miRNA function is inherently conditional on the cellular transcript pool, future isoTar releases will aim to better connect predicted interactions with molecular context by enabling optional integration of expression or differential expression results (when available) to filter or rank predicted targets and sponge candidates in a sample- or state-aware manner. Together, these expansions—more species, more prediction engines, explicit support for composite miRNA variants, and ceRNA/lncRNA-focused functionality—will strengthen isoTar as a versatile, next-generation resource for interpreting canonical and modified miRNAs within the broader RNA regulatory landscape.

Acknowledgments

This work was supported by the European Union's NextGenerationEU initiative under the Italian Ministry of University and Research as part of the PNRR – M4C2.-I1.3 PE00000019 "HEAL ITALIA," CUP B73C22001250006, awarded to S.D.B. F.O. was supported by the European Union under the Italian Ministry of University and Research as part of the Project: Health Extended ALliance for Innovative Therapies, Advanced Lab research, and Integrated Approaches of Precision Medicine (HEAL ITALIA) Spoke 3 – Task 1.2 CUP B73C22001250006. G.N. was supported in part by National Cancer Institute (National Institute of Health) grant (R21CA287180).

Conflict of Interest The authors have no conflicts of interest to declare that are relevant to the content of this chapter.

References

1. Treiber T, Treiber N, Meister G (2019) Regulation of microRNA biogenesis and its crosstalk with other cellular pathways. Nat Rev Mol Cell Biol 20:5–20. https://doi.org/10.1038/s41580-018-0059-1
2. Gebert LFR, Macrae IJ (2018) Regulation of microRNA function in animals. Nature Rev Mol Cell Biol 21:743. https://doi.org/10.1038/s41580-018-0045-7
3. Yi R, Qin Y, Macara IG, Cullen BR (2003) Exportin-5 mediates the nuclear export of pre-microRNAs and short hairpin RNAs. Genes Dev 17:3011–3016. https://doi.org/10.1101/gad.1158803
4. Bohnsack MT, Czaplinski K, Gorlich D (2004) Exportin 5 is a RanGTP-dependent dsRNA-binding protein that mediates nuclear export of pre-miRNAs. RNA (New York, NY) 10:185–191. https://doi.org/10.1261/rna.5167604
5. Ha M, Kim VN (2014) Regulation of microRNA biogenesis. Nat Rev Mol Cell Biol 15:509–524. https://doi.org/10.1038/nrm3838
6. Nigita G, Marceca GP, Tomasello L et al (2019) ncRNA editing: functional characterization and computational resources. Methods Mol Biol 1912:133–174. https://doi.org/10.1007/978-1-4939-8982-9_6
7. Tomasello L, Distefano R, Nigita G, Croce CM (2021) The microRNA family gets wider: the isomiRs classification and role. Front Cell Dev Biol 9:668648. https://doi.org/10.3389/fcell.2021.668648
8. Nishikura K (2016) A-to-I editing of coding and non-coding RNAs by ADARs. Nat

Rev Mol Cell Biol 17:83–96. https://doi.org/10.1038/nrm.2015.4
9. Nigita G, Veneziano D, Ferro A (2015) A-to-I RNA editing: current knowledge sources and computational approaches with special emphasis on non-coding RNA molecules. Front Bioeng Biotechnol 3:37. https://doi.org/10.3389/fbioe.2015.00037
10. Zhou H, Arcila ML, Li Z et al (2012) Deep annotation of mouse iso-miR and iso-moR variation. Nucleic Acids Res 40:5864–5875. https://doi.org/10.1093/nar/gks247
11. Neilsen CT, Goodall GJ, Bracken CP (2012) IsomiRs – the overlooked repertoire in the dynamic microRNAome. Trends in Genet 28:544–549. https://doi.org/10.1016/j.tig.2012.07.005
12. Magee RG, Telonis AG, Loher P et al (2018) Profiles of miRNA isoforms and tRNA fragments in prostate cancer. Sci Rep 8:5314. https://doi.org/10.1038/s41598-018-22488-2
13. Kozomara A, Birgaoanu M, Griffiths-Jones S (2019) miRBase: from microRNA sequences to function. Nucleic Acids Res 47:D155–D162. https://doi.org/10.1093/nar/gky1141
14. Yang W, Chendrimada TP, Wang Q et al (2006) Modulation of microRNA processing and expression through RNA editing by ADAR deaminases. Nature Struct & Mol Biol 13:13–21. https://doi.org/10.1038/nsmb1041
15. Kawahara Y, Zinshteyn B, Sethupathy P et al (2007) Redirection of silencing targets by adenosine-to-inosine editing of miRNAs. Science 315:1137–1140. https://doi.org/10.1126/science.1138050
16. Nigita G, Acunzo M, Romano G et al (2016) microRNA editing in seed region aligns with cellular changes in hypoxic conditions. Nucleic Acids Res 44:6298–6308. https://doi.org/10.1093/nar/gkw532
17. Distefano R, Nigita G, Veneziano D et al (2019) isoTar: consensus target prediction with enrichment analysis for microRNAs harboring editing sites and other variations. MicroRNA TI 1970:211–235. https://doi.org/10.1007/978-1-4939-9207-2_12
18. Merkel D (2014) Docker: lightweight Linux containers for consistent development and deployment. Linux J 2014(239):2
19. Spurlock J (2013) Bootstrap: responsive web development. O'Reilly Media, Inc.
20. Volder KD (2006) JQuery: A generic code browser with a declarative configuration language. In: Van Hentenryck P (eds) Practical Aspects of Declarative Languages. PADL (2006). Lecture Notes in Computer Science, vol 3819. Springer, Heidelberg, pp 88–102. https://doi.org/10.1007/11603023_7
21. Bardou P, Mariette J, Escudié F et al (2014) jvenn: an interactive Venn diagram viewer. BMC Bioinform 15:293. https://doi.org/10.1186/1471-2105-15-293
22. Kertesz M, Iovino N, Unnerstall U et al (2007) The role of site accessibility in microRNA target recognition. Nature Genet 39:1278–1284. https://doi.org/10.1038/ng2135
23. John B, Enright AJ, Aravin A et al (2004) Human microRNA targets. PLOS Biol 2:e363. https://doi.org/10.1371/journal.pbio.0020363
24. Heyne S, Costa F, Rose D, Backofen R (2012) GraphClust: alignment-free structural clustering of local RNA secondary structures. Bioinformatics 28:i224–i232. https://doi.org/10.1093/bioinformatics/bts224
25. Vejnar CE, Zdobnov EM (2012) MiRmap: comprehensive prediction of microRNA target repression strength. Nucleic Acids Res 40:11673–11683. https://doi.org/10.1093/nar/gks901
26. Lewis BP, Burge CB, Bartel DP (2005) Conserved seed pairing, often flanked by adenosines, indicates that thousands of human genes are microRNA targets. Cell 120:15–20. https://doi.org/10.1016/j.cell.2004.12.035
27. Marín RM, Vanícek J (2011) Efficient use of accessibility in microRNA target prediction. Nucleic Acids Res 39:19–29. https://doi.org/10.1093/nar/gkq768
28. Sherry ST, Ward M-H, Kholodov M et al (2001) dbSNP: the NCBI database of genetic variation. Nucleic Acids Res 29:308–311. https://doi.org/10.1093/nar/29.1.308
29. Harris MA, Clark J, Ireland A et al (2004) The gene ontology (GO) database and informatics resource. Nucleic Acids Res 32:D258–61. https://doi.org/10.1093/nar/gkh036
30. Fabregat A, Sidiropoulos K, Garapati P et al (2016) The reactome pathway knowledgebase. Nucleic Acids Res 44:D481–D487. https://doi.org/10.1093/nar/gkv1351
31. Croft D, Mundo AF, Haw R et al (2014) The reactome pathway knowledgebase. Nucleic Acids Res 42:D472–7. https://doi.org/10.1093/nar/gkt1102
32. Fabregat A, Jupe S, Matthews L et al (2018) The reactome pathway knowledgebase. Nucleic Acids Res 46:D649–D655. https://doi.org/10.1093/nar/gkx1132
33. Kanehisa M, Goto S (2000) KEGG: kyoto encyclopedia of genes and genomes. Nucleic Acids Res 28:27–30. https://doi.org/10.1093/nar/28.1.27
34. Klopfenstein DV, Zhang L, Pedersen BS et al (2018) GOATOOLS: a python library for gene ontology analyses. Sci Rep 8:10872. https://doi.org/10.1038/s41598-018-28948-z

35. Schröder MS, Gusenleitner D, Quackenbush J et al (2013) RamiGO: an R/Bioconductor package providing an AmiGO Visualize interface. Bioinformatics 29:666–668. https://doi.org/10.1093/bioinformatics/bts708

36. Gentleman RC, Carey VJ, Bates DM et al (2004) Bioconductor: open software development for computational biology and bioinformatics. Genome Biol 5:R80. https://doi.org/10.1186/gb-2004-5-10-r80

Check for updates

Chapter 6

MiREDiBase: A Comprehensive Resource for miRNA Editing Events

Ilaria Cosentini, Gioacchino P. Marceca, Rosario Distefano, Shaopeng Gu, Shaohong Feng, Eswar Shankar, Mario Acunzo, Giulia Romano, Alessandro Laganà, Ernesto Picardi, Christian Rolfo, Qin Ma and Giovanni Nigita

Abstract

MicroRNA (miRNA) editing introduces a new layer of complexity in post-transcriptional gene regulation, with potential implications for miRNA biogenesis, target specificity, and function. However, the lack of a dedicated repository to comprehensively catalog miRNA editing events has hindered advances in the field. MiREDiBase addresses this gap by offering a meticulously curated database focused exclusively on validated, high-confidence putative miRNA editing events, with emphasis on A-to-I (adenosine-to-inosine) and C-to-U (cytidine-to-uridine) editing. This resource provides extensive annotations, including genomic locations, structural changes, and biological context, alongside tools for comparative analysis and predictive modeling of miRNA editing effects. MiREDiBase integrates data from experimental validation and high-throughput sequencing, delivering insights into the functional and pathological significance of miRNA editing across species. Unique features include a user-friendly interface for searching and comparing editing patterns, access to RESTful APIs for advanced data retrieval, and integration with external databases to enhance usability in epitranscriptomics research. By bridging the gap between RNA editing and miRNA research, MiREDiBase fosters innovative exploration of miRNA biology, advancing our understanding of its role in health and disease.

Keywords A-to-I RNA editing, C-to-U RNA editing, Epitranscriptomics, microRNA, database

1 Introduction

1.1 Background on miRNA and RNA Editing

MicroRNAs (miRNAs) are pivotal regulators of gene expression, playing crucial roles in various biological processes such as development, cell differentiation, and response to environmental stimuli [1]. These small, non-coding RNA molecules, typically 19–24 nucleotides in length, function by binding to complementary sequences in target messenger RNAs (mRNAs) [2]. This binding results in mRNA

Alessandro Laganà and Giovanni Nigita (eds.), *MicroRNA Target Identification*, Methods in Molecular Biology, vol. 3056,
https://doi.org/10.1007/978-1-0716-5392-0_6,

degradation or translational repression, thereby finely regulating gene expression [3, 4]. The specificity of miRNA-mRNA interactions allows miRNAs to modulate complex regulatory networks, influencing entire pathways involved in cellular functioning [5]. Furthermore, miRNA dysregulation has been implicated in numerous diseases, including cancer, cardiovascular diseases, and neurological disorders, making them potential non-invasive biomarkers for early diagnosis and therapeutic targets of several pathologies [6–8]. Research continues to explore the distinct roles different miRNAs play, contributing to the growing understanding of their importance in both normal conditions and disease states.

Complementing the regulatory functions of miRNAs, RNA editing—particularly A-to-I (adenosine-to-inosine) and C-to-U (cytidine-to-uridine) editing—introduces additional layers of complexity in post-transcriptional regulation [9–11]. A-to-I editing, predominantly mediated by adenosine deaminases acting on RNA (ADARs), can alter mRNA coding sequences, leading to changes in protein function and the generation of diverse protein isoforms from a single gene [12, 13]. This event can affect various aspects of cellular function, such as synaptic plasticity in neurons and immune responses [14, 15]. Similarly, C-to-U editing plays a significant role in regulating gene expression dynamics as well, influencing mRNA stability and the translation process [16–19]. Understanding how miRNAs and RNA editing interact is essential for elucidating their roles in health and disease [20–26], paving the way for the development of innovative therapeutic strategies.

1.2 The Need for a Dedicated miRNA Editing Database

Existing RNA editing databases, such as REDIportal [27], REIA [28], TCEA [29], and GPEdit [30], primarily catalog RNA editing events across the transcriptome, focusing mainly on mRNAs and long non-coding RNAs (lncRNAs). However, these resources provide limited insights into miRNA editing, which is a crucial post-transcriptional modification that can significantly affect miRNA biogenesis, targeting, and function. In the aforementioned databases, miRNA editing events are often underrepresented, poorly annotated, or embedded within broader datasets, making it difficult for researchers to retrieve specific information on miRNA modifications. Additionally, these platforms rarely offer crucial details, such as the functional impact of miRNA editing, structural changes to pre- and mature miRNAs, or specific biological contexts—elements essential for understanding how these modifications influence miRNA activity and their potential roles in diseases, including cancer.

Given these limitations, the development of a specialized database of miRNA editing sites was essential to fill the gap between RNA editing and miRNA research [31]. MiREDiBase offers a dedicated, manually curated catalog focused exclusively on validated and putative miRNA editing events, accessible via a user-friendly platform. Unlike general-purpose RNA editing resources, it provides comprehensive annotations, including editing positions in pre- and mature miRNAs, structural changes, predictions of miRNA re-targeting, and enrichment analyses. Such a platform empowers researchers to explore how editing alters the target repertoire of miRNAs, thereby impacting biological processes and facilitating the discovery of diagnostic and prognostic biomarkers. MiREDiBase also integrates tools for comparing editing patterns between normal and pathological conditions, enabling studies on the regulatory roles of miRNA editing in various tissues and disease models. By addressing the unmet need for an edited-miRNA-centric database, MiREDiBase aims to advance the understanding of epitranscriptomic regulation and foster new insights into miRNA biology and its biomedical applications.

2 Overview of MiREDiBase

2.1 Development and Curation Process

The development of MiREDiBase involved a meticulous, manual curation process to ensure a reliable and comprehensive catalog of miRNA editing events. The curation team systematically reviewed the scientific literature by searching for keywords such as "miRNA editing" and "microRNA editing" and narrowed the time range to studies published between 2000 and 2019. Relevant articles were manually filtered to retain only those reporting validated or high-confidence miRNA editing events. Editing events confirmed through targeted experiments, such as Sanger sequencing or ADAR enzyme perturbation, were classified as validated modifications. In contrast, events identified by high-throughput sequencing (HTS) methods were categorized as putative, provided they met statistical significance criteria.

Additionally, existing databases, including DARNED [32], RADAR [33], and REDIportal [27], were screened for miRNA-specific editing events. The curation process excluded misassigned miRNAs, dead entries, and incorrectly annotated editing sites to maintain the highest data integrity. Genomic coordinates from previous assemblies, such as hg19, were converted to the current hg38 assembly using the UCSC LiftOver tool to ensure alignment with updated genomic references. This rigorous curation strategy allowed MiREDiBase to collect both experimental and computationally inferred data, establishing a valuable resource for studying miRNA editing.

2.2 Database Structure and Content

MiREDiBase is designed to provide comprehensive coverage of miRNA editing, containing both validated and putative A-to-I and C-to-U editing events. The database includes information on the genomic location, the position within pre-miRNA or mature miRNA transcripts, and the detection method used. For validated editing events, details of the experimental validation approach are provided, while putative events include metadata on detection techniques and statistical

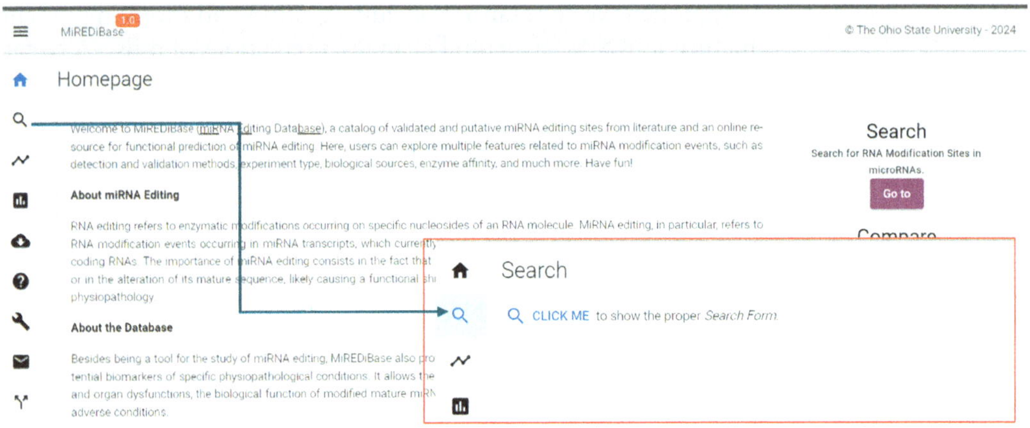

Fig. 1 MiREDiBase homepage with quick access to key modules, including the "Search" feature for detailed queries

thresholds. MiREDiBase goes beyond raw editing data by incorporating biological annotations, such as tissue-specific editing patterns, target prediction analyses, and thermodynamic comparisons between edited and non-edited miRNAs. This allows researchers to explore the structural and functional impact of miRNA editing events and compare their distribution in healthy versus pathological conditions.

Statistically, the current version of MiREDiBase catalogs over 3000 unique editing events from humans and three non-human primates (*Macaca mulatta*, *Pan troglodytes*, and *Gorilla gorilla*), distributed across 626 pre-miRNAs. The human dataset includes 2989 A-to-I and C-to-U editing events, with approximately 38.6% occurring within mature miRNAs, 31.7% within precursor miRNAs, and 29.7% in regions outside of pre-miRNAs. Although a substantial portion of the reported events remain functionally uncharacterized, MiREDiBase highlights 24 sites with experimentally validated functional implications. These include editing-induced changes that influence miRNA biogenesis or re-targeting potential, providing insights into their regulatory roles. By providing both genomic and functional data, MiREDiBase offers a comprehensive platform for studying the biological relevance of miRNA editing and its potential impact on disease pathways.

3 Using MiREDiBase

3.1 Navigating the User Interface

MiREDiBase is an online, open-source platform with a user-friendly interface, available at https://ncrnaome.osumc.edu/miredibase/. The homepage provides a brief explanation of the database and quick access to its sections, visually displayed on both sides of the home screen (Fig. 1). It includes essential links leading to the consultation

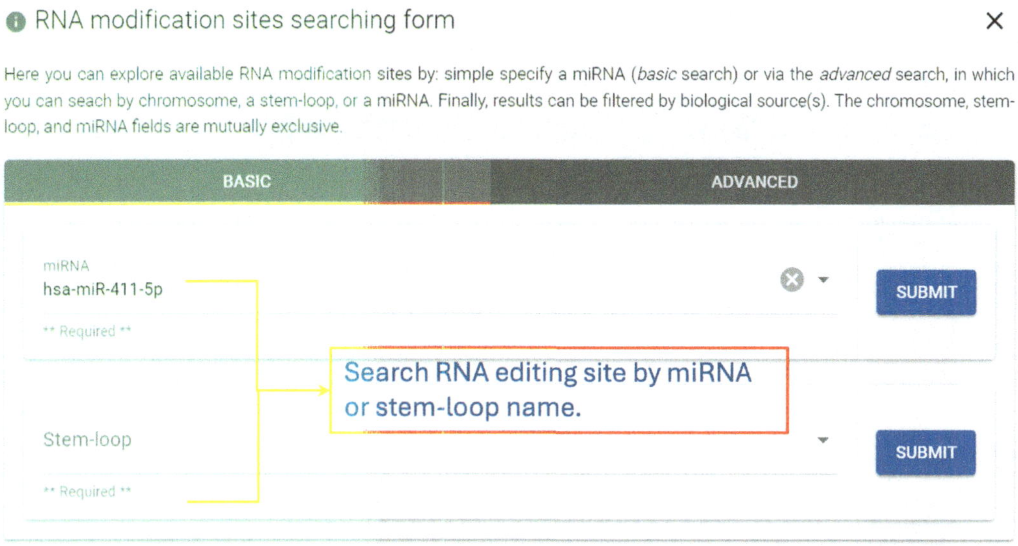

Fig. 2 The "Search" module with "Basic" and "Advanced" forms for retrieving miRNA editing sites by keywords and filtering by species or editing type

modules of the database: "Search," which allows to conduct detailed queries; "Compare," which allows comparative analysis between physiological and pathological conditions; "Statistics," showing a report about miRNA editing sites subdivided by species; "Download," that allows to obtain the complete list of unique A-to-I and C-to-U miRNA modification sites included in the current version; "Help," containing a detailed guide to navigating the MiREDiBase website; "Contact Us" for reaching out to the MiREDiBase team. Finally, the "Changelog" section highlights newly added data, significant findings, and recent database updates, keeping users informed about the latest developments in miRNA editing research. The most recent update consists of API and UI extensions for data retrieval.

3.2 The Search module

The "Search" module allows the retrieval of validated and putative miRNA editing sites by providing specific keywords (e.g., miRNA name or gene symbol) in the "Search form" and the filtering of results by species or editing type. The "Search form" includes two distinct sections, i.e., "Basic" and "Advanced." The first one offers a straightforward way to search for RNA modifications falling into entire mature miRNAs or stem-loops (Fig. 2). The second one (Fig. 3) allows for more refined searches, enabling the user to filter results by species (indicated as "organism"—optionally selecting a specific RNA modification type) (Step 1) or by genomic region (Step 2). The search by genomic region envisages three distinct fields: chromosome (allowing the retrieval of miRNA editing events mapped to whole chromosomes or restricting the search to specific chromosome regions and DNA strand), stem-loop or mature miRNA.

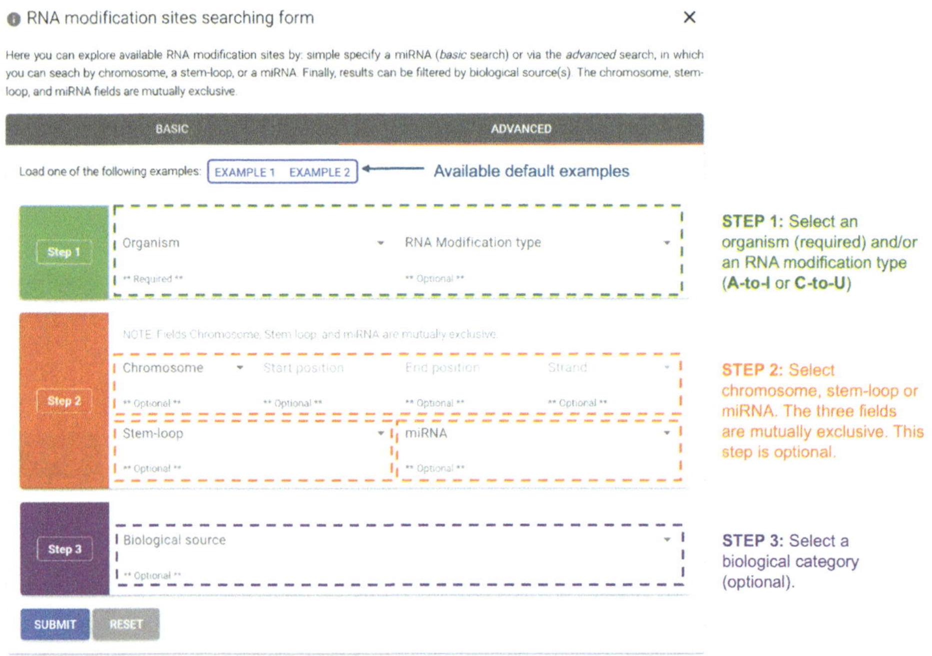

Fig. 3 Advanced search module for filtering miRNA editing sites by species, RNA modification type, genomic region, or biological source, with options for detailed refinement and default examples for guidance

These three fields are mutually exclusive. Once one of the fields is filled, the remaining will be automatically deactivated.

Additionally, the user can specify one or more biological sources (e.g., cell lines or tissues) to further refine the search (Step 3). Two distinct default examples can be loaded by the user in order to understand the advanced search functionality better. After clicking the "Submit" button, the query result will be displayed.

The search results are shown in a table listing the unique miRNA modification sites retrieved. The table comprises four main parts (Fig. 4A). The first six columns (white) report basic information about the selected editing site, including organism (species), modification type, genomic details (chromosome, strand, UCSC position), and conservation across species. The second part of the table (orange columns) includes details concerning the stem-loop that the editing site falls into. The third part (blue columns) provides information on how many times the selected editing site was detected by specific detection methods (HTS methods, enzyme perturbation, and targeted methods) in the literature. Finally, the green columns report whether the editing site is putative (Yes) or experimentally validated by targeted methods (No).

By clicking the left-sided red icon, the user will open a new page reporting two tables: the "RNA modification summary" table (Fig. 4B, upper side), which is an overview of the selected row, and the "RNA modification site studies" table (Fig. 4B, lower side),

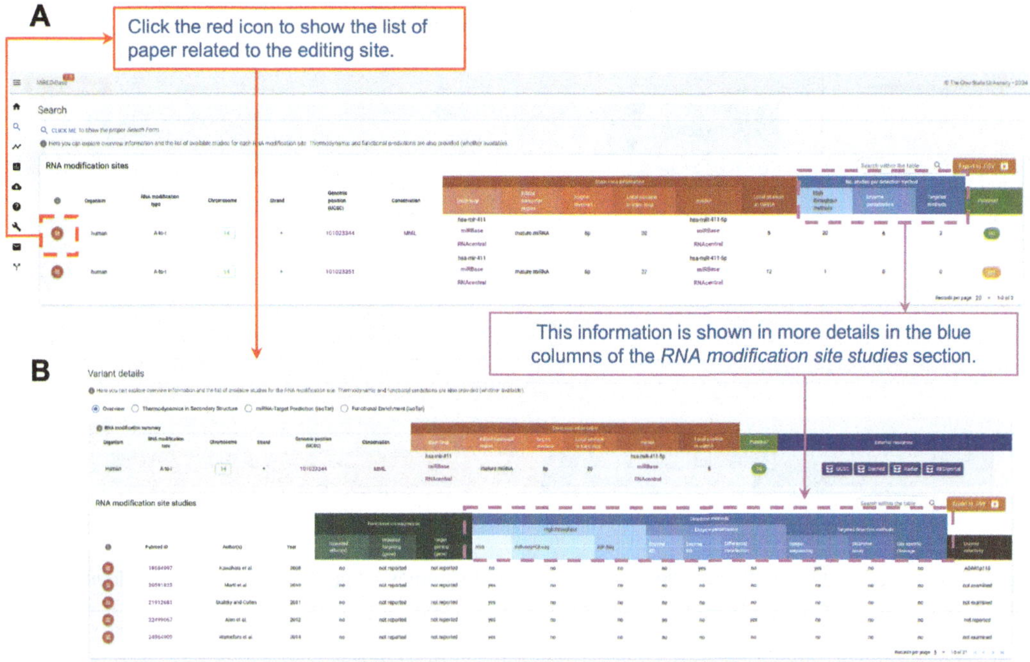

Fig. 4 General information and available studies for each RNA modification site

displaying the list of published original papers for the selected record. More precisely, the latter shows, on the left, information about each scientific paper, i.e., PubMed ID, authors, and year of publication. By shifting to the right, the table reports whether the authors assessed any functional consequences (green columns). The table then reports more detailed information about the experimental strategies used—alone or in combination—to detect the selected miRNA editing site (blue columns), as a major detail of the aforementioned blue columns in the previous table. Finally, on the far right, the table shows whether an enzyme affinity was assessed for the selected editing site by the authors.

The left-red icon in the "RNA modification site studies" table opens and additional section, named "Additional info" containing complementary data about authors' findings for the selected record (Fig. 5, lower side)—from left to right: information on biological samples (grey columns), experiment type (*in vitro, ex vivo, in vivo*, blue columns), editing levels in the primary (pri-), premature (pre-) and mature miRNA transcript (orange-to-red coluns), and eventual comparison between physiological and pathological conditions (green column).

In all of the mentioned sections, results can be filtered by typing appropriate keywords or numerical values in the search bars. All tables can be downloaded as a CSV (comma-separated data) file by clicking the "Export" button located in the top-right side of each

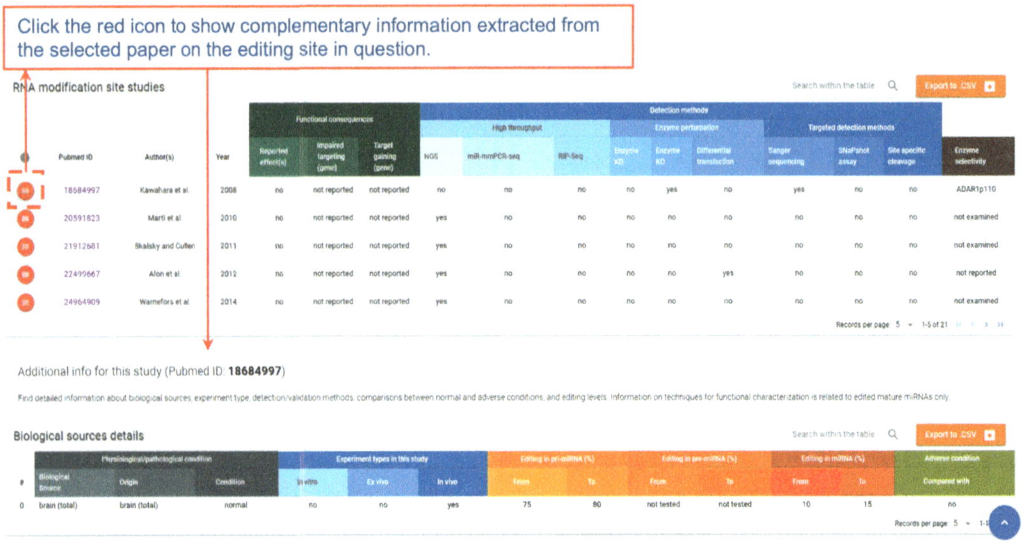

Fig. 5 The additional information section details biological samples, experiment types, editing levels across miRNA transcripts, and comparisons of physiological and pathological conditions

table. Moreover, by clicking the purple inscriptions, users can access to external resources.

3.3 MiREDiBase Predictive Functions

The tables and information mentioned so far are from the MiREDiBase "Overview" section, where users can visualize the factual data extracted from original papers. However, the platform also provides three additional sections devoted to predictions. Users can switch from one section to another by clicking the corresponding selection box posed at the top of the screen (Fig. 6).

The first predictive section is devoted to thermodynamic predictions on double-stranded RNA structures. Such predictions were carried out by the employment of the RNAfold tool, included in the ViennaRNA package (v2.0) [34] with default settings. The aim of this section is to infer whether a single nucleotide substitution can impact the secondary structure of the RNA stem-loop, potentially causing variations in molecule stability or binding efficiency. Minimum free energy (MFE) structures have been generated for all stem-loops subjected to editing events, as well as for their wild-type (WT) counterparts. The two versions of the stem-loop (WT and edited) are then compared as shown in Fig. 6. A standardized color scale represents the positional entropy for both transcript versions, with the color legend at the bottom left of the MFE structure. Finally, the calculated free energy and maximum expected accuracy (MEA) are provided at the bottom of the panel. The former, in particular, is used to give a quick estimate of changes in structural stability. For instance, in the case shown in Fig. 6, RNAfold reports a free energy value of −30.6 for the WT hsa-mir-411 version versus −37.2 for the transcript edited

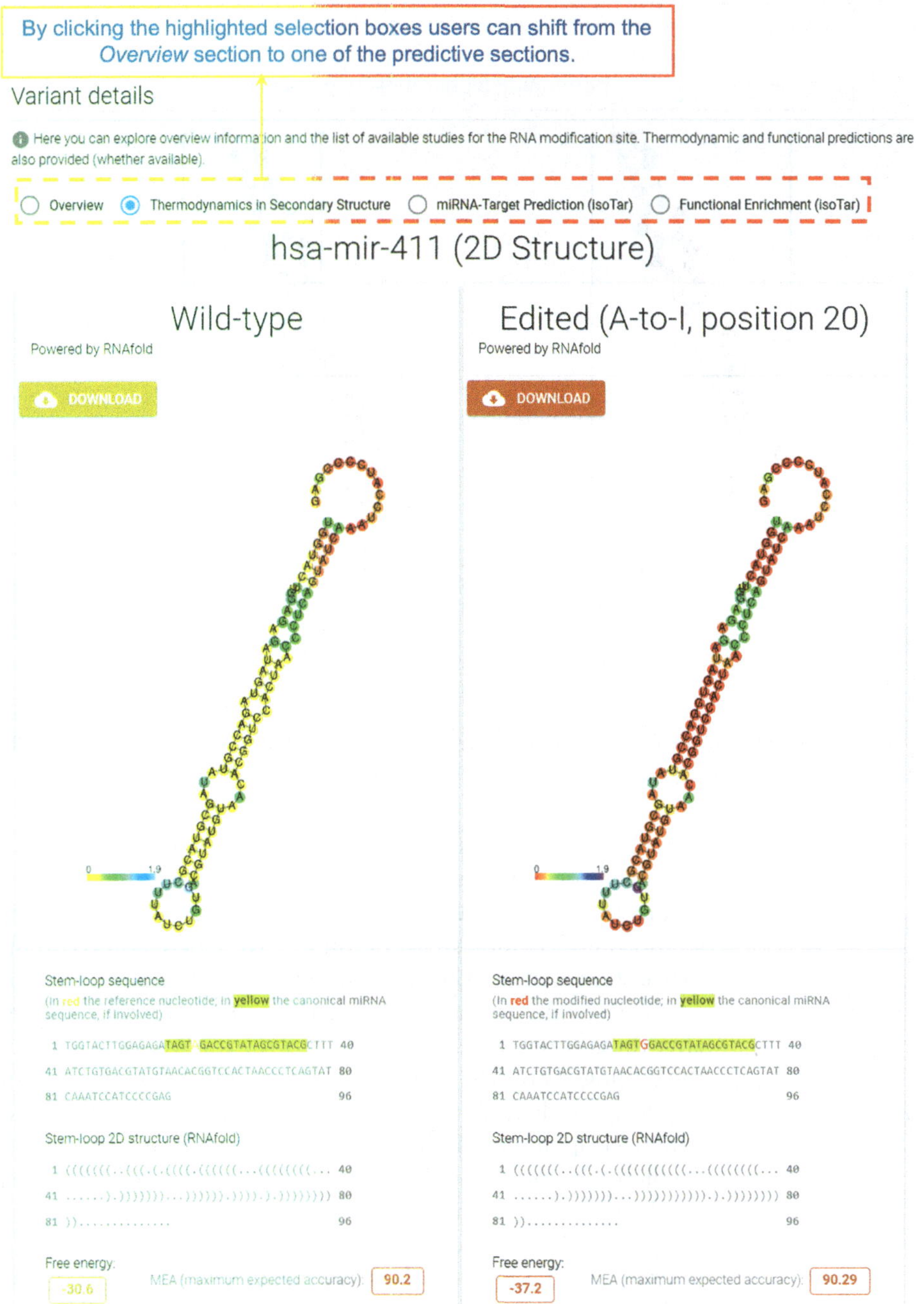

Fig. 6 The predictive module displays thermodynamic comparisons of RNA stem-loops (wild-type vs. edited) using RNAfold, including positional entropy, free energy, and structural stability estimates

at position 20 of the pre-miRNA sequence (corresponding to position 5 of the hsa-miR-411-5p), with an almost overlapping MEA estimate between the two predictions. This value indicates that the edited hsa-mir-411 stem-loop is assumed to be more stable than the

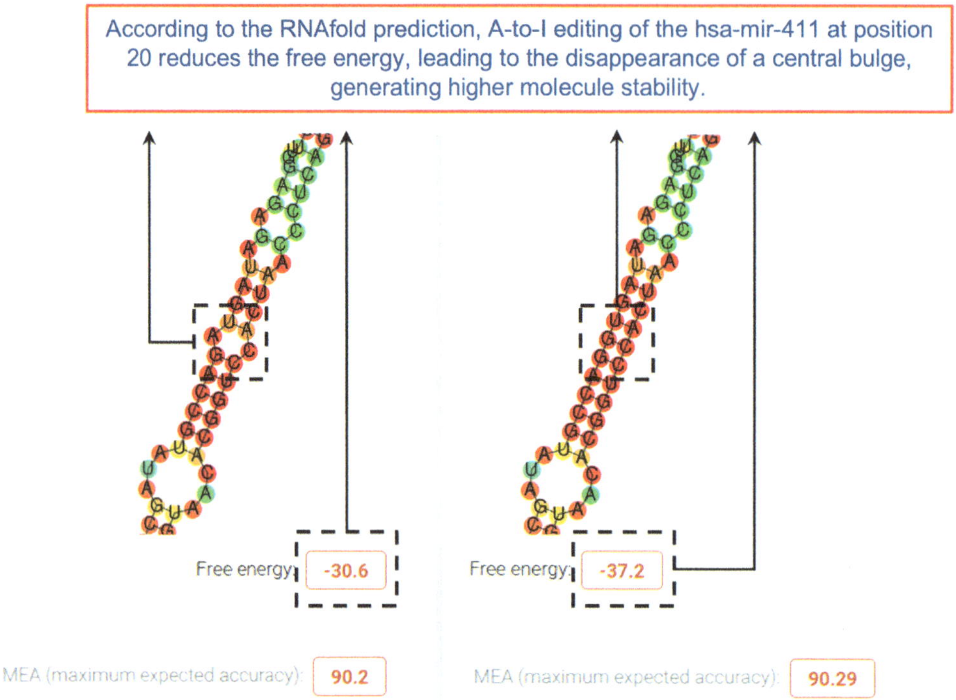

Fig. 7 Highlights of the comparison between the wild-type (left) and edited (right) versions of hsa-miR-411, focusing on the thermodynamic properties of the stem-loop in the predictive module

WT version. In fact, the tool predicts the disappearance of a bulge in the middle of the stem region, as an A-to-I editing at position 20 would result in a perfect match with the nucleotides on the opposite strand (Fig. 7). Such an event might have repercussions on miRNA maturation steps and/or miRNA-mRNA binding.

The second predictive section infers changes in miRNA targeting after the editing event. Obviously, this section will be available only if the editing event falls within the mature miRNA sequence. miRNA targeting predictions were performed by using our web-based containerized application isoTar [35]. By clicking the blue button ("Load prediction results") on the left side of the screen (Fig. 8), five tables will be shown. The three tables at the top of this section report the unique set of predicted WT miRNA targets (blue table, left side), the unique set of predicted edited miRNA targets (red table, right side), and the overlap set of predicted targets for both the miRNA versions (green table, in the middle), respectively (Fig. 8A). This setting is intended to represent a Venn diagram in a tabular format. The other two tables at the bottom list the predicted miRNA targets for both versions, with more detailed information. In particular, users can note the five prediction tools employed in IsoTar and the outcomes returned by each tool, with a consensus score shown in the far-right column (Fig. 8B). A threshold score of 3 was set as the minimum consensus in MiREDiBase.

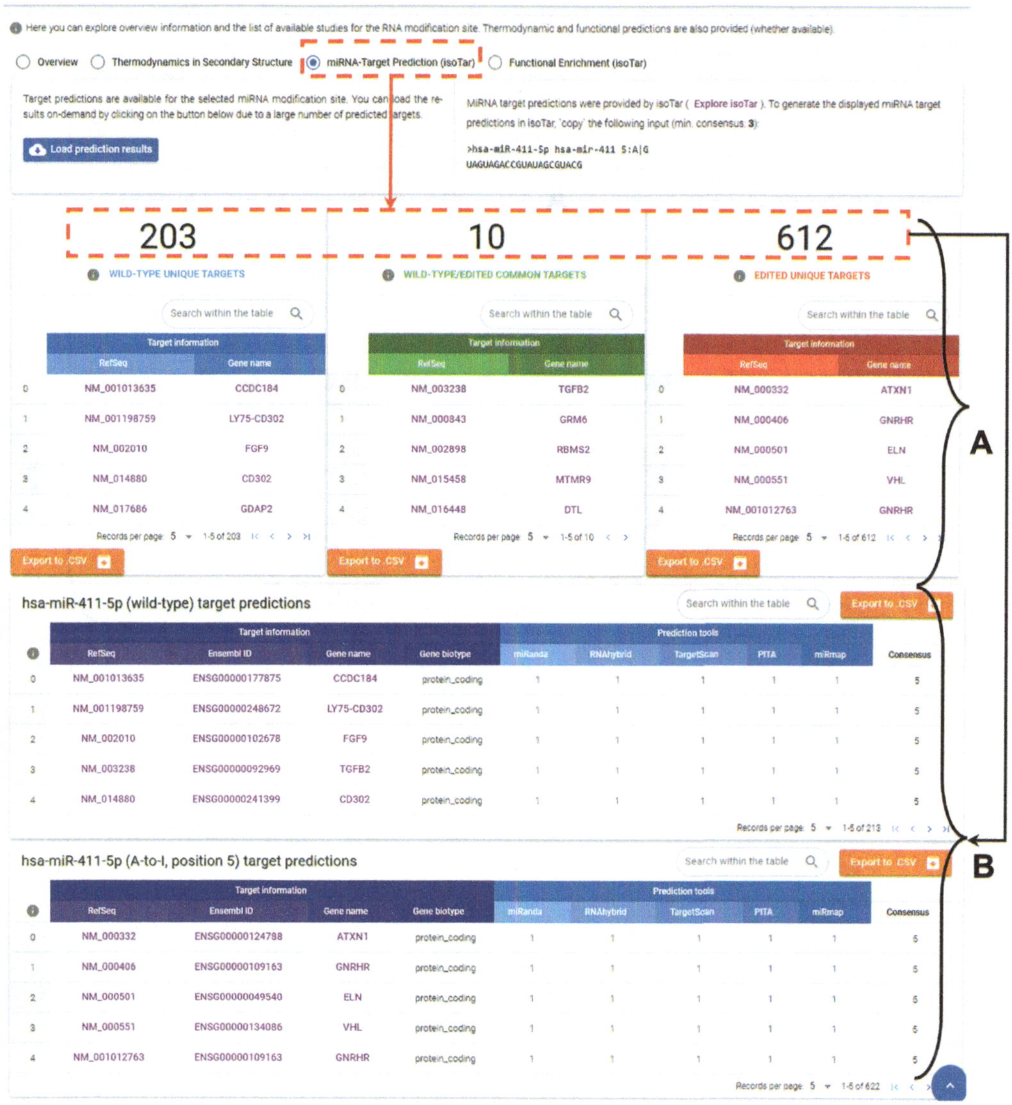

Fig. 8 Predictive module for miRNA targeting shifts post-editing, displaying unique and overlapping targets for wild-type and edited miRNAs, along with detailed predictions from five tools and consensus scores

The third predictive section is dedicated to the functional enrichment analysis performed by IsoTar. The analysis is based on the previous miRNA-target prediction analysis. Results are then ranked by p-value and adjusted p-value. By clicking the blue button ("Load enrichment results") on the left side of the screen, results will be displayed in a tabular format. At the top of the screen, users will visualize the GO panel, reporting results (GO terms) obtained by running IsoTar on the Gene Ontology (GO) database [36] (Fig. 9A). Likewise the "miRNA-Target prediction" section, here GO terms are firstly shown in three distinct tables: one table reporting the unique

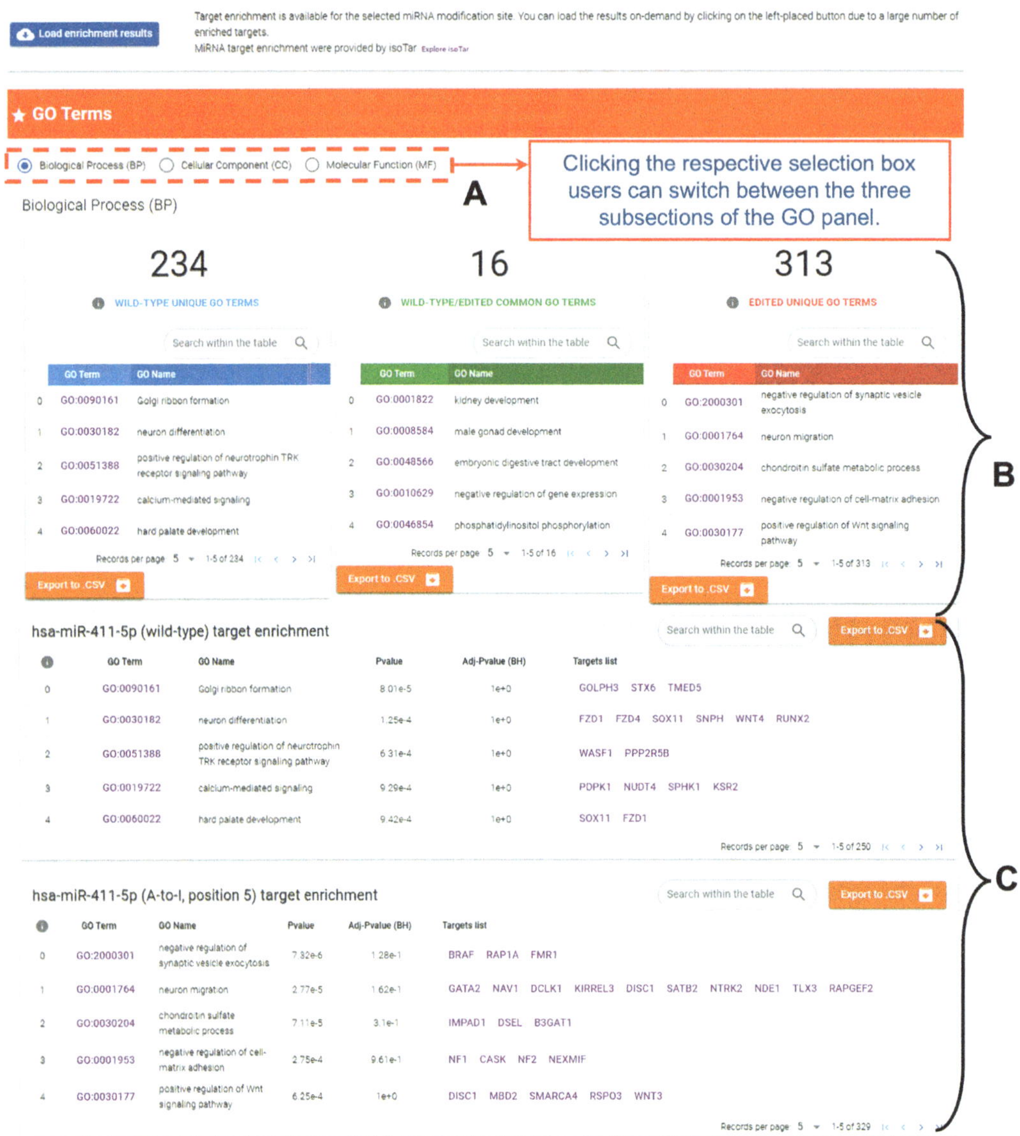

Fig. 9 Functional enrichment analysis module displaying gene ontology (GO) terms for wild-type and edited miRNAs, including unique and overlapping terms across biological processes, cellular components, and molecular functions. Results are ranked by p-value and adjusted *p*-value

set of GO terms where the WT miRNA is involved in (blue table, left side), another one listing the unique set of GO terms impacted by the edited miRNA (red table, right side) and the last one listing common GO terms for the two miRNA versions (green table, in the middle), (Fig. 9B) respectively. Right at the bottom of these three tables, users will see two additional tables listing, respectively, the predicted GO terms influenced by the two miRNA versions, with more detailed information (Fig. 9C). Importantly, the GO panel is

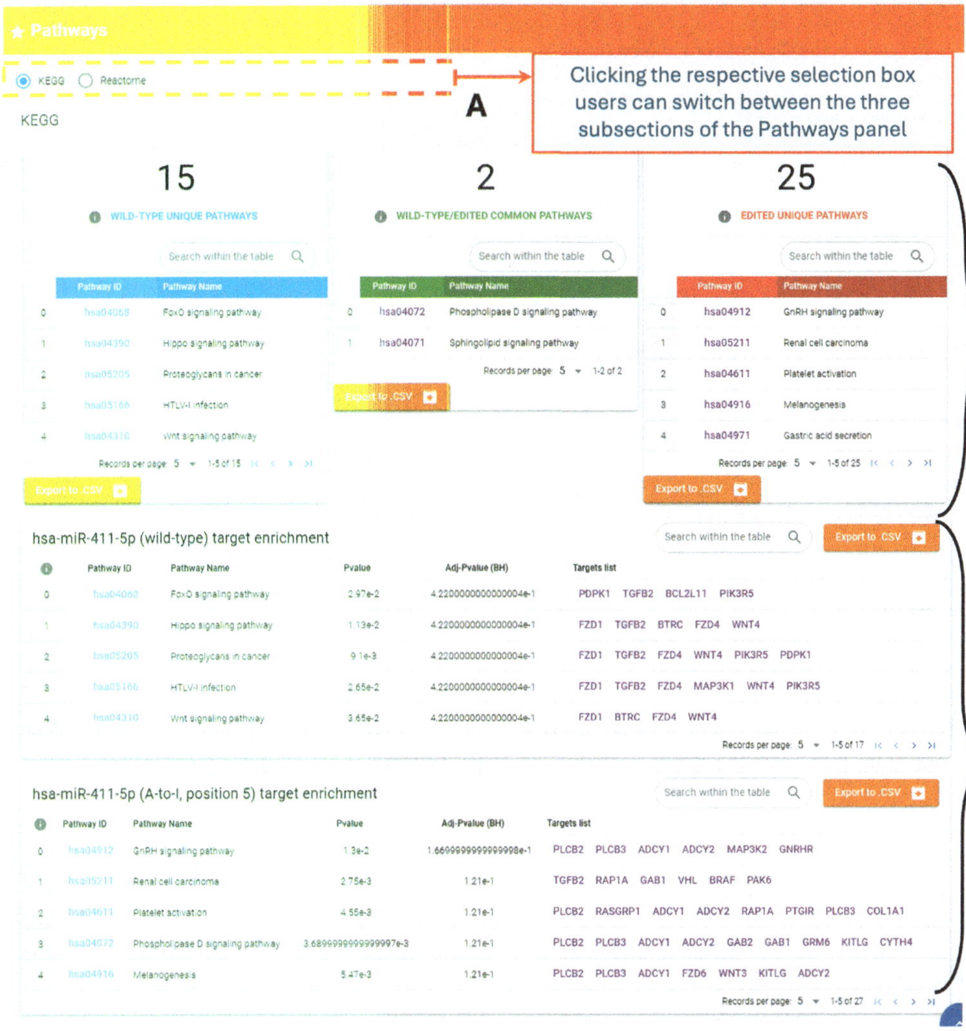

Fig. 10 Pathways panel displaying probabilistically impacted pathways for wild-type and edited miRNAs across KEGG and Reactome subsections, with results presented in tabular format for unique and overlapping pathways

composed of three subsections: "Biological processes," "Cellular components," and "Molecular function." Users can switch the subsections by clicking the respective selecting box at the top of the panel (Fig. 9A). For each subsection, the page will display the same tabular setting. Scrolling below the GO panel, users can visualize the Pathways panel (Fig. 10A). The setting of this panel mirrors that of the previous one, except that in the Pathways panel, the summary tables report pathways that are probabilistically impaired or influenced by the two miRNA versions (Fig. 10B, C). The Pathways panel consists of two subsections: "KEGG" [37] and "Reactome" [38]. Users can switch the subsections by clicking the respective selecting box at the top of the panel (Fig. 10A).

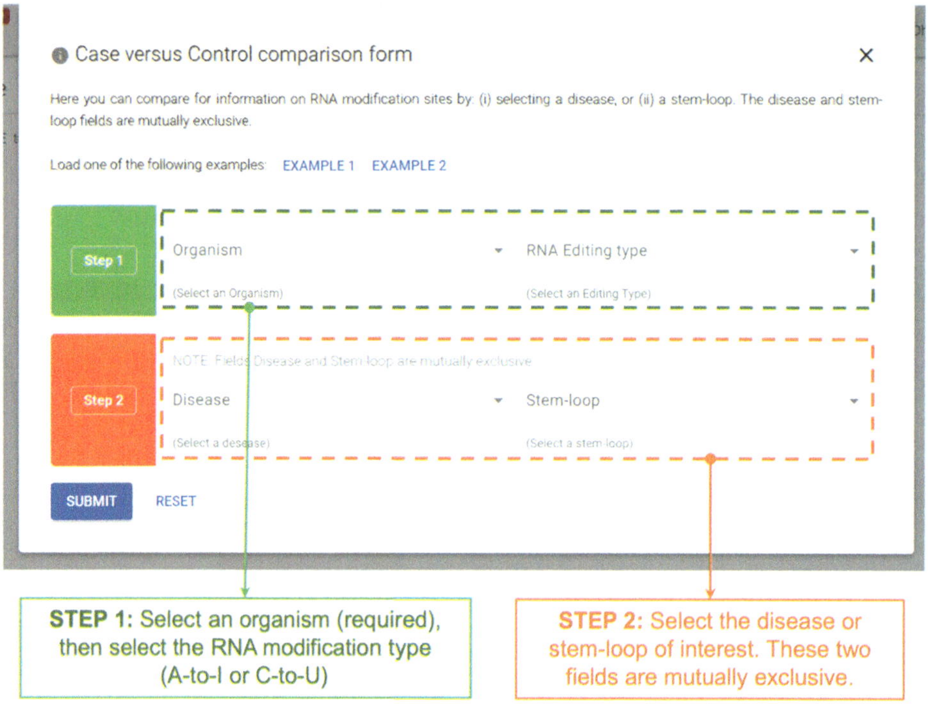

Fig. 11 The compare module interface allows users to select an organism, editing type, and either a disease condition or a stem-loop name to retrieve query results. Default examples are provided for guidance

As for results from the "Overview" section, results from predictive sections can be filtered by using the search bars, and all tables can be downloaded as CSV files by clicking the "Export" button. Purple inscriptions redirect users to external resources.

3.4 Advanced Features

Exploring the Comparison Module in MiREDiBase

Since RNA editing sites are often differentially edited under adverse versus normal conditions, our tool allows comparison of miRNA editing levels at specific editing sites across conditions. Thus, MiREDiBase is provided with a dedicated section, named "Compare," that allows for the evidence trends in miRNA editing levels under specific conditions. The "Compare" section icon, located right under the "Search" section icon in the left menu, allows users to retrieve intra-study comparisons of editing levels for the selected miRNA editing sites. The decision to restrict comparisons to site-specific editing levels within the same study, rather than allowing inter-study comparisons, was made to avoid bias from differences in experimental conditions, detection methods, and/or bioinformatic procedures.

After the Compare form appears (Fig. 11), users must select an organism (currently only human) and specify the editing type (A-to-I or C-to-U in the current version) (Step 1). Secondly, users must select either a disease condition or a stem-loop name from the

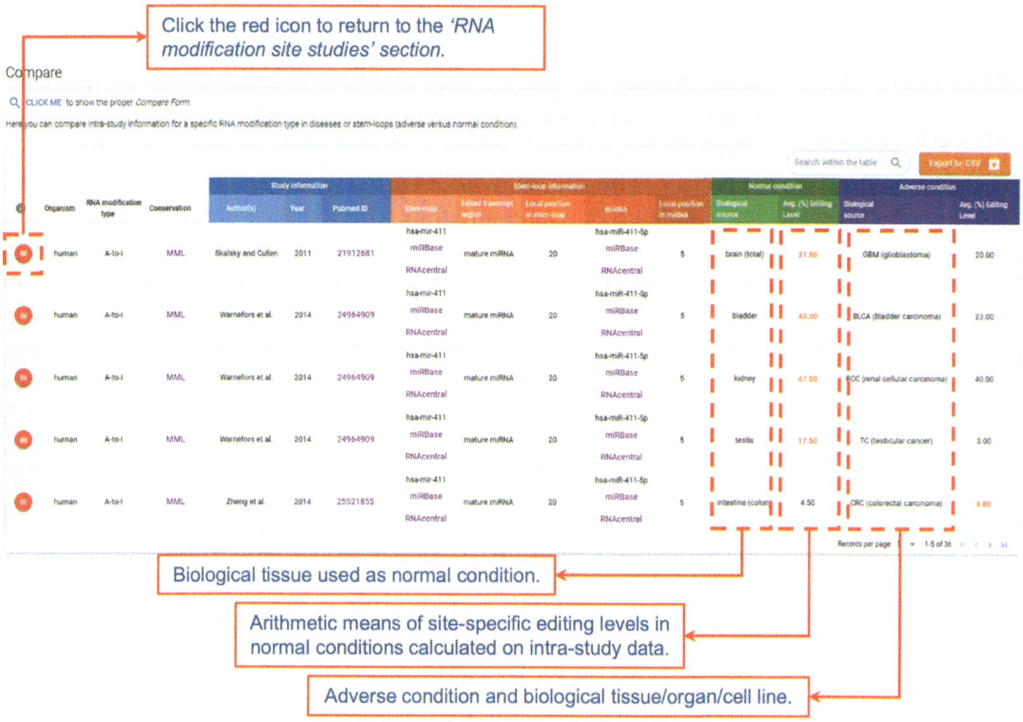

Fig. 12 Comparison table showing editing levels for specific sites across conditions, with arithmetic means highlighted, matched-control data, and detailed biological sources. Rows link to the "RNA modification site studies" module for further exploration

respective dropdown menu (Step 2). Finally, clicking the "Submit" button displays the query results. Note that the two fields of Step 2 are mutually exclusive. Users can load two distinct default examples to understand the Compare module's functionality better.

MiREDiBase will return an exportable table in CSV format that shows comparisons of editing levels for single editing sites. If the query is based on a pathological condition, comparisons will include editing sites retrieved from the literature for the specified condition, grouped by different pre-miRNAs. If the query is made using a pre-miRNA stem-loop, comparisons will consist only of editing sites within the selected stem-loops retrieved under various conditions. In any case, editing levels will also be reported for the matched control condition in the same study, and biological sources are always explicated (Fig. 12). Comparisons are made by taking into account the arithmetic means of editing levels reported for each condition. Within the comparison table, the arithmetic means are reported under the "Avg. (%) Editing Level" columns (Fig. 12) and were extracted from the respective studies. Row by row, the highest numerical value is colored in orange, while the lowest one is colored in black. By clicking the red button on the left side of each row, the system will open the "RNA modification site studies" module under the "Overview" section, as described previously (see Sect. 3.2).

Access to RESTful API for Advanced Data Retrieval and Integration

MiREDiBase provides seamless access to its data via a dedicated RESTful API [39], designed for advanced users seeking efficient data retrieval and integration. The API provides a high-performance interface that allows researchers to query and extract specific datasets directly from the platform without relying on the graphical user interface. This capability facilitates programmatic access to the database, enabling users to integrate MiREDiBase data into their custom workflows, bioinformatics pipelines, or external analytical tools. The API supports a variety of filtering parameters, including organism, modification type (e.g., A-to-I or C-to-U editing), genomic region, biological condition, and editing validation status.

The RESTful API is built using modern web technologies, including FastAPI, to ensure secure, fast, and scalable operations. Users can embed API calls into their code to retrieve structured information, including validated and putative miRNA editing events, target predictions, and enrichment analyses, along with metadata on experimental conditions. Furthermore, MiREDiBase's microservice architecture ensures continuous delivery and reliability, making the API a dependable tool for both exploratory research and large-scale data mining. Through this interface, users have the flexibility to automate data extraction processes and perform comparative analyses across normal and pathological conditions, significantly enhancing the usability of the database for researchers in bioinformatics and epitranscriptomics.

4 MiREDiBase Current Applications

Since the identification of disease-related miRNAs is highly significant for studying disease pathogenesis and drug development, MiREDiBase represents a valuable tool for cell biology and biomedical research. One of its kind, our tool serves as a facilitator for researchers working in epitranscriptomics, as retrieving information on miRNA editing from the literature can be laborious and time-consuming. Moreover, thanks to its Comparison section, it can quickly summarize correlations between several diseases and site-specific editing levels occurring in miRNAs and pre-miRNAs. As such, MiREDiBase has been implemented as an internal resource of miRTarBase [40, 41]—a manually curated database that serves as one of the main repositories of miRNA-target interactions—adding a new layer of complexity alongside other specialized resources. Data from MiREDiBase have also been integrated into the recently developed Non-Coding variant Annotation Database (NCAD [42]), a tool that aims to help researchers and clinicians interpret the biological roles of non-coding variants in various diseases. Data from the Comparison section and miRNA-disease correlations were found to be suitable for developing a Karnel Bayesian model that integrates multi-network information to improve inference performance in predicting miRNA-disease associations [43]. The same data might be suitable for developing certain machine learning approaches [44].

In a recent study investigating the impact of miRNA Isoforms in lung cancer [45], Distefano et al. employed data from MiREDiBase to annotate single-nucleotide variants in miRNA sequences detected by a high-throughput method. In a distinct study published in the same year, Distefano et al. [24] exploited IsoTar-derived predictions included in MiREDiBase to identify differentially regulated miRNA-targets in breast cancer. Here, after measuring the A-to-I RNA editing abundance across several cancer cohorts and tissues, the edited miR-381–3p (A-to-I RNA editing at position 4) resulted in one of the most diffused dysregulated edited miRNAs in cancers, including breast cancer. By intersecting the sets of significantly dysregulated genes with those identified by IsoTar, the authors obtained two reduced sets of potential direct targets for both versions of the selected miRNA (WT miR-381–3p and miR-381–3p_4_A_G). Among the predicted target genes, the authors validated UBE2C as a target of the wild-type miRNA and SYT13 as a target of the edited miRNA using *in vitro* experiments.

MiREDiBase might also be used for other biology-related purposes than biomedical research. An example is the study of the conservation of editing events across species as an evolutionary aspect. Indeed, while investigating the molecular mechanism of porcine fat deposition, Jiayu Lv et al. [46] very recently used data on miRNA editing from the human MiREDiBase dataset to assess the conservation of miRNA editing sites between humans and pigs. This demonstrates the versatility of our tool as an open online resource for clinical, biological, and molecular investigation.

5 Conclusion and Future Directions

MiREDiBase represents an advancement in RNA biology, providing the first comprehensive, dedicated resource for miRNA editing events. By integrating validated and putative editing data with advanced analytical tools, the database offers researchers unparalleled insights into how miRNA editing influences gene regulation and disease pathways. Its user-friendly interface, predictive functionalities, and robust integration with external resources make MiREDiBase an indispensable platform for epitranscriptomics research. Through its curated datasets and tools for comparative analysis, MiREDiBase has facilitated the identification of editing-driven miRNA re-targeting and highlighted the regulatory significance of editing in health and disease contexts. These contributions highlight the resource's potential to accelerate discoveries in biomarker research and therapeutic development.

MiREDiBase is set to become an even more versatile tool, helping researchers better understand the complex role of miRNA editing in biology and health science. Planned updates include adding the ability to edit data for new species, such as mice and Drosophila, enabling comparative evolutionary studies and expanding the database's applicability to diverse research models. Concurrently, efforts are underway to integrate newly published evidence, ensuring that MiREDiBase remains current and comprehensive. Functional annotations will also be expanded to include detailed analyses of how editing affects miRNA-mRNA interactions and structural stability. These enhancements aim to provide a richer context for understanding the biological consequences of miRNA editing. With these advancements, MiREDiBase is set to become an even more versatile tool, allowing researchers to uncover the intricate mechanisms of miRNA editing and its far-reaching implications in the biomedical field.

Acknowledgments

Giovanni Nigita was supported in part by the National Cancer Institute (National Institute of Health) grant (R21CA287180).

Conflict of Interest The authors have no conflicts of interest to declare that are relevant to the content of this chapter.

References

1. Kozomara A, Birgaoanu M, Griffiths-Jones S (2019) miRBase: from microRNA sequences to function. Nucleic Acids Res 47:D155–D162. https://doi.org/10.1093/nar/gky1141
2. Gebert LFR, Macrae IJ (2018) Regulation of microRNA function in animals. Nat Rev Mol Cell Biol 21:743. https://doi.org/10.1038/s41580-018-0045-7
3. Ameres SL, Zamore PD (2013) Diversifying microRNA sequence and function. Nat Rev Mol Cell Bio 14:475–488. https://doi.org/10.1038/nrm3611
4. Bartel DP (2009) MicroRNAs: target recognition and regulatory functions. Cell 136:215–233. https://doi.org/10.1016/j.cell.2009.01.002
5. Nam J-W, Rissland OS, Koppstein D et al (2014) Global analyses of the effect of different cellular contexts on MicroRNA targeting. Mol Cell 53:1031–1043. https://doi.org/10.1016/j.molcel.2014.02.013
6. Croce CM (2009) Causes and consequences of microRNA dysregulation in cancer. Nat Rev Genet 10:704–714. https://doi.org/10.1038/nrg2634
7. Condrat CE, Thompson DC, Barbu MG et al (2020) miRNAs as biomarkers in disease: latest findings regarding their role in diagnosis and prognosis. Cells 9:276. https://doi.org/10.3390/cells9020276
8. Rupaimoole R, Slack FJ (2017) MicroRNA therapeutics: towards a new era for the management of cancer and other diseases. Nat Rev Drug Discov 16:203–222. https://doi.org/10.1038/nrd.2016.246
9. Nishikura K (2016) A-to-I editing of coding and non-coding RNAs by ADARs. Nat Rev Mol Cell Biol 17:83–96. https://doi.org/10.1038/nrm.2015.4
10. Nigita G, Veneziano D, Ferro A (2015) A-to-I RNA editing: current knowledge sources and computational approaches with special emphasis on non-coding RNA molecules. Front Bioeng Biotechnol 3:37. https://doi.org/10.3389/fbioe.2015.00037

11. Nigita G, Marceca GP, Tomasello L et al (2019) ncRNA editing: functional characterization and computational resources. Methods Mol Biology 1912:133–174. https://doi.org/10.1007/978-1-4939-8982-9_6
12. Nishikura K (2006) Editor meets silencer: crosstalk between RNA editing and RNA interference. Nat Rev Mol Cell Biol 7:919–931. https://doi.org/10.1038/nrm2061
13. Nishikura K (2016) A-to-I editing of coding and non-coding RNAs by ADARs. Nat Rev Mol Cell Biol 17:83–96. https://doi.org/10.1038/nrm.2015.4
14. Liddicoat BJ, Chalk AM, Walkley CR (2016) ADAR1, inosine and the immune sensing system: distinguishing self from non-self. Wiley Interdiscip Rev RNA 7:157–172. https://doi.org/10.1002/wrna.1322
15. Ahmad S, Mu X, Hur S (2021) The role of RNA editing in the immune response. Methods Mol Biology 2181:287–307. https://doi.org/10.1007/978-1-0716-0787-9_17
16. Blanc V, Davidson NO (2003) C-to-U RNA editing: mechanisms leading to genetic diversity. J Biol Chem 278:1395–1398. https://doi.org/10.1074/jbc.r200024200
17. Gu T, Buaas FW, Simons AK et al (2012) Canonical A-to-I and C-to-U RNA editing Is enriched at 3′ UTRs and microRNA target sites in multiple mouse tissues. PLOS One 7:e33720. https://doi.org/10.1371/journal.pone.0033720.g003
18. Bertotti S, Fleming I, Cámara M, de los M et al (2022) Characterization of ADAT2/3 molecules in Trypanosoma cruzi and regulation of mucin gene expression by tRNA editing. Biochem J 479:561–580. https://doi.org/10.1042/bcj20210850
19. Norden MV, Falls Z, Mandloi S et al (2024) The implications of APOBEC3-mediated C-to-U RNA editing for human disease. Commun Biol 7:529. https://doi.org/10.1038/s42003-024-06239-w
20. Paul D, Sinha AN, Ray A et al (2017) A-to-I editing in human miRNAs is enriched in seed sequence, influenced by sequence contexts and significantly hypoedited in glioblastoma multiforme. Scientific Rep 7:2466. https://doi.org/10.1038/s41598-017-02397-6
21. Cesarini V, Silvestris DA, Tassinari V et al (2018) ADAR2/miR-589-3p axis controls glioblastoma cell migration/invasion. Nucleic Acids Res 46:2045–2059. https://doi.org/10.1093/nar/gkx1257
22. Liu W-H, Chen C-H, Yeh K-H et al (2013) ADAR2-mediated editing of miR-214 and miR-122 precursor and antisense RNA transcripts in liver cancers. PLOS One 8:e81922. https://doi.org/10.1371/journal.pone.0081922.s002
23. Kim HS, Na MJ, Son KH et al (2023) ADAR1-dependent miR-3144-3p editing simultaneously induces MSI2 expression and suppresses SLC38A4 expression in liver cancer. Exp Mol Med 55:95–107. https://doi.org/10.1038/s12276-022-00916-8
24. Distefano R, Tomasello L, Vinciguerra GLR et al (2022) Pan-cancer analysis of canonical and modified miRNAs enhances the resolution of the functional miRNAome in cancer. Cancer Res 82:3687–3700. https://doi.org/10.1158/0008-5472.can-22-0240
25. Nigita G, Distefano R, Veneziano D et al (2018) Tissue and exosomal miRNA editing in non-small cell lung cancer. Scientific Rep 8: 10222. https://doi.org/10.1038/s41598-018-28528-1
26. Pinto Y, Buchumenski I, Levanon EY, Eisenberg E (2018) Human cancer tissues exhibit reduced A-to-I editing of miRNAs coupled with elevated editing of their targets. Nucleic Acids Res 46:71–82. https://doi.org/10.1093/nar/gkx1176
27. Picardi E, D'Erchia AM, Giudice CL, Pesole G (2017) REDIportal: a comprehensive database of A-to-I RNA editing events in humans. Nucleic Acids Res 45:D750–D757. https://doi.org/10.1093/nar/gkw767
28. Zhu H, Huang L, Liu S et al (2022) REIA: a database for cancer A-to-I RNA editing with interactive analysis. Int J Biol Sci 18:2472–2483. https://doi.org/10.7150/ijbs.69458
29. Lin C-H, Chen SC-C (2019) The cancer editome atlas: a resource for exploratory analysis of the adenosine-to-inosine RNA editome in cancer. Cancer Res 3501:2018. https://doi.org/10.1158/0008-5472.can-18-3501
30. Ruan H, Li Q, Liu Y et al (2021) GPEdit: the genetic and pharmacogenomic landscape of A-to-I RNA editing in cancers. Nucleic Acids Res 50:D1231–D1237. https://doi.org/10.1093/nar/gkab810
31. Marceca GP, Distefano R, Tomasello L et al (2021) MiREDiBase, a manually curated database of validated and putative editing events in microRNAs. Sci Data 8:199. https://doi.org/10.1038/s41597-021-00979-8
32. Kiran AM, O'Mahony JJ, Sanjeev K, Baranov PV (2013) Darned in 2013: inclusion of model organisms and linking with Wikipedia 41:D258–D261. https://doi.org/10.1093/nar/gks961
33. Ramaswami G, Li JB (2014) RADAR: a rigorously annotated database of A-to-I RNA editing. Nucleic Acids Res 42:D109–D113. https://doi.org/10.1093/nar/gkt996
34. Lorenz R, Bernhart SH, Siederdissen CHzu et al (2011) ViennaRNA package 2.0. Algorithms Mol Biol 6:1–14. https://doi.org/10.1186/1748-7188-6-26

35. Distefano R, Nigita G, Veneziano D et al (2019) isoTar: consensus target prediction with enrichment analysis for MicroRNAs harboring editing sites and other variations. MicroRNA TI 1970:211–235. https://doi.org/10.1007/978-1-4939-9207-2_12
36. Harris MA, Clark J, Ireland A et al (2004) The gene ontology (GO) database and informatics resource. Nucleic Acids Res 32:D258–D261. https://doi.org/10.1093/nar/gkh036
37. Kanehisa M, Furumichi M, Tanabe M et al (2017) KEGG: new perspectives on genomes, pathways, diseases and drugs. Nucleic Acids Res 45:D353–D361. https://doi.org/10.1093/nar/gkw1092
38. Fabregat A, Jupe S, Matthews L et al (2018) The reactome pathway knowledgebase. Nucleic Acids Res 46:D649–D655. https://doi.org/10.1093/nar/gkx1132
39. Fielding RT, Taylor RN (2002) Principled design of the modern web architecture. ACM Trans Internet Technol 2:115–150. https://doi.org/10.1145/514183.514185
40. Huang H-Y, Lin Y-C-D, Cui S et al (2021) miRTarBase update 2022: an informative resource for experimentally validated miRNA–target interactions. Nucleic Acids Res 50:D222–D230. https://doi.org/10.1093/nar/gkab1079
41. Cui S, Yu S, Huang H-Y et al (2024) miRTarBase 2025: updates to the collection of experimentally validated microRNA–target interactions. Nucleic Acids Res 53:D147–D156. https://doi.org/10.1093/nar/gkae1072
42. Feng X, Liu S, Li K et al (2024) NCAD v1.0: a database for non-coding variant annotation and interpretation. J Genet Genom 51:230–242. https://doi.org/10.1016/j.jgg.2023.12.005
43. Ma Y, Ma Y (2024) Kernel Bayesian logistic tensor decomposition with automatic rank determination for predicting multiple types of miRNA-disease associations. PLOS Comput Biol 20:e1012287. https://doi.org/10.1371/journal.pcbi.1012287
44. Thomas SD, Vijayakumar K, John L et al (2024) Machine learning strategies in microRNA research: bridging genome to phenome. OMICS A J Integr Biol 28:213–233. https://doi.org/10.1089/omi.2024.0047
45. Distefano R, Nigita G, Le P et al (2022) Disparities in lung cancer: miRNA isoform characterization in lung adenocarcinoma. Cancers 14:773. https://doi.org/10.3390/cancers14030773
46. Lv J, Yang F, Li Y et al (2024) Characterization and function analysis of miRNA editing during fat deposition in Chinese indigenous Ningxiang pigs. Vet Sci 11:183. https://doi.org/10.3390/vetsci11040183

Chapter 7

Construction and Analysis of miRNA Regulatory Networks

Simone Avesani, Simone Caligola and Rosalba Giugno

Abstract

Unveiling the role of miRNAs in regulating gene expression is crucial for developing the most effective drugs and therapies, for example, in cancer. To this end, the complex interplay between miRNAs and transcription factors, ncRNAs, genetic/epigenetic events, diseases, and drugs must be taken into consideration. The aim of this chapter is to illustrate the usage of specialized web platforms that aim to dissect this complexity through the construction and analysis of miRNA regulatory networks.

Keywords miRNA regulatory networks, miRNA-target networks, Network analysis, Web applications

1 Introduction

miRNAs are small non-coding RNAs (~22 nucleotides) which have been found to be deeply involved in the process of gene expression regulation [1]. A miRNA can target several genes, and a single gene can be targeted by more than one miRNA. miRNAs and target genes are inversely correlated in terms of expression due to the mechanism of gene repression mediated by miRNAs. Therefore, considering the expression values in the network topology can be useful to understand the dynamics of the gene regulation system. The regulatory activity of miRNAs has also been linked to other ncRNAs, transcription factors (TFs), epigenetic events, diseases, drugs, and phenotypes that can be viewed as additional nodes of the network. The combination of such interactions forms complex heterogeneous systems that can be analyzed using Systems Biology approaches. For example, gene regulatory networks, composed of TF-gene interactions, can be enriched with miRNA targeting information that adds another layer of complexity [2]. This representation allowed, for example, the investigation of complex biological phenomena that were linked to cell fate [3].

This chapter illustrates the usage of several web tools designed for the miRNA regulatory network construction and analysis:

Alessandro Laganà and Giovanni Nigita (eds.), *MicroRNA Target Identification*, Methods in Molecular Biology, vol. 3056,
https://doi.org/10.1007/978-1-0716-5392-0_7,

miRNet [4], MIENTURNET [5], miRTargetLink 2.0 [6], FFLtool [7], miRViz [8], and Arena-Idb [9]. In the Methods section, all the different software tools are described in-depth, and all their features are shown with a case study using gene expression data from lung adenocarcinoma of the Cancer Genome Atlas (TCGA-LUAD). This chapter provides an overview of cutting-edge web platforms for exploring miRNA regulatory networks that do not require specialized computational expertise. However, the chapter is accompanied by some lines of R code that were used for data download and processing.

2 Materials

R (version 4.4.1) scripts have been executed on a machine with CPU 2.6 GHz 6-Core Intel Core i7, RAM 16 GB 2667 MHz DDR5 with macOS Sonoma 14.5. The web platforms have been used with Google Chrome (version 129.0.6668.91).

3 Methods

Web platforms are useful tools for studying miRNA regulatory networks. These tools differ for input type (e.g., miRNA, gene), type of target interactions (experimentally validated or predicted), network interactions (e.g., miRNA-gene, miRNA-pathway), number of organisms supported, and possibility to perform functional enrichment and network-based analyses (Table 1). miRNet is an all-in-one tool for exploring miRNA functions. It allows users to study different types of miRNA interactions, such as miRNA-gene, miRNA-disease, and miRNA-SNPs in 28 different organisms, with the possibility to create a heterogeneous network. In addition, it allows pathway enrichment using well-known biological databases and network analysis to discover central nodes and modules. MIENTURNET is a web platform for studying miRNA-gene interactions from 6 different organisms starting from a list of miRNAs or genes. It also allows users to perform pathway enrichment and network analysis giving information about node properties such as degree, closeness, and betweenness. miRTargetLink 2.0 takes in input miRNAs, genes, or pathways from 3 organisms to obtain predicted and validated interactions. It enables the creation of simple (e.g., miRNA-gene) or heterogeneous networks (i.e., miRNA-gene-pathway) and performs enrichment analysis using target genes or miRNAs. FFLtool is a web application designed to explore interactions among miRNAs, genes, and transcription factors (TFs). Specifically, it identifies regulatory feed-forward loops (FFLs) within TF-miRNA-gene networks in the human context. miRViz is an intuitive tool designed for exploring miRNA–miRNA interactions that does not require computational

Table 1
Web platforms for the construction and analysis of miRNA regulatory networks and their key features

Web application	Input				Targets		Network interactions	Number of organisms	Pathway enrichment	Network properties
	miRNA	Gene	Expression data	Other input nodes	Experimentally validated	Pred-icted				
miRNet 2.0 [4]	x	x	x	Drug, epigenetic, ncRNA, disease, SNPs, xeno, TFs	x	x	Heterog-eneous	28	x	x
MIENTURNET [5]	x	x	–	–	x	x	miRNA-gene	6	x	x
miRTargetLink 2.0 [6]	x	x	–	Pathway	x	x	Heterog-eneous	3	x	–
FFLtool [7]	x	x	–	TFs	x	–	Heterog-eneous	1	x	x
miRViz [8]	x	–	x	–	–	–	miRNA-miRNA	11	–	x
Arena-Idb [9]	x	x	–	ncRNA, pseudogene, phenotype	x	–	Heterog-eneous	1	–	–

"Heterogeneous" network type, here, means that "other input nodes" can be added to the canonical miRNA-gene network skills. It enables the analysis of miRNA interactions across various network models, where miRNAs are linked based on shared target genes or proximity within specific genomic regions. Lastly, Arena-Idb is a web platform designed to explore non-coding RNA (ncRNA) regulatory networks, which allows users to explore interactions between miRNAs, ncRNA, genes, pseudogenes, and phenotypes.

Gene expression data of lung adenocarcinoma coming from the Cancer Genome Atlas (TCGA-LUAD) (https://portal.gdc.cancer.gov/projects/TCGA-LUAD) were used as a case study to demonstrate the functionalities of the aforementioned tools. RNA-seq data from tumor and adjacent normal tissue of 58 TCGA-LUAD patients were downloaded using the TCGABiolinks [10] R/Bioconductor package (*see* **Note 1**). The case study data were downloaded and analyzed using the version of the web application available in November 2024. As the tools may have been updated since then, analysis results obtained in subsequent months may differ slightly.

3.1 mirNet 2.0

miRNet 2.0 (https://www.mirnet.ca/miRNet/home.xhtml) is the updated version of the former web-based platform introduced in 2016 [11]. miRNet is a comprehensive analytics tool for exploring miRNA functions using a network-based approach. miRNet differs from the other web platforms for its possibility to integrate different molecular, functional, and pathological information about miRNA interactions from different tissues and species. All this information can be integrated into different networks or into a single network. In particular, miRNet takes as input three types of data: expression data, databases of miRNA interactions, and user query lists. Expression data from RNA-seq, microarray, proteomics, and RT-qPCR experiments can be uploaded in the form of text files to perform differential expression and time series/dose response analysis. miRNet 2.0 integrates the latest version of miRBase [12], miRTarBase [13], and TarBase [14]. Further, miRNet allows to explore miRNA-disease interactions from miR2Disease [15], HMDD 4.0 [16], and PhenomiR [17], miRNA-ncRNA interactions from starBase [18], and Xeno-miRs interactions from 15 studies. In addition, miRNet 2.0 has added support for transcription factors integrating TransmiR [19], TRRUST [20], RegNetwork [21], ENCODE [22], JASPAR [23], and ChEA3 [24] databases and single-nucleotide polymorphisms integrating ADmiRe [25], PolymiRTS [26], and SNP2TFBS [27]. miRNet also supports tissue-specific interactions integrating TSmiR [28] and IMOTA [29] databases. The different input types available in miRNet with the corresponding analysis modules are shown in Fig. 1.

In the next sections, we will show how to construct and analyze miRNA regulatory networks starting from gene expression of TCGA-LUAD, exploiting different modules of miRNet to upload data, perform enrichment analysis, retrieve miRNA-gene and miRNA-drug interactions, and network analysis.

Import and Analysis of Expression Data

From the miRNet modules shown in Fig. 1, the user can select "Expression table" to upload an expression matrix. Of note, the expression module of miRNet is based on ExpressAnalyst [30], i.e., a complete platform for gene expression analysis. From the upload page, the user must specify different information: 1 of 28 different organisms, analysis type (Differential expression, time series/dose response), data type (RNA-seq, microarray, proteomics), and the ID type of the features (e.g., ENTREZ, SYMBOL). The user can also upload a custom library of gene IDs using the section "(Optional) Custom library." In that case, the section "ID type" must be left "not specified" (Fig. 2).

To retrieve differentially expressed genes in LUAD compared to the normal adjacent tissue, we select "H. sapiens" as organism, "Differential Expression" as analysis type, "Counts (bulk RNA-seq)" as data type, and "Official Gene Symbol" as ID type. Next, the count

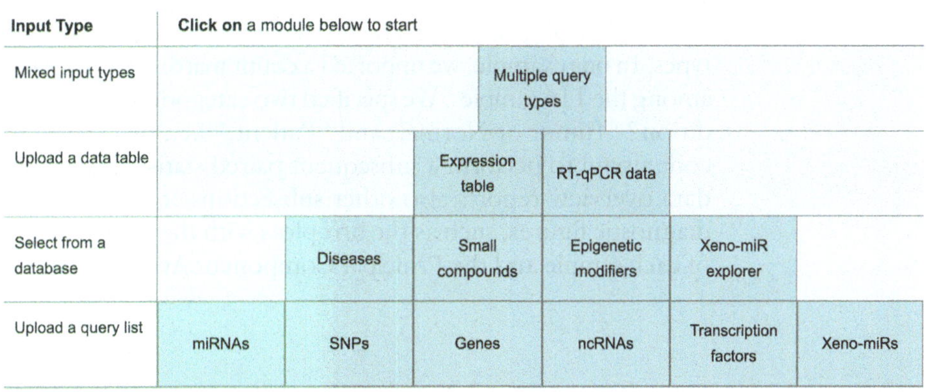

Fig. 1 Input types and analysis modules of miRNet

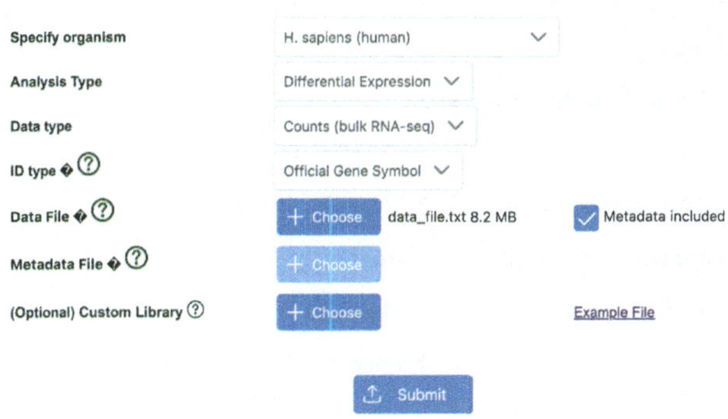

Fig. 2 Upload of expression data in miRNet

matrix previously obtained (*see* **Note 1**) is uploaded in the section "Data File." The count matrix must be a tab-delimited text file with a proper format. The first row comprises the sample names in the format *#NAME, Sample1, Sample2, ..., SampleN*. Starting from the second line, the user can, optionally, specify several metadata associated with the samples in the format *#CLASS, Meta1, Meta2, ..., MetaN*. After metadata, each row must have an ID type and the expression intensities for each sample. If the expression data includes metadata, the user must select the proper checkbox before submitting the page. Of note, metadata can be uploaded as a separate file using the section "Metadata File." After submitting the upload module and clicking the button "Proceed" at the bottom right of the page, miRNet will show the data quality check page. This page includes 2 main sections: omics data overview and metadata overview.

Omics data overview summarizes useful information about the count matrix, such as the total number of features, the number of samples, and the average counts per sample (Fig. 3). It also reports the number of grouping factors specified in the metadata and their types. In our example, we imported a count matrix with 19,048 genes among the 116 samples. We specified two categorical variables, "Condition" (Tumor or Normal) and "Patient," i.e., the TCGA patient code useful to perform a subsequent paired statistical analysis. The data overview reports also other subsections containing common diagnostic figures, such as the box plots with the count distribution of each sample and the Principal Component Analysis (PCA).

Data type:	RNA count table
Total feature number:	19048
Matched feature number:	18697 (98%)
Sample number:	116
Number of experimental factors:	2
Total read counts:	4.52e+09
Average counts per sample:	3.89e+07
Maximum counts per sample:	1.01e+08
Minimum counts per sample:	1.68e+07
Group names:	2 discrete factors: Condition, Patient;

Fig. 3 Summary of the uploaded RNA-seq data in ExpressAnalyst

Metadata overview reports a table with the metadata associated with the samples. In Fig. 4, we report page 1 of 20. The table can be modified using the icons in the last two columns to change sample names, metadata variable type (discrete or continuous), or remove samples. After quality control, clicking the button "Proceed" will show the section "Data Filtering and Normalization" (Fig. 5). Unannotated, low-abundant features (i.e., low number of counts) and features with a slight variation among the samples can be discarded from the analysis using this section. miRNet provides different normalization strategies: log2-counts per million (logCPM) transformation, upper quantile normalization (UQ), trimmed mean of M-values (TMM) normalization, and relative log expression (RLE) normalization. In addition, the user can choose not to normalize data by selecting "None." After keeping the default

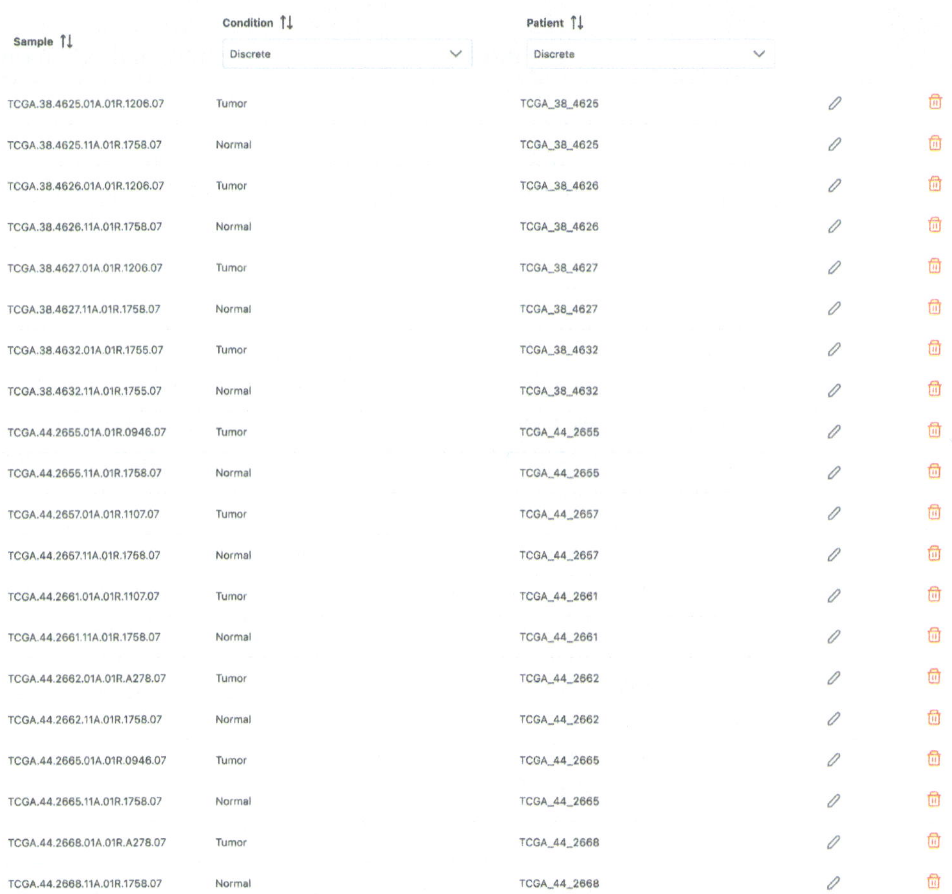

Sample ⇅	Condition ⇅ (Discrete)	Patient ⇅ (Discrete)
TCGA.38.4625.01A.01R.1206.07	Tumor	TCGA_38_4625
TCGA.38.4625.11A.01R.1758.07	Normal	TCGA_38_4625
TCGA.38.4626.01A.01R.1206.07	Tumor	TCGA_38_4626
TCGA.38.4626.11A.01R.1758.07	Normal	TCGA_38_4626
TCGA.38.4627.01A.01R.1206.07	Tumor	TCGA_38_4627
TCGA.38.4627.11A.01R.1758.07	Normal	TCGA_38_4627
TCGA.38.4632.01A.01R.1755.07	Tumor	TCGA_38_4632
TCGA.38.4632.11A.01R.1755.07	Normal	TCGA_38_4632
TCGA.44.2655.01A.01R.0946.07	Tumor	TCGA_44_2655
TCGA.44.2655.11A.01R.1758.07	Normal	TCGA_44_2655
TCGA.44.2657.01A.01R.1107.07	Tumor	TCGA_44_2657
TCGA.44.2657.11A.01R.1758.07	Normal	TCGA_44_2657
TCGA.44.2661.01A.01R.1107.07	Tumor	TCGA_44_2661
TCGA.44.2661.11A.01R.1758.07	Normal	TCGA_44_2661
TCGA.44.2662.01A.01R.A278.07	Tumor	TCGA_44_2662
TCGA.44.2662.11A.01R.1758.07	Normal	TCGA_44_2662
TCGA.44.2665.01A.01R.0946.07	Tumor	TCGA_44_2665
TCGA.44.2665.11A.01R.1758.07	Normal	TCGA_44_2665
TCGA.44.2668.01A.01R.A278.07	Tumor	TCGA_44_2668
TCGA.44.2668.11A.01R.1758.07	Normal	TCGA_44_2668

Fig. 4 Metadata of the uploaded RNA-seq samples

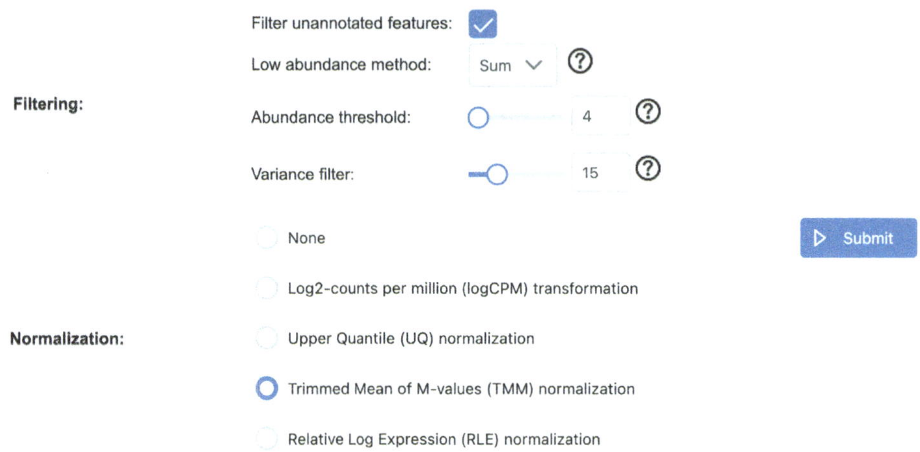

Fig. 5 Data filtering and normalization of RNA-seq data

filtering parameters and selecting the normalization method, in our case, TMM normalization, the configuration is saved by clicking the submit button. Diagnostic figures (e.g., box plot, PCA) similar to the ones shown in the quality control page will be shown after the application of filtering and normalization. A mean-variance plot will also be shown to check the dependency between counts and variance in the experiment.

After submitting the page, the next section is the "Differential Expression Analysis". Two types of analysis are possible for simple or more complex experimental designs using the two tabs "Simple Metadata" and "Complex Metadata" (Fig. 6). In the Simple Metadata tab, the user can select between 3 statistical methods: limma [31], EdgeR [32], and DESeq2 [33] with the possibility to adjust the prior variance using a robust intensity-dependent trend. In the study design, the user can specify the primary factor of interest and a secondary factor that can be a blocking factor. In the section "Comparison of Interest," the user can specify the comparisons of interest and the proper controls. In the Complex Metadata tab, the user can choose the primary and control groups or perform all comparisons in an ANOVA-like fashion. The user can also specify different covariates to control for and a blocking factor. In our example, we specify a simple design using limma and select "Condition" as primary factor and "Patient" as secondary blocking factor. After selecting "Tumor vs Normal" comparison and clicking Submit and Proceed, the page with the significant features is shown (Fig. 7). In this page, the user can change the *p*-value and log2 fold change thresholds and download the significant results using the "Download" button. The differentially expressed features are shown as a table in which it is possible to sort based on the different statistical indicators, search for specific features, and download a figure for each feature. In our

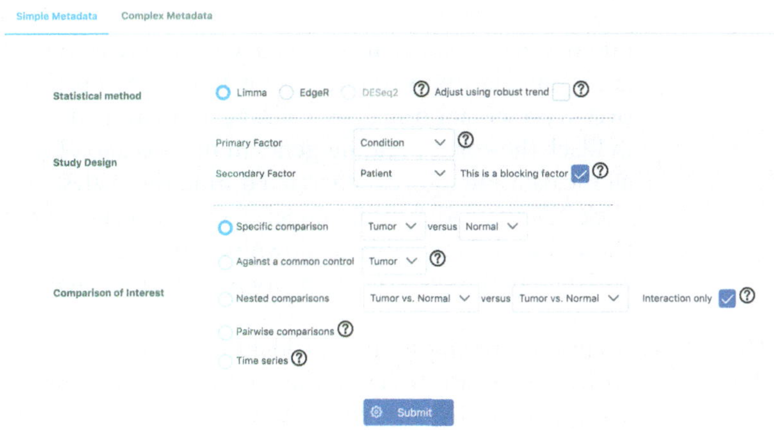

Fig. 6 Experimental design of RNA-seq data

Fig. 7 Differentially expressed genes of RNA-seq data

example, we select only genes up/down-regulated with a log2 fold change of 1 and *p*-value cutoff of 0.05. With these parameters, 3829 differentially expressed genes are obtained. After differential analysis, several graphical and functional analyses can be performed in the next section (Fig. 8A). For example, selecting the section "Volcano Plot," an interactive scatter plot will be shown with up-, down-regulated, and not significant genes (Fig. 8B). In addition, on the left of the page, an enrichment analysis is shown with the possibility to choose different databases such as KEGG, GO, and Reactome. Selecting the latter and clicking submit, we obtain enriched pathways following over-representation analysis. Clicking a specific pathway will highlight in black the corresponding genes in the volcano plot. To download all the data and figures generated during the analysis, it is possible to click "Downloads" in the upper bar of the page (Fig. 8B). Among these files, it is possible to download the CSV file containing the enriched pathways that can be used for further analyses.

Construction and Analysis of miRNA Regulatory Networks

An interesting question is the identification of miRNAs that control pathways in cancer. Two important pathways enriched in tumor samples of LUAD are "G2/M Checkpoints" and "Cell junction organization." The G2/M checkpoints pathway is fundamental for cell cycle and its dysregulation may cause tumorigenesis [34]. Cell junction organization is critical to maintain intercellular adhesion and integrity of epithelial tissue, and the down-modulation of its proteins is essential for epithelial-mesenchymal transition (EMT), a hallmark of cancer progression and metastasis [35]. From the CSV file of enriched pathways downloaded in Sect. 3.1.1, it is possible, using the R programming language, to select the differentially expressed genes involved in these biological pathways and save them into TXT files:

```
## Load pathways enriched in Reactome
$ pathways <- as.data.frame(fread("mirnet_
enrichment_reactome_1.csv"))
## Select the genes of "G2/M Checkpoints" and "Cell
junction organization"
$ to_keep <- c("G2/M Checkpoints", "Cell junction
organization")
$ pathways <- pathways[pathways$Pathway %in%
to_keep,]
$ g2m_genes <
-strsplit(pathways$Genes[pathways$Pathway == to_
keep[1]], split=",")
$ g2m_genes <- unlist(g2m_genes)
$ cell_junction_genes <
-strsplit(pathways$Genes[pathways$Pathway == to_
keep[2]], split=",")
$ cell_junction_genes <
- unlist(cell_junction_genes)
## Save gene SYMBOLS into files
$ write.table(g2m_genes, file = "g2m_genes.txt",
quote = F, row.names = F, col.names = F)
$ write.table(cell_junction_genes, file = "cell_
junction_genes.txt", quote = F, row.names = F, col.
names = F)
```

To construct a miRNA-gene regulatory network using the genes involved in "G2/M checkpoints" and "cell junction organization", we copy and paste the genes saved in the files "g2m_genes.txt" and

Fig. 8 (**A**) Graphical and functional analyses of RNA-seq data. (**B**) Volcano plot with the obtained up- and down-regulated genes in tumor vs adjacent normal tissue

"cell_junction_genes.txt" as a query list in the Genes module of miRNet (Fig. 1). The user must specify the organism, the ID type of the genes, and (optionally) the tissue. Besides miRNA-gene interactions, the module allows users to add gene-TF, protein-protein, and TF-miRNA interactions using the aforementioned databases. After submitting the information shown in Fig. 9 and clicking the proceed button, the page with the downloadable interaction tables and networks is shown. The network created is composed of 96 miRNAs, 83 TFs, and 49 genes with 1368 edges. Before proceeding with the creation and visualization of the network, several filters can be applied to reduce the number of network nodes using the "Network Tools" menu on the right of the page. The filters include methods based on network properties such as degree, betweenness centrality, and shortest paths. The user can also simplify a dense network using algorithms (minimum network and Steiner Forest Network) to construct a minimally connected network containing the input nodes (seeds). Here, we apply the minimum network algorithm and

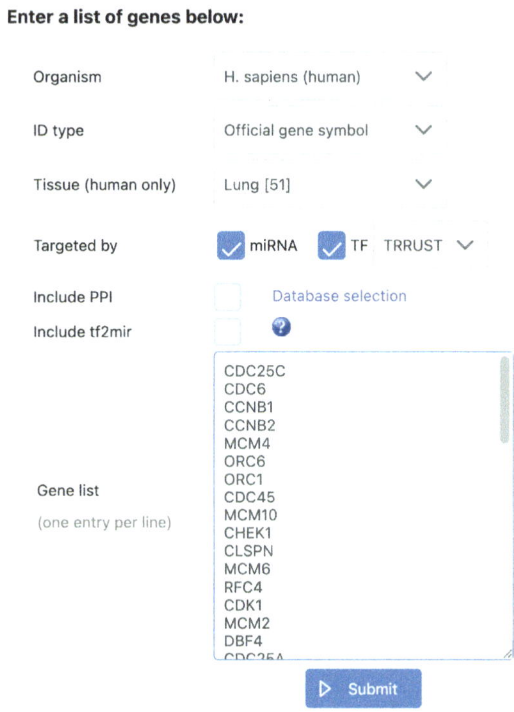

Fig. 9 Gene module of miRNet with the genes involved in "G2/M Checkpoints" and "Cell junction organization"

update the current network. The resulting network has 31 miRNAs, 83 TFs, and 49 genes with 793 edges. Clicking the proceed button, the "Network Viewer" of miRNet is shown.

This page is composed by a central interactive view of the network with a gray bar in the upper left corner composed of several graphical controls (e.g., zoom, move up, move down), a menu on the top in which the user can change several visualization options and other seven main panels (Fig. 10). In the "Global Node Styles" panel, the user can change the size and color of the different types of nodes (e.g., miRNAs, genes). In the "Node Explorer", all the nodes of the network are reported in a multi-page table. The user can select, search, and delete nodes from the network. This panel also shows network properties, such as degree and betweenness centrality, that the user can use to sort the table. In our example, Cadherin-1 (*CDH1*), is the node with highest degree and betweenness centrality. The nodes selected in the "Node explorer" or in the interactive network are reported in the "Current Selection" panel. In the "Function Explorer," miRNet allows functional enrichment on different genes or miRNAs databases using hypergeometric test or empirical sampling. The results are shown in the same panel in the form of a table reporting the enriched terms, found nodes, and p-values. In the module explorer, the user can execute algorithms to find network modules, i.e., tightly connected subnetworks that might perform a specific biological function. The "Batch Selection"

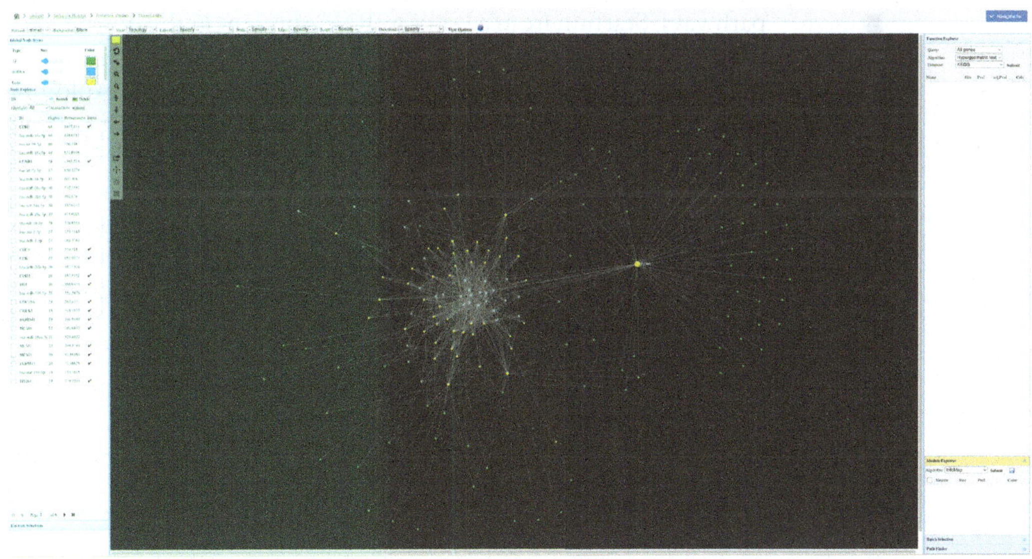

Fig. 10 Interactive view of the miRNA-gene-TF network in miRNet

and "Path Finder" modules allow, respectively, to highlight or delete nodes from an input list and to highlight in the network view the path between two nodes of the network.

To obtain the modules of our network, we select the Walktrap algorithm and click submit. miRNet also allows us to use InfoMap and Label propagation algorithms. As a result, we obtain two modules (0 and 1). Of note, module 1 is built around *CDH1*, an important adhesion molecule dysregulated in cancer and tightly regulated by miRNAs and transcription factors [36]. The user can highlight each module with a single color in the network view or use distinct colors. To obtain the latter, for each module, the user has to (i) select a color in the gray bar in the upper left corner of the network view and (ii) select a module with the appropriate checkbox in the Module Explorer (Fig. 10).

To download the figure of the network, the user can select the format (e.g., PNG) in the drop-down menu of the top bar and save it on their own computer (Fig. 10). To download all the files associated with the analysis, the user can click the "Downloads" link in the upper top menu and save specific files or all the files in a zip format (Fig. 10). Downloadable files comprise an XML file of the network, CSV files with miRNA-gene and TF-gene interactions, node and edge lists, module table and the R history script.

Starting from a list of miRNAs implicated in a specific disease or condition, an interesting question is how such miRNAs are affected by drugs or how to target them. We previously identified a set of miRNAs potentially implicated in the regulation of cancer-associated pathways. Starting from a list of miRNAs present in the downloaded file "orig_node_list.csv" containing all the nodes of the network, miRNet allows users to create a miRNA-drug network exploiting SM2mir [37]. SM2mir is a curated database containing interactions between small bioactive molecules and miRNAs. Selecting the miRNA module of miRNet (Fig. 1), the user can copy and paste a list of miRNAs and the type of targets after selecting the organism, ID type, and tissue. After selecting "small molecules" in the targets drop-down menu, the user can submit the choice and proceed with the page showing the Network Builder and, finally, the Network Viewer with the miRNA-drug network (Fig. 11). Sorting the table in the Node Explorer, it appears, for example, that 5-fluorouracil has the highest degree centrality among the small molecules, emphasizing its potential in affecting many of the miRNAs present in the network. Of note, 5-fluorouracil has been a widely used drug for cancer treatment [38].

3.2 MIENTURNET

MIENTURNET (http://userver.bio.uniroma1.it/apps/mienturnet/) is a web platform for studying miRNA-gene interactions using enrichment and network-based analyses. It takes as inputs a list of genes or miRNAs of 6 different organisms using TargetScan [39] and miRTarBase databases to select miRNA-target interactions.

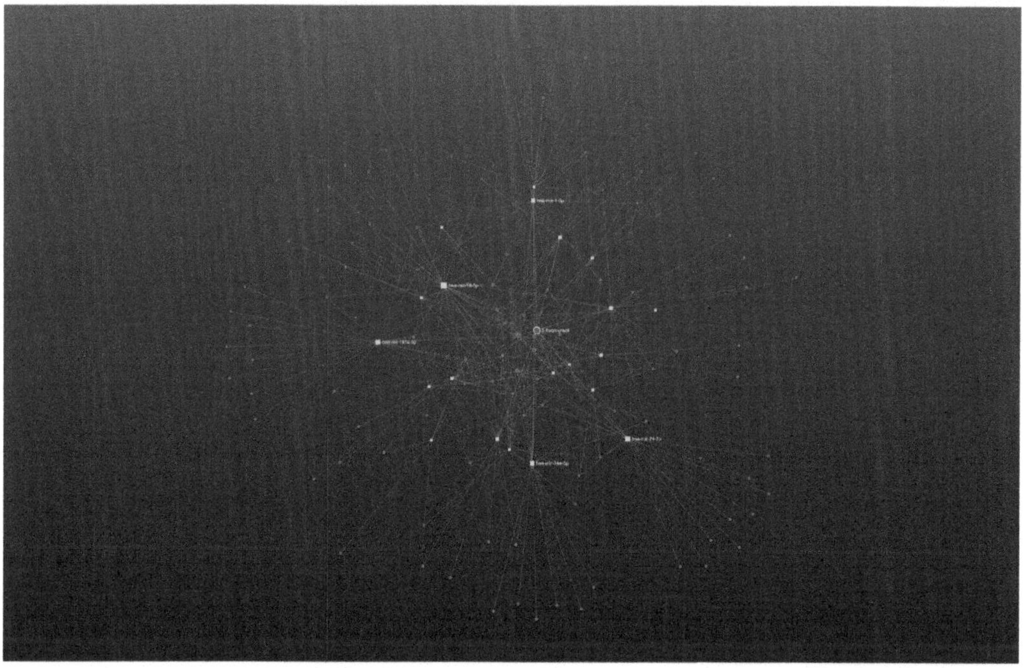

Fig. 11 Interactive view of the miRNA-drug network in miRNet

The latter are obtained by performing enrichment analysis and then used for the creation of the network. MIENTURNET also allows functional enrichment, network analyses, and export of the results. Through the navigation bar placed on top of the web page, the user can access the modules of MIENTURNET.

In the next section, we will see how to create and analyze a miRNA-target network starting from a list of miRNAs previously identified as regulators of lung cancer pathways.

Construction and Analysis of miRNA-Target Networks

In the "Data Upload" section (Fig. 12), the user can upload or paste a list of genes (i.e., SYMBOLS) or miRNAs (i.e., miRBase IDs) of a specific organism among Homo sapiens, Mus musculus, Rattus norvegicus, Drosophila melanogaster, Danio rerio, and Caenorhabditis elegans. Of note, MIENTURNET requires mature miRNA IDs and does not recognize miRNA precursors, while miRNet deals with both through automatic conversions. For example, "hsa-mir-29c" is not recognized by MIENTURNET while miRNet automatically converts it into "hsa-miR-29c-3p" and "hsa-miR-29c-5p." Submitting the input list of the mature miRNAs obtained from Sect. 3.1.2, a multi-page summary table is shown with information about microRNA name, miRBase accession, and mature miRNA sequence (Fig. 13). The user can search specific miRNAs and download the table in CSV or Excel formats with the appropriate buttons.

In the "miRNA-target Enrichment" section, the user can set a threshold for the minimum number of miRNA-target interactions

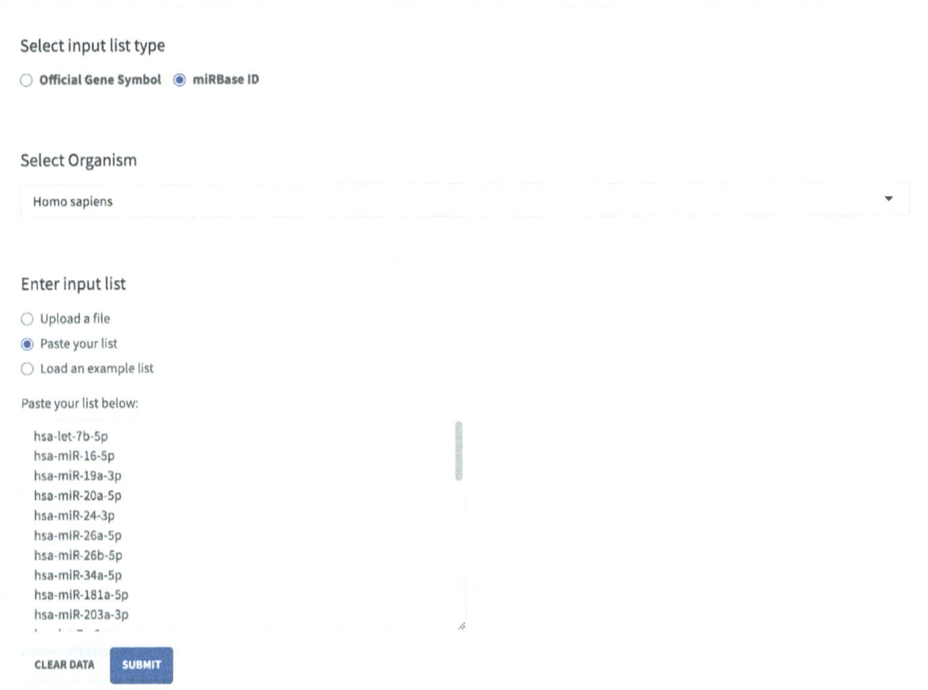

Fig. 12 Data upload page of MIENTURNET

Fig. 13 Table with information about the miRNAs uploaded

and the adjusted p-value cutoff (False Discovery Rate, FDR) of the enrichment test. A hypergeometric test is performed to test the significance of the miRNA input list enrichment in each list of miRNAs targeting a gene in TargetScan or miRTarBase. Selecting an FDR cutoff of 0.05, the enrichment results are shown in two main tabs, one for each database. Figure 14 reports the results obtained on miRTarBase. Each tab reports a table with the enriched genes in the rows (Fig. 14, top). For each gene, statistical significance (p-value and

FDR), odds ratio, and number of interactions are reported. The other columns of the table differ between TargetScan and miRTarBase. In the “TargetScan” tab, a column for each miRNA family enriched is reported since TargetScan predicts targets based on seed regions that are shared between miRNA families. In “miRTarBase” tab, a column for each enriched miRNA is reported since the interactions are experimentally validated individually for each miRNA. In addition to the table, a bar plot showing the ten most targeted genes is shown (Fig. 14, bottom). The miRNA-target network is obtained clicking the “Network” section in the navigation bar and selecting the database. Selecting miRTarBase, a page composed of two main sections, is shown. On the left part (Fig. 15), the user can apply several filters and customization on the network that is visualized in the right part of the page (Fig. 16). In particular, the user can change the minimum number of miRNA-target interactions, the FDR threshold, the confidence of the interactions (strong, weak or both), and several graphical features such as the layout of the network and size, color, and shape of the nodes. In the right panel, there are three main tabs. In the “Network” tab, the user can interact with the network, moving or zooming the entire network using the proper button controls and selecting/moving single nodes. The user can also save the network

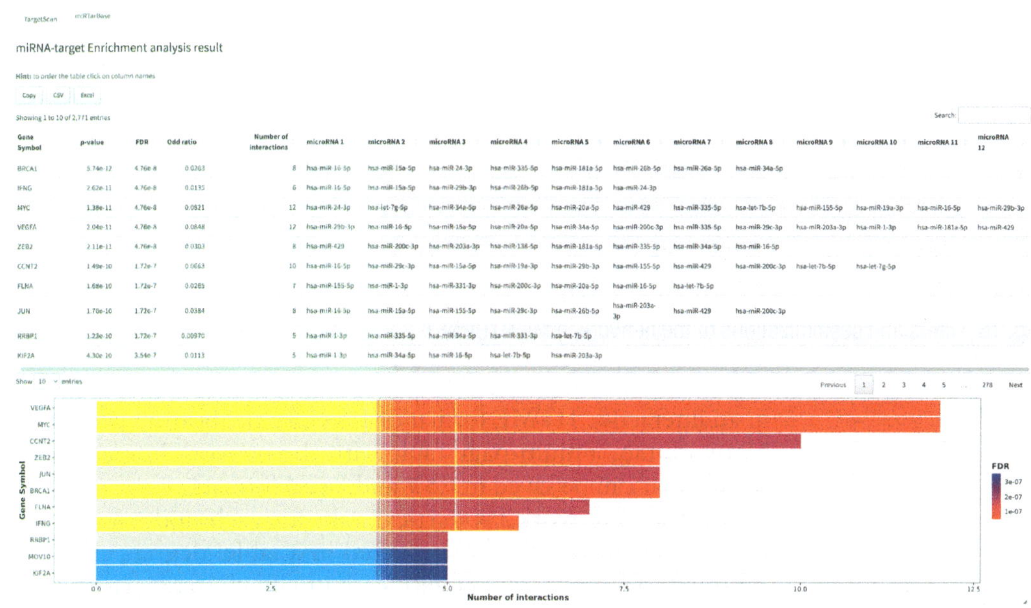

Fig. 14 Enrichment results of the target genes obtained in miRTarBase (top) and bar plot with 10 most targeted genes (bottom)

Fig. 15 Filters and customizations of the network in MIENTURNET

in an interactive HTML format or PNG. In the bottom part of the Network tab, a multi-page table with the network interactions is shown that can be saved in CSV and Excel formats.

In the "Network Properties" tab, for each node, a series of network characteristics are reported in a table: degree, closeness centrality, betweenness centrality, eccentricity, clustering coefficient, and average shortest path length (Fig. 17). Clicking the small arrows in the column names will sort the table in ascending or descending order. In our example, hsa-miR-155-5p is the most central node in the network in terms of degree, closeness, and betweenness centrality. Of note, elevated levels of hsa-miR-155 were reported as an indicator of poor prognosis in patients with lung adenocarcinoma

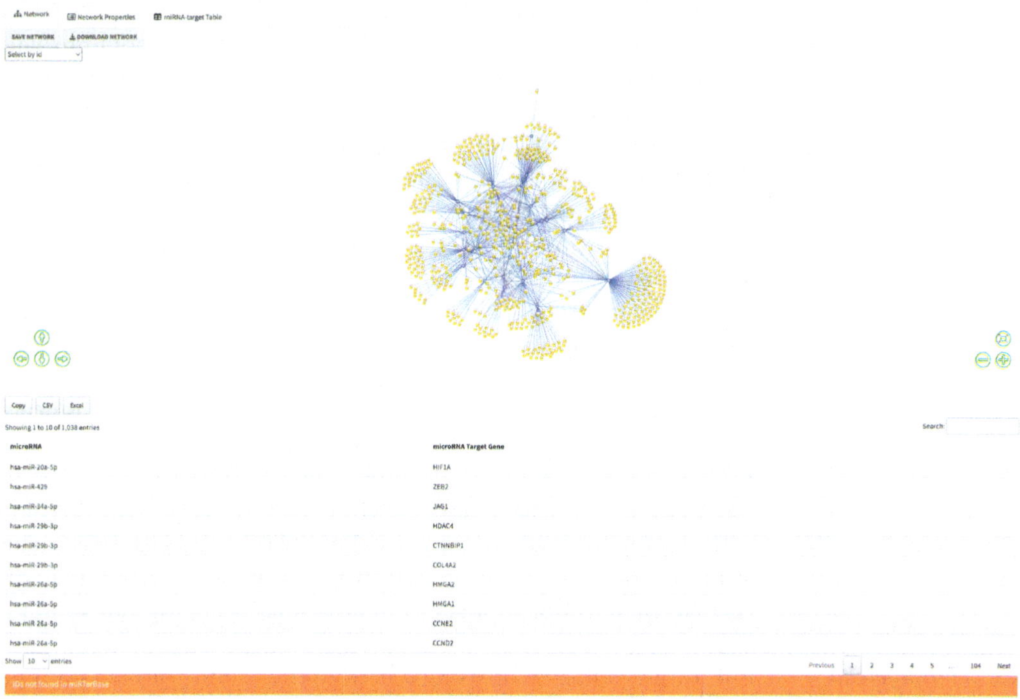

Fig. 16 Network visualization in MIENTURNET

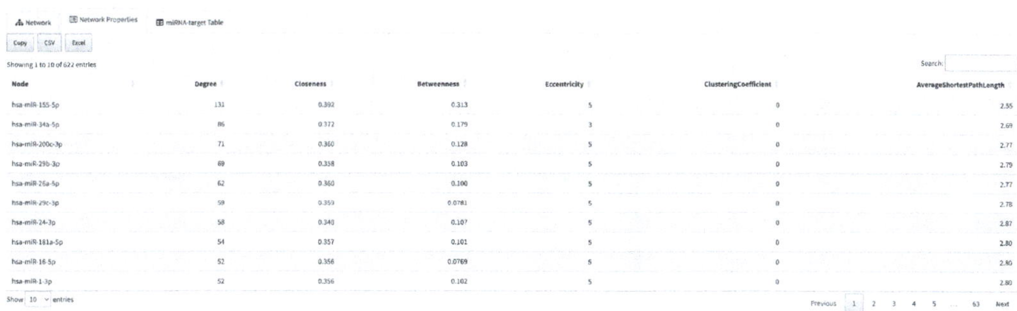

Fig. 17 Table with the network properties of the 10 most central (based on degree) miRNAs of the network

[40]. The table with the network properties can be downloaded in CSV and Excel files clicking the buttons on top of the table. Below the table, MIENTURNET reports three network degree plots that can be saved in PDF format: 2 bar plots with the degree of the top 30 (based on the degree) genes (Fig. 18A) and miRNAs (Fig. 18B), and a plot showing the degree distribution of the network with a

best-fitting curve (Fig. 18C). The "miRNA-target Table" tab displays a downloadable table with the miRNAs in the columns with the corresponding gene targets found in the selected database, in this case, miRTarBase. The last section that the user can select from the navigation bar of the web platform is the "Functional Enrichment". With this module, the user can select one or more miRNAs and perform functional enrichment of the target genes on KEGG, Reactome, WikiPathways, and Disease Ontology databases. The user can select TargetScan or miRTarBase interactions using the tabs in the right part of the page. MIENTURNET will show the miRNAs selected and, for each database, a table with the following enrichment results: miRNA, category ID, description, p-value, FDR, target genes, and gene count. Further, a dot plot with the enriched category for each miRNA/miRNA family showing the gene ratio as point size and FDR as color gradient. All the tables and the figures can be saved in Excel and PDF formats, respectively. Figure 19 reports the dot plots with the top five pathways enriched in the different databases using as input hsa-miR-let-7g-5p, a miRNA of the let-7 family that was reported as a tumor suppressor in non-small-cell lung cancer [41].

3.3 miRTargetLink 2.0

miRTargetLink 2.0 (https://ccb-compute.cs.uni-saarland.de/mirtargetlink2/) is the updated version of the previous interactive tool for the construction and analysis of miRNA regulatory networks [42]. It allows to query miRNAs, target genes, and pathways to retrieve predicted and validated miRNA interactions using, respectively, mirDIP/miRDB [43, 44] and miRTarBase. miRTargetLink 2.0 supports 2 different modalities of search, unidirectional and bidirectional. In the unidirectional modality, the user can perform a miRNA-centric or a target-centric search. In the miRNA-centric section, a single miRNA, multiple miRNAs, and a miRNA set from miEAA [45] can be specified. Similarly, for the target-centric section,

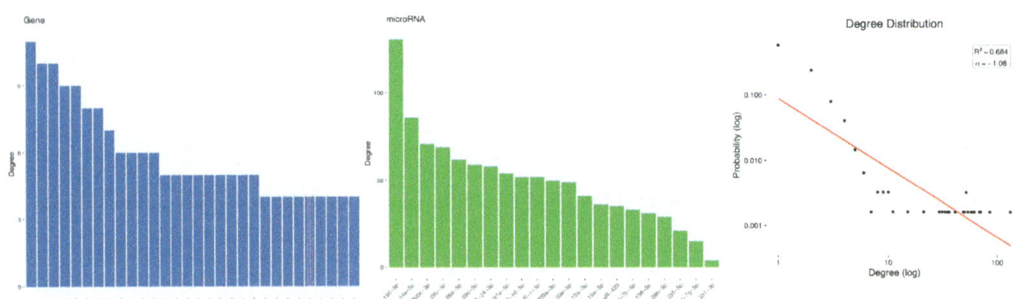

Fig. 18 Bar plot showing the degree of the top connected genes (**A**) and miRNAs (**B**), and scatter plot with the degree distribution of the network with a best fitting curve (**C**)

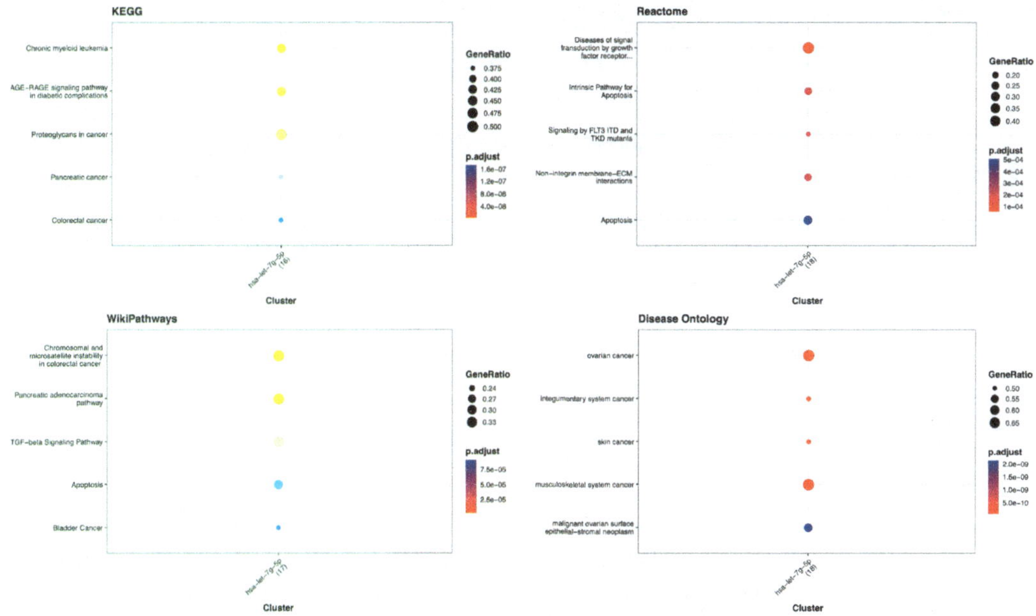

Fig. 19 Dot plots with the top 5 pathways enriched in the different databases using the miRNA hsa-miR-let-7g-5p

the user can specify single or multiple genes or a pathway from mirPathDB [46] and GeneTrail [47]. In the bidirectional section, the user can specify a list of miRNAs and a list of target genes to include in the network.

Construction of miRNA-Pathway Network

Different from the other web platforms, miRTargetLink allows users to create a regulatory network starting from a biological pathway. This operation can be performed by typing the name of the pathway in the "Quickstart" form on the homepage or using the "Unidirectional search" in the menu on the upper left part of the page (Fig. 20). Typing "G2/M Checkpoints" in the quickstart form, an autocomplete drop-down menu is shown with the pathways that match the user key words. Selecting "G2/M Checkpoints [Homo sapiens] - Pathway" from the drop-down menu and clicking the button "Create network" a page with different sections is shown. In the "Interaction graph" section, a panel with the interactive network is shown with graphical commands to move and zoom it. In the "Edit network" section, the user can select the type of miRNA-target interactions (predicted, weak, or strong validated), whether to include pathways using weak and strong validated targets, the minimum number of shared targets, and different network layouts for visualization, such

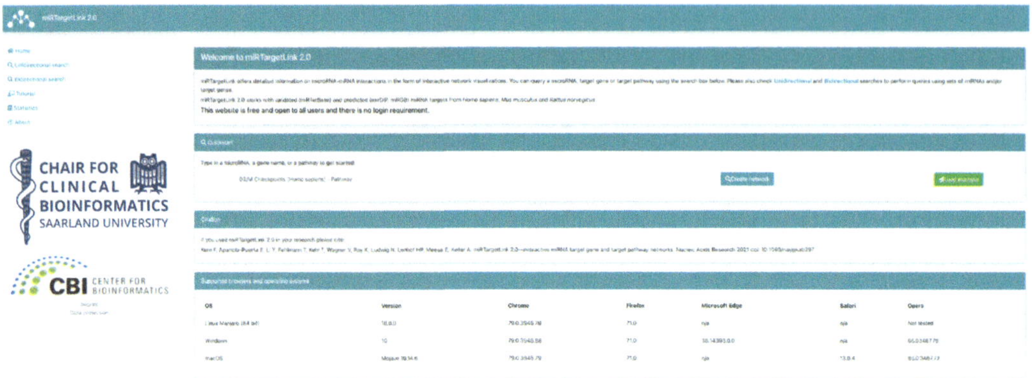

Fig. 20 Homepage of miRTargetLink

as concentric, circle, and random. Selecting "miRNA target pathways" and clicking the update button, the network will be enriched with those miRNAs that target the selected pathway (Fig. 21). In the "Interaction table" section, a table with the following information about miRNA-pathway interactions is shown: miRNA, target pathway, confidence of the interaction, database used, experiment, and reference of the study (PubMed ID). In the "Node annotation table," a table with the miRNA categories annotated is shown. In general, in this section, annotations about miRNAs and genes are reported, respectively, based on miEAA and GeneTrail. All the tables are exportable in CSV and Excel formats. In the right part of the page, the user can obtain additional information about the nodes, remove nodes, export the network to PNG and JPG, and run an over-representation analysis using miEAA and GeneTrail. The results of the enrichment will be provided as downloadable files.

Construction of miRNA-Target Network Using a Bidirectional Search

miRTargetLink allows users to construct a miRNA-regulatory network specifying both miRNAs and target genes that the user wants to include in the network. Selecting "Bidirectional search" in the main menu on the right, the user can select the species and paste a list of miRNAs and target genes. Using miRNAs and target genes of the network obtained in Sect. 3.1.2 and clicking the button "Go," a page with a summary of the nodes uploaded will be shown. In this page, the user can refine each node if it is necessary to disambiguate some miRNA or gene names. After submitting the refinement page, the network is shown. Selecting only strong validated interactions, deselecting the "additional filters" checkbox, and selecting the "Euler" network layout, the network in Fig. 22 will be shown.

3.4 FFLtool

FFLtool (https://guolab.wchscu.cn/FFLtool//#!/) is a web application designed to investigate interactions between two key regulators of gene expression: miRNAs and TFs, along with their target

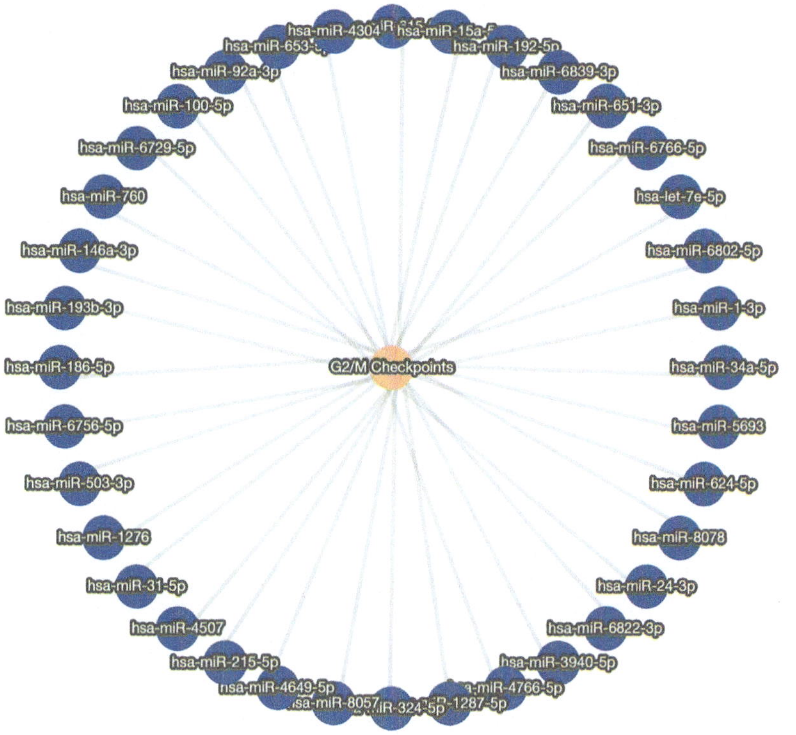

Fig. 21 miRNA-pathway network constructed with miRTargetLink using as input the pathway "G2/M Checkpoints [Homo sapiens]"

genes. This tool facilitates the identification of potential FFLs within TF-miRNA-target gene networks in the human context, offering valuable insights into complex gene regulatory mechanisms. FFL-tool integrates transcription factor (TF)-target interactions from the hTFtarget [48] database and miRNA-target interactions from six experimentally verified databases, including miR2Disease, miR-TarBase 2018 [49], miRecords [50], DIANA-TarBase V8 [51], miRWalk V3 [52], and an interaction database curated by London's Global University (UCL), as well as two predictive tools, TargetScan and miRmap [53]. When the user uploads a specific gene list, FFLtool explores five types of regulatory interactions: TF-TF, TF-gene, TF-miRNA, miRNA-gene, and miRNA-TF to construct the TF-miRNA-target gene network and identify FFLs. Additionally, the tool provides three methods to validate these interactions. The first is an expression validation approach performed by checking if each molecule composing the FFL is expressed inside the tissue of interest. The second approach is based on the correlation metric to ensure that each pair of molecules in the FFL exhibits a concordant

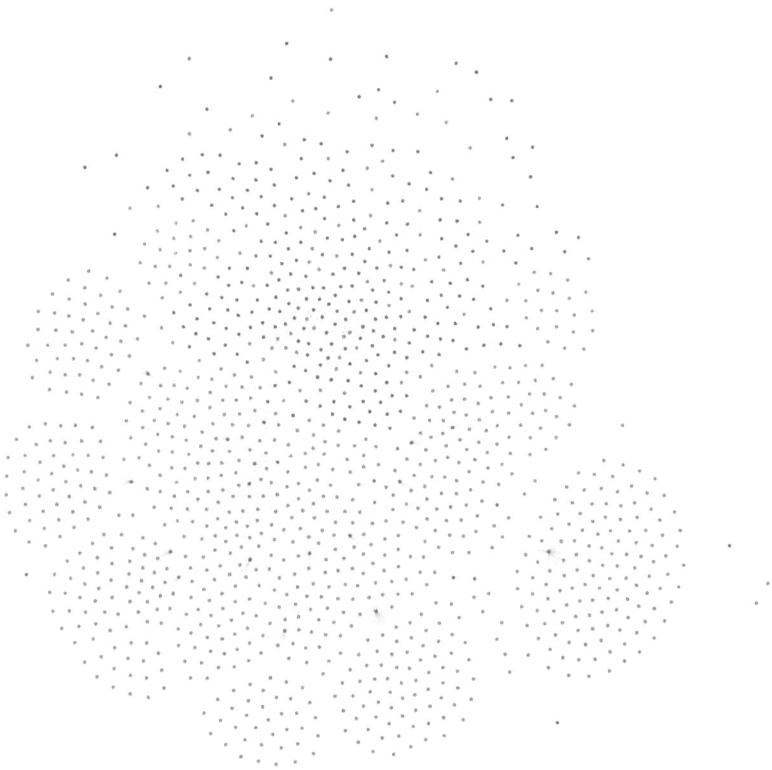

Fig. 22 miRNA-target network obtained using the bidirectional search of miRTargetLink. Blue points represent miRNAs, while green points represent target genes

correlation pattern that aligns with their regulatory relationship. Finally, the tool verifies whether the FFL has been directly or indirectly validated by existing literature. Furthermore, through the R package clusterProfiler [54], FFLtool performs enrichment analysis on the target genes of TFs and miRNAs, helping to uncover the biological significance of the identified regulatory networks.

FFLtool includes two distinct modules. The first module allows users to browse each FFL stored in the tool database and associated with a specific type of cancer without the need to submit external data as input. The second module allows users to upload a personalized list of TFs, miRNAs, and genes to check for the presence of FFLs associated with the input list of molecules. FFLtool features are presented in the next sections using a case study on lung cancer. This case study utilizes the same list of miRNAs, transcription factors, and target genes previously analyzed with miRNet 2.0 in Sect. 3.1.2.

Browse FFLs

The first module is composed of two different panels. The first panel includes three sequential steps (Fig. 23). In the first step, the user can select the cancer type of interest along with a subset of top-ranked differentially expressed genes of the corresponding tissue type, while,

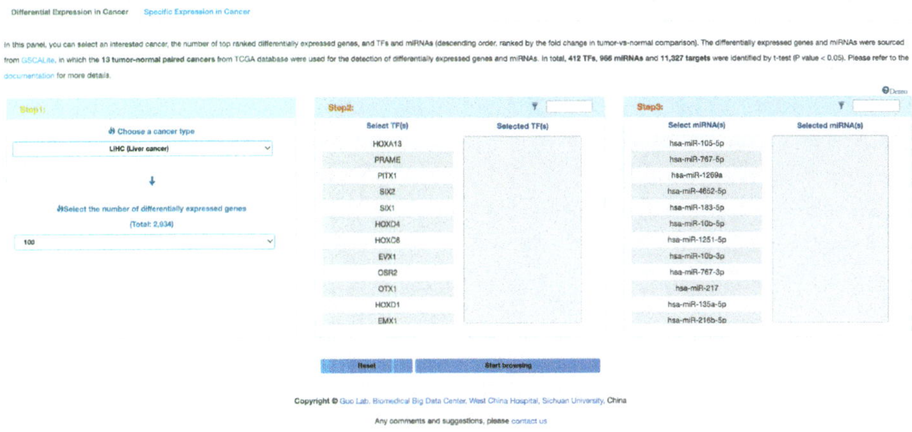

Fig. 23 Overview of the first panel in the FFL browse module

in the second and third steps, the user can select respectively a subset of TFs and a subset of miRNAs of interest. Upon completing these steps, the tool generates potential FFLs based on the selected molecules. This approach enables researchers to explore pre-existing cancer-associated FFLs or identify novel FFLs from a custom selection of molecules. The second panel (Fig. 24) allows to explore FFLs in the context of a high or low level of expression of the target genes in cancer data from GDC [55] and GSCA [56]. In the first step, users specify the cancer type, target gene expression level (high or low), and the number of targets to analyze. In the second and third steps, the user can select respectively the TFs and miRNAs of interest. Once the analysis is submitted, the tool reconstructs feed-forward loops involving the selected molecules. All the FFLs can be downloaded as images showing the different belonging molecules as nodes of a network, or in a textual format where all the molecule identifiers are listed.

FFL Personalized Analysis

The second module of FFLtool enables the study of FFLs inside a personalized list of molecules. In the previous section, we tested the module on the lung cancer regulator molecule list used in the Sect. 3.1.2 of this chapter, containing respectively 83 TFs, 31 miRNAs, and 49 genes. The same list of molecules can be uploaded copying and pasting the identifiers, separated by line break, in the appropriate box or uploading a textual file from a local folder (Fig. 25). Once the molecules are uploaded, the user can select a tissue or a group of tissues of interest and submit the analysis. In our case study, we select the lung tissue. Once the analysis is completed, the web app opens a new page where the results can be explored. At the top of the page,

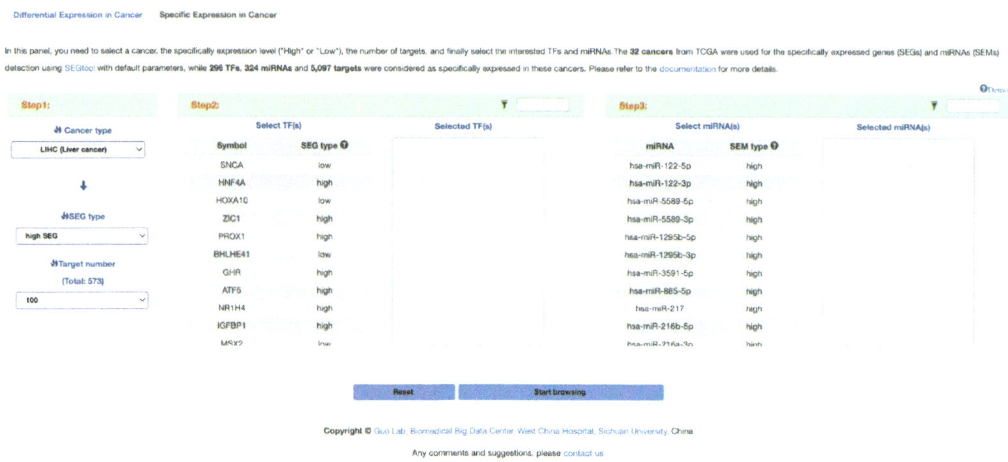

Fig. 24 Overview of the second panel in the FFL browse module

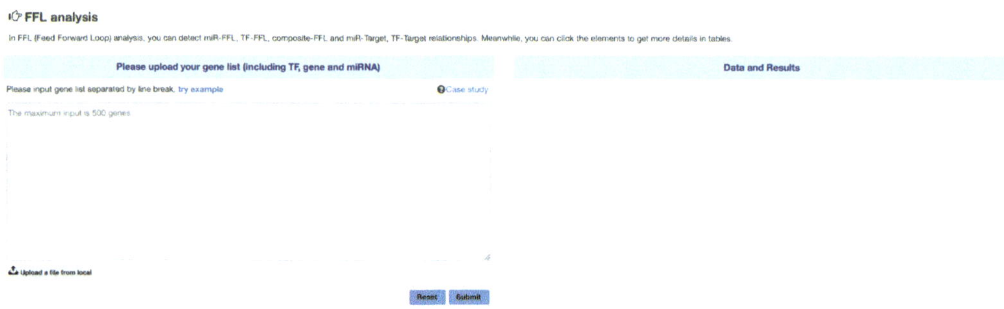

Fig. 25 Data upload module for FFL analysis

miRNA-TF-gene regulatory networks can be visualized individually or aggregated in a single network (Fig. 26). Below the graphic section, two tables are shown (Fig. 27). In the left table, the full list of FFLs classified with respect to the FFL type (miR-FFL, TF-FFL, composite-FFL) is stored together with the names of the TFs and miRNAs involved and the number of target genes. When the user selects an FFL entry inside the left table, the graphical representation of the regulative network is dynamically updated according to the regulative interaction selected. Simultaneously, the right table provides detailed information about the chosen FFL. Specifically, the table on the right shows each single interaction between TF, miRNA, and gene inside the FFL selected, the reference tissue, and three columns, namely "Expr.," "Cor.," and "Validation." These columns confirm, respectively, if each node of the FFL is expressed in

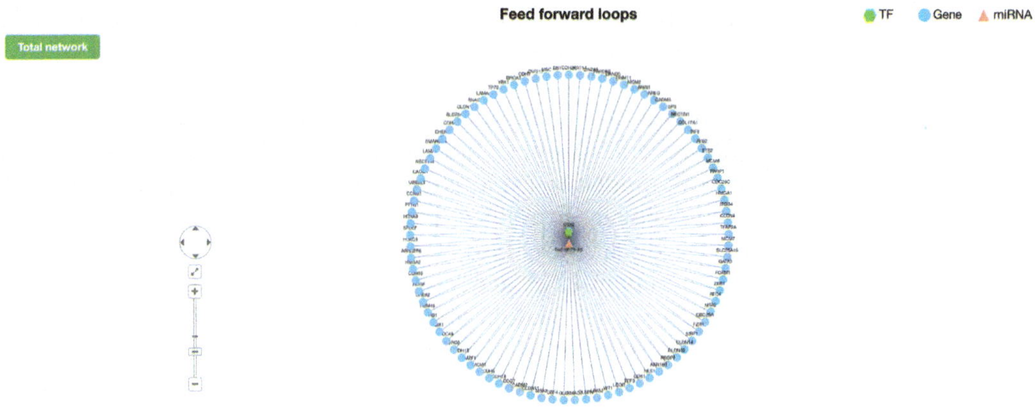

Fig. 26 miRNA-Gene-TF interaction network resulting from the case study

Fig. 27 FFLtool results tables and download interface for the case study

the reference tumor dataset, each pair within the FFL is supported by a Pearson expression correlation in the reference expression dataset, and a miRNA-target interaction is directly or indirectly supported by literature. The table on the left allows also to perform enrichment analysis on the list of target genes of each FFL if the number of target genes is greater than 50. Finally, all FFLs can be downloaded as images or in textual format for further analysis.

3.5 miRViz

miRViz (http://mirviz.prabi.fr/) is a web application developed in Java and JavaScript designed to facilitate the exploration, visualization, and interpretation of miRNA datasets without requiring specialized computational skills. While miRViz does not support comprehensive computational analysis, such as data preprocessing or miRNA differential expression analysis, which must be conducted prior to its use, it can be used in combination with other tools, such as miRNet, to initially identify miRNAs of interest and to explore miRNA-miRNA interaction networks. miRViz leverages multiple network models to explore miRNA relationships. The "Seed2_7" network visualizes miRNA families by linking nodes that share the same seed sequence. Two "Genomic_Distance" networks connect neighboring miRNA genes on the genome: the "2k" version links miRNAs within 2 kilobases (kb), while the "50k" version uses a 50 kb threshold to account for larger genomic reorganizations such as those of tumors. Three co-regulation networks connect miRNAs based on shared mRNA targets: "Diana50" links miRNAs sharing more than 50% of mRNA targets, predicted by Diana MicroT v3 [57], "TargetScan54" connects miRNAs sharing over 54% of targets, predicted by TargetScan v6.2 [58], and "DianaTarBase50" links miRNAs sharing more than 50% experimentally validated targets from Diana-TarBase V8. Finally, to simplify visualization, two additional pruned networks were developed, including "Genomic_Distance_50k_clusters_3," which focuses on the 820 miRNAs in clusters of three or more members, and "TargetScan54_degree_10," which retains the layout of TargetScan54 but excludes nodes with fewer than 10 connections. These networks provide diverse views for analyzing miRNA interactions. miRViz also includes three pre-loaded datasets: "hsa_miRNAmine_cells", "hsa_miRNAmine_tissues" collected from miRmine [59], and "hsa_TissueAtlas" collected from TissueAtlas [60] that can be used to make comparisons with the user datasets. miRViz offers powerful visualization capabilities for exploring miRNA aggregations within families or genomic clusters. Additionally, it provides both *local* and *global significance scores*, which evaluate the significance of miRNA hits and their connectivity within networks.

The next section demonstrates the features of miRViz using a simple case study. For this, we employ the same list of miRNAs, transcription factors, and genes for lung cancer used in the previous sections.

Construction of miRNA–miRNA Interaction Networks

The miRViz web application has a user-friendly and intuitive interface designed for easy navigation (Fig. 28). At the top of the page, there is an interactive toolbar that provides quick access to the tool manual and the network datasets included in the application. The interface is divided into two main sections. On the left part, an interactive menu allows users to select various functionalities, including uploading data, selecting network features, and highlighting specific miRNA nodes inside the network. The right part contains four visualization tabs where different output networks can be displayed. The default functionality of miRViz is the network exploration, which is the "Network" option in the top-left menu section. This feature allows

Fig. 28 miRViz web interface

users to visualize miRNA networks and customize several parameters, including the choice of one of the 11 species available, the selection of one of the 8 graph models available, and the configuration of node colors. The latter can be chosen according to various attributes such as miRNAs from pre-loaded datasets, the tissue of interest, the miRNA seed, and color scales. Additionally, users can highlight specific nodes within the network using a dedicated text box. Another miRViz key feature is the possibility to load custom data. The "Load data" function enables users to upload their own miRNA datasets in the CSV format. By clicking the button "Load CSV," a pop-up window opens that allows users to drag and drop their custom file opens (Fig. 29). The file should include a list of miRNAs, and it could contain their average expression values under different study conditions. Data in the file can be separated by tabs, commas, or semicolons. After uploading the file, a summary of the data is displayed in the left menu. This summary includes the proportion of miRNAs uploaded that were found in each of the eight network types and the number of miRNAs not present in any network. In our case study, a list of 31 miRNAs was uploaded, of which 12 were not recognized by the tool, probably because they were not involved in any relationship. Once the data is uploaded, users can move back to the "Network" panel and select their dataset from the "miR data table" section to visualize the involvement of their miRNAs in the 8 network models. Together with the "Network" and "Load data" sections, "Display/

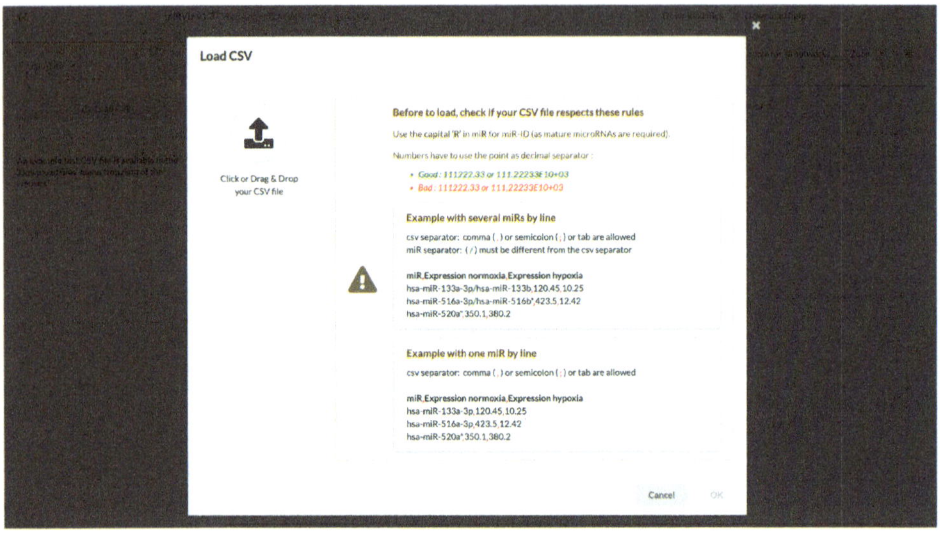

Fig. 29 miRViz data loading interface

Hide" and "Highlight" functionalities are provided. They can be used to highlight the miRNA nodes in bold or to hide nodes and connections that are not present in the uploaded dataset. For example, selecting the "TargetScan54_degree10+" network and hiding nodes not included in the uploaded miRNA list revealed three pairwise miRNA interactions with a degree greater than nine. These interactions involved miRNAs associated with lung cancer, either as tumor suppressors or tumor promoters. These findings suggest potential regulatory relationships between miRNA pairs and provide insights into the regulation of proliferative processes in lung cancer cells. Lastly, miRViz provides the functionalities to visualize different networks in four different boxes (Fig. 30) and to download the networks in different image and file formats, including PNG, TIFF, and PDF, making it easy to share visualizations or integrate them into presentations and reports.

3.6 Arena-Idb

Arena-Idb (http://ncrnadb.scienze.univr.it/sites/arenaidb/) is a comprehensive web resource dedicated to non-coding RNAs (ncRNAs) and their interactions. By integrating multiple sources, it provides a non-redundant collection of ncRNA objects and their associations with other biological entities, including miRNAs. The web interface allows users to search for specific ncRNAs, miRNAs, genes, or diseases, serving as a starting point to explore ncRNA properties and retrieve networks of interactions involving the searched entities. The platform is organized into 4 main modules. The first

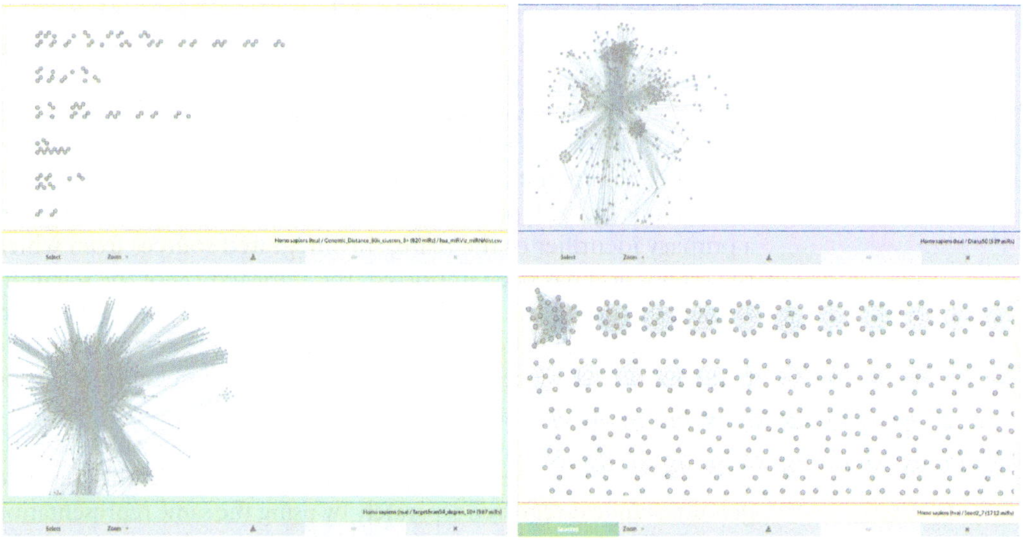

Fig. 30 miRViz network visualization boxes

module (i.e., search module) enables users to query and investigate the database content. The second module allows users to browse the entire collection of interactions between ncRNAs and other biological entities. Additionally, a download section provides access to raw data in TSV (Tab-Separated Values) format, making easy to export and analyze the information offline. Lastly, a network visualization module is available that can be accessed after selecting an entity from the search interface. This module offers an interactive way to visualize the relationships and interactions within the dataset.

Retrieving and Integration of Network Data

The search interface allows three different ways of searching. The user can search an ncRNA, its gene, or a miRNA by specifying the name, an alias, or any given identifier with which the entity is labeled. Alternatively, ncRNAs can be searched by their genomic location. The user selects a human chromosome and specifies a genomic range. All the ncRNAs located within the genomic range are returned as a result of the query. Lastly, the resource can be queried by specifying a disease to study ncRNAs that can be implicated in it.

The search returns the set of entities associated with the identifiers. Each entity is reported separately as a card, where summary information is displayed. A detailed description of the entity is available by clicking on the corresponding information button (a circled "i").Figure 31 shows a detailed description of the miRNA hsa-miR-26b-3p, which was selected from the list of miRNAs previously identified in LUAD. The description is composed of 4 sections reporting the aliases of the entity, its genomic locations, its sequence, and the set of interactions in which the miRNA is involved. Aliases can be in the form of generic names or symbols, or they can be identifiers defined in an external resource, for example, "MIMAT0004500" is the identifier assigned in miRBase to "hsa-miR-26b-3p."

Locations are described by reporting the human chromosome, the strand (+ or −), and the absolute start and end genomic coordinates. The interaction sections provide a summary, grouping the interactions by the type of interacting mate. Each interaction reports a primary identifier of the mate, the integrated sources from which the interaction has been imported, the PubMed ID of the scientific articles reporting the relation, and supplementary support information that usually contains the sentence in the scientific article that supports the interaction. In addition, the interactions can be downloaded in CSV and XLSX.

The search by genomic coordinates presents the results similarly to the search by name. The set of ncRNAs located in the specified genomic range is reported separately by using the same representation as cards. This type of search is useful in retrieving clusters of spatially close miRNAs.

The search by disease name is another useful tool to investigate ncRNAs that are specific for a given phenotype. After assigning a name to the disease, the set of disease labeled with the given name is reported by using the same concept of entity cards used for the

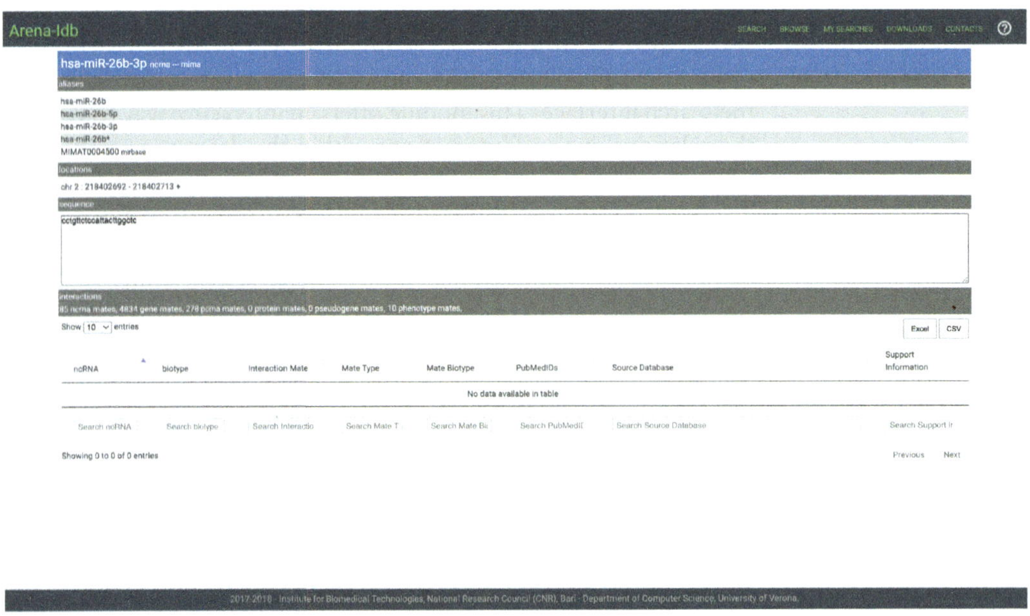

Fig. 31 Arena-Idb description panel for the miRNA hsa-miR-26b-3p

ncRNAs. Each card represents a specific disease, and it is an entry point to visualize the detailed description of it. The description reports the set of aliases associated with the disease and the set of (functional) interactions linking the disease and the ncRNAs registered in the resource.

Visualization of the Network

On the entity card, the button represented as a downward arrow allows the user to download, as textual files, both the same information reported in the detailed description and the interactions involving the given object. A graphical visualization of the network is accessible via the checkerboard pattern button placed on the entity card. The module interface allows the user to select which type of mate has to be included in the visualization (Fig. 32). Each type of entity is drawn in the resultant visualization as ellipses filled with the same color used for the labels of the checkboxes in the selection panel.

Figure 33 gives an example of a network representing the interactions between a queried miRNA, the "hsa-miR-26b-3p," and its mate of type "ncRNAs". Entities are labeled with their representative name, and the complete set of aliases assigned to them is displayed by clicking on the entity node. Clicking on a specific node also highlights the direct neighborhood of the node and makes opaque the rest of the network. Edges are colored from gray to black depending on the amount of evidence that Arena-Idb stores for the specific connection. The higher the amount of evidence, the darker the edge color. By default, only interactions between queried entities and their direct mates are displayed. The network can contain ncRNAs that have not

Fig. 32 Arena-Idb interaction network for the miRNA hsa-miR-26b-3p

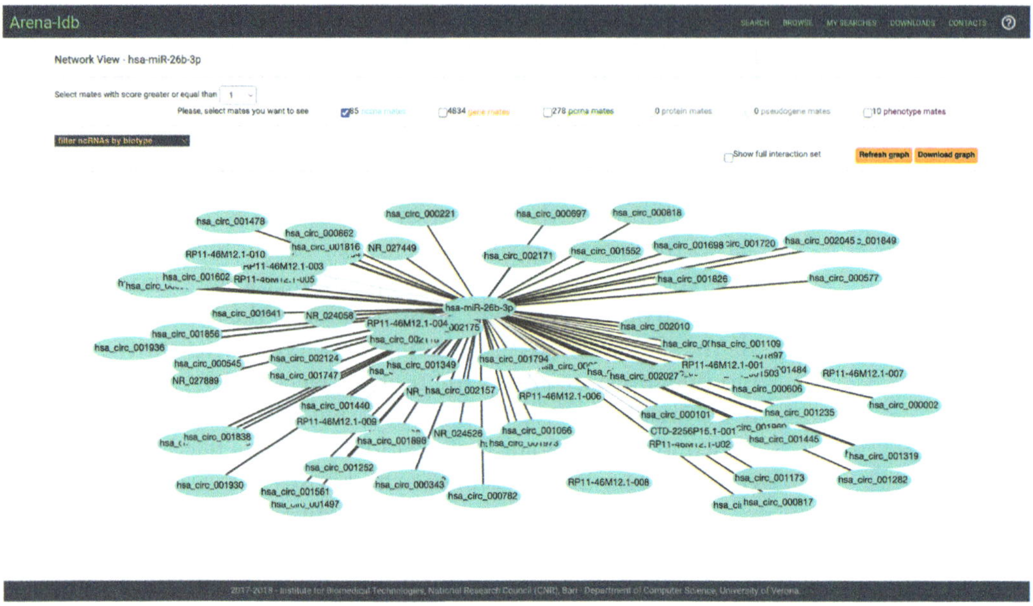

Fig. 33 Arena-Idb interaction network between hsa-miR-26b-3p and ncRNA molecules

been queried but that are just neighbors of the searched ones. The interactions involving these unqueried ncRNAs, but not the queried entities, are not shown in the visualization. However, the "Show full interaction set" checkbox on the selection panel allows the user to retrieve and visualize these types of secondary relations.

4 Notes

1. The following script was used to download and save the count matrix of TCGA-LUAD paired tumor-normal samples and to extract genes on enriched pathways obtained from miRNet:

```
## Installation of the required packages
install.packages("data.table")
if (!require("BiocManager", quietly = TRUE))
    install.packages("BiocManager")
BiocManager::install("SummarizedExperiment",
"biomaRt", "TCGAbiolinks")
## Import of the packages
library(data.table)
library(SummarizedExperiment)
```

```
library(TCGAbiolinks)
library(biomaRt)
## Gene expression aligned against hg38
query <- GDCquery(
  project = "TCGA-LUAD",
  data.category = "Transcriptome Profiling",
  data.type = "Gene Expression Quantification",
  workflow.type = "STAR - Counts",
  sample.type = c("Primary Tumor", "Solid Tissue
Normal"))
GDCdownload(query = query)
data <- GDCprepare(query = query)
## Clinical information
tab <- data.table(
  as.data.frame(colData(data)),
  options = list(scrollX = TRUE, keys = TRUE,
pageLength = 5),
  rownames = FALSE)
## Matrix of counts
counts <- assay(data)
## Remove samples fixed in formalin and embedded in
paraffin
not_ffpe <- which(!tab$is_ffpe)
tab <- tab[not_ffpe,]
counts <- counts[, not_ffpe]
identical(colnames(counts), tab$barcode)
## Paired tumor-normal samples
normal <- sort(tab$patient[tab$definition ==
"Solid Tissue Normal"])
tumor <- sort(tab$patient[tab$definition ==
"Primary solid Tumor"])
int <- intersect(normal, tumor)
tab <- tab[tab$patient %in% int,]
tab <- tab[!duplicated(tab$sample),]
counts <- counts[, colnames(counts) %in%
tab$barcode]
tab <- tab[order(tab$sample_type_id),]
tab <- tab[order(tab$patient),]
counts <- counts[, tab$barcode]
to_keep <- apply(counts, 1, function(row){
  return(sum(row) > 0)
})
counts <- counts[to_keep,]
counts[1:10, 1:10]
ids <- sapply(strsplit(rownames(counts),"[.]"),
`[`, 1)
mart <- useEnsembl(biomart = "ensembl",
                   dataset =
"hsapiens_gene_ensembl",
                   verbose = T)
ann <- getBM(c("hgnc_symbol","ensembl_gene_id",
"gene_biotype"), "ensembl_gene_id", ids, mart)
## Uncharacterized genes
ann <- ann[ann$hgnc_symbol!= "",]
```

```
## Only protein coding genes
ann <- ann[ann$gene_biotype == "protein_coding",]
counts_df <- data.frame(ENSEMBL = ids, counts)
rownames(counts_df) <- ids
counts_df[1:10, 1:10]
## Remove 2 duplicates
ann <- ann[!duplicated(ann$hgnc_symbol),]
colnames(ann) <- c("SYMBOL", "ENSEMBL", "BIOTYPE")
ann <- ann[, -3]
counts <- merge(counts_df, ann, by = "ENSEMBL")
genes <- counts$SYMBOL
counts <- counts[, -which(colnames(counts) %in%
c("ENSEMBL", "SYMBOL"))]
rownames(counts) <- genes
col <- c("#CLASS:Condition", "#CLASS:Patient",
rownames(counts))
row1 <- ifelse(tab$sample_type_id == "01",
"Tumor", "Normal")
row2 <- tab$patient
data_file <- data.frame(cbind(col, rbind(row1,
row2, counts)))
colnames(data_file) <- c("#NAME", colnames(data_
file)[-1])
metadata_file <- data.frame(Name =  colnames(data_
file)[-1], Condition = row1, Patient = row2)
colnames(metadata_file)[1] <- "#Name"
write.table(data_file, file = "data_file.txt",
quote = F, row.names = F, sep = "\t")
write.table(metadata_file, file = "metadata_file.
txt", quote = F, row.names = F, sep = "\t")
## Read DE genes LUAD (Tumor vs Normal)
de_genes <- as.data.frame(fread("DE_GENES_LUAD.
csv"))
de_genes <- de_genes[de_genes$adj.P.Val < 0.05 &
abs(de_genes$logFC) > 1,]
de_genes <- de_genes[order(de_genes$logFC,
decreasing = T),]
## Pathways enriched
pathways <- as.data.frame(fread("mirnet_enrichment_
reactome_1.csv"))
to_keep <- c("G2/M Checkpoints", "Cell junction
organization")
pathways <- pathways[pathways$Pathway %in%
to_keep,]
g2m_genes <
- unlist(strsplit(pathways$Genes[pathways$Pathway =
= to_keep[1]], split = ","))
cell_junction_genes <
- unlist(strsplit(pathways$Genes[pathways$Pathway =
= to_keep[2]], split = ","))
write.table(g2m_genes, file = "g2m_genes.txt",
quote = F, row.names = F, col.names = F)
write.table(cell_junction_genes, file = "cell_
junction_genes.txt", quote = F, row.names = F, col.
names = F)
```

Conflict of Interest The authors have no conflicts of interest to declare that are relevant to the content of this chapter.

References

1. Flynt AS, Lai EC (2008) Biological principles of microRNA-mediated regulation: shared themes amid diversity. Nat Rev Genet 9:831–842
2. Gurtan AM, Sharp PA (2013) The role of miRNAs in regulating gene expression networks. J Mol Biol 425:3582–3600
3. Johnston RJ, Chang S, Etchberger JF et al (2005) MicroRNAs acting in a double-negative feedback loop to control a neuronal cell fate decision. Proc Natl Acad Sci USA 102:12449–12454
4. Chang L, Zhou G, Soufan O et al (2020) miRNet 2.0: network-based visual analytics for miRNA functional analysis and systems biology. Nucleic Acids Res 48:W244–W251
5. Licursi V, Conte F, Fiscon G et al (2019) MIENTURNET: an interactive web tool for microRNA-target enrichment and network-based analysis. BMC Bioinform 20:545
6. Kern F, Aparicio-Puerta E, Li Y et al (2021) miRTargetLink 2.0-interactive miRNA target gene and target pathway networks. Nucleic Acids Res 49:W409–W416
7. Xie G-Y, Xia M, Miao Y-R et al (2020) FFLtool: a web server for transcription factor and miRNA feed forward loop analysis in human. Bioinformatics 36:2605–2607
8. Giroux P, Bhajun R, Segard S et al (2020) miRViz: a novel webserver application to visualize and interpret microRNA datasets. Nucleic Acids Res 48:W252–W261
9. Bonnici V, Caro GDe, Constantino G et al (2018) Arena-Idb: a platform to build human non-coding RNA interaction networks. BMC Bioinformatics 19:350
10. Colaprico A, Silva TC, Olsen C et al (2016) TCGAbiolinks: an R/bioconductor package for integrative analysis of TCGA data. Nucleic Acids Res 44:e71
11. Fan Y, Siklenka K, Arora SK et al (2016) miRNet - dissecting miRNA-target interactions and functional associations through network-based visual analysis. Nucleic Acids Res 44:W135–W141
12. Kozomara A, Birgaoanu M, Griffiths-Jones S (2019) miRBase: from microRNA sequences to function. Nucleic Acids Res 47:D155–D162
13. Huang H-Y, Lin Y-C-D, Cui S et al (2022) miRTarBase update 2022: an informative resource for experimentally validated miRNA-target interactions. Nucleic Acids Res 50:D222–D230
14. Skoufos G, Kakoulidis P, Tastsoglou S et al (2024) TarBase-v9.0 extends experimentally supported miRNA-gene interactions to cell-types and virally encoded miRNAs. Nucleic Acids Res 52:D304–D310
15. Jiang Q, Wang Y, Hao Y et al (2009) miR2Disease: a manually curated database for microRNA deregulation in human disease. Nucleic Acids Res 37:D98–104
16. Cui C, Zhong B, Fan R et al (2024) HMDD v4.0: a database for experimentally supported human microRNA-disease associations. Nucleic Acids Res 52:D1327–D1332
17. Ruepp A, Kowarsch A, Schmidl D et al (2010) PhenomiR: a knowledgebase for microRNA expression in diseases and biological processes. Genome Biol 11:R6
18. Li J-H, Liu S, Zhou H et al (2014) starBase v2.0: decoding miRNA-ceRNA, miRNA-ncRNA and protein-RNA interaction networks from large-scale CLIP-Seq data. Nucleic Acids Res 42:D92–D97
19. Tong Z, Cui Q, Wang J et al (2019) TransmiR v2.0: an updated transcription factor-microRNA regulation database. Nucleic Acids Res 47:D253–D258
20. Han H, Cho J-W, Lee S et al (2018) TRRUST v2: an expanded reference database of human and mouse transcriptional regulatory interactions. Nucleic Acids Res 46:D380–D386
21. Liu Z-P, Wu C, Miao H et al (2015) RegNetwork: an integrated database of transcriptional and post-transcriptional regulatory networks in human and mouse. Database (Oxford) 2015:bav095
22. ENCODE Project Consortium (2004) The ENCODE (ENCyclopedia Of DNA elements) project. Science 306:636–640
23. Fornes O, Castro-Mondragon JA, Khan A et al (2020) JASPAR 2020: update of the open-access database of transcription factor binding profiles. Nucleic Acids Res 48:D87–D92
24. Keenan AB, Torre D, Lachmann A et al (2019) ChEA3: transcription factor enrichment analysis by orthogonal omics integration. Nucleic Acids Res 47:W212–W224
25. Oak N, Ghosh R, Huang K-L et al (2019) Framework for microRNA variant annotation and prioritization using human population and disease datasets. Hum Mutat 40:73–89
26. Bhattacharya A, Ziebarth JD, Cui Y (2014) PolymiRTS Database 3.0: linking

polymorphisms in microRNAs and their target sites with human diseases and biological pathways. Nucleic Acids Res 42:D86–D91
27. Kumar S, Ambrosini G, Bucher P (2017) SNP2TFBS – a database of regulatory SNPs affecting predicted transcription factor binding site affinity. Nucleic Acids Res 45:D139–D144
28. Guo Z, Maki M, Ding R et al (2014) Genome-wide survey of tissue-specific microRNA and transcription factor regulatory networks in 12 tissues. Sci Rep 4:5150
29. Palmieri V, Backes C, Ludwig N et al (2018) IMOTA: an interactive multi-omics tissue atlas for the analysis of human miRNA-target interactions. Nucleic Acids Res 46:D770–D775
30. Liu P, Ewald J, Pang Z et al (2023) ExpressAnalyst: a unified platform for RNA-sequencing analysis in non-model species. Nat Commun 14:2995
31. Ritchie ME, Phipson B, Wu D et al (2015) limma powers differential expression analyses for RNA-sequencing and microarray studies. Nucleic Acids Res 43:e47
32. Robinson MD, McCarthy DJ, Smyth GK (2010) edgeR: a bioconductor package for differential expression analysis of digital gene expression data. Bioinformatics 26:139–140
33. Love MI, Huber W, Anders S (2014) Moderated estimation of fold change and dispersion for RNA-seq data with DESeq2. Genome Biol 15:550
34. Löbrich M, Jeggo PA (2007) The impact of a negligent G2/M checkpoint on genomic instability and cancer induction. Nat Rev Cancer 7:861–869
35. Brabletz T, Kalluri R, Nieto MA et al (2018) EMT in cancer. Nat Rev Cancer 18:128–134
36. Polyak K, Weinberg RA (2009) Transitions between epithelial and mesenchymal states: acquisition of malignant and stem cell traits. Nat Rev Cancer 9:265–273
37. Liu X, Wang S, Meng F et al (2013) SM2miR: a database of the experimentally validated small molecules' effects on microRNA expression. Bioinformatics 29:409–411
38. Longley DB, Harkin DP, Johnston PG (2003) 5-fluorouracil: mechanisms of action and clinical strategies. Nat Rev Cancer 3:330–338
39. Agarwal V, Bell GW, Nam J-W et al (2015) Predicting effective microRNA target sites in mammalian mRNAs. Elife 4:e05005
40. Yanaihara N, Caplen N, Bowman E et al (2006) Unique microRNA molecular profiles in lung cancer diagnosis and prognosis. Cancer Cell 9:189–198
41. Kumar MS, Erkeland SJ, Pester RE et al (2008) Suppression of non-small cell lung tumor development by the let-7 microRNA family. Proc Natl Acad Sci U S A 105:3903–3908
42. Hamberg M, Backes C, Fehlmann T et al (2016) MiRTargetLink–miRNAs, genes and interaction networks. Int J Mol Sci 17:564
43. Tokar T, Pastrello C, Rossos AEM et al (2018) mirDIP 4.1-integrative database of human microRNA target predictions. Nucleic Acids Res 46:D360–D370
44. Wong N, Wang X (2015) miRDB: an online resource for microRNA target prediction and functional annotations. Nucleic Acids Res 43:D146–D152
45. Kern F, Fehlmann T, Solomon J et al (2020) miEAA 2.0: integrating multi-species microRNA enrichment analysis and workflow management systems. Nucleic Acids Res 48:W521–W528
46. Kehl T, Kern F, Backes C et al (2020) miRPathDB 2.0: a novel release of the miRNA Pathway Dictionary Database. Nucleic Acids Res 48:D142–D147
47. Gerstner N, Kehl T, Lenhof K et al (2020) GeneTrail 3: advanced high-throughput enrichment analysis. Nucleic Acids Res 48:W515–W520
48. Zhang Q, Liu W, Zhang H-M et al (2020) hTFtarget: a comprehensive database for regulations of human transcription factors and their targets. Genom Proteom Bioinform 18:120–128
49. Chou C-H, Shrestha S, Yang C-D et al (2018) miRTarBase update 2018: a resource for experimentally validated microRNA-target interactions. Nucleic Acids Res 46:D296–D302
50. Xiao F, Zuo Z, Cai G et al (2009) miRecords: an integrated resource for microRNA-target interactions. Nucleic Acids Res 37:D105–D110
51. Karagkouni D, Paraskevopoulou MD, Chatzopoulos S et al (2018) DIANA-TarBase v8: a decade-long collection of experimentally supported miRNA-gene interactions. Nucleic Acids Res 46:D239–D245
52. Sticht C, La Torre CDe, Parveen A et al (2018) miRWalk: an online resource for prediction of microRNA binding sites. PLOS One 13:e0206239
53. Vejnar CE, Blum M, Zdobnov EM (2013) miRmap web: comprehensive microRNA target prediction online. Nucleic Acids Res 41:W165–W168
54. Yu G, Wang L-G, Han Y et al (2012) clusterProfiler: an R package for comparing biological themes among gene clusters. OMICS 16:284–287
55. Jensen MA, Ferretti V, Grossman RL et al (2017) The NCI genomic data commons as an engine for precision medicine. Blood 130:453–459

56. Liu C-J, Hu F-F, Xia M-X et al (2018) GSCALite: a web server for gene set cancer analysis. Bioinformatics 34:3771–3772
57. Maragkakis M, Reczko M, Simossis VA et al (2009) DIANA-microT web server: elucidating microRNA functions through target prediction. Nucleic Acids Res 37:W273–W276
58. Lewis BP, Burge CB, Bartel DP (2005) Conserved seed pairing, often flanked by adenosines, indicates that thousands of human genes are microRNA targets. Cell 120:15–20
59. Panwar B, Omenn GS, Guan Y (2017) miRmine: a database of human miRNA expression profiles. Bioinformatics 33:1554–1560
60. Ludwig N, Leidinger P, Becker K et al (2016) Distribution of miRNA expression across human tissues. Nucleic Acids Res 44:3865–3877

Check for updates

Chapter 8

Computational Methods for Pathway and Network Analysis with MicroRNAs

Salvatore Alaimo and Alfredo Pulvirenti

Abstract

Pathway analysis is a key computational framework for interpreting high-throughput molecular data and contextualizing biological alterations within known cellular processes. The integration of microRNAs (miRNAs), key post-transcriptional regulators of gene expression, has substantially expanded the ability of pathway-based approaches to capture complex regulatory mechanisms underlying human diseases. Although methodologies such as Over-Representation Analysis (ORA), Functional Class Scoring (FCS), and Pathway Topology (PT)-based methods established the field's foundation, recent advances have profoundly reshaped both its methodological landscape and its biological scope. This chapter reviews the emerging computational paradigms driving the next generation of miRNA-integrated pathway analysis. In particular, it highlights three major developments: the rise of multi-omics integration, which enables the joint analysis of heterogeneous molecular layers such as transcriptomics, methylomics, and proteomics; the growing adoption of deep learning approaches, especially Graph Neural Networks (GNNs), for modeling complex and non-linear dependencies in biological networks; and the advent of single-cell sequencing, which provides unprecedented resolution for investigating cellular heterogeneity, dynamic regulatory programs, and causal relationships along pseudotemporal trajectories. The chapter also discusses recent advances in subpathway and network module analysis, which shift attention from individual pathways to coordinated network behavior and higher-order regulatory organization. Overall, these developments reflect a broader transition toward more integrative, dynamic, and mechanistically informative models of miRNA-mediated regulation, expanding our capacity to characterize complex biological systems.

Keywords Pathway analysis, MiRNA-integrated analysis, Multi-omics integration, Graph neural networks, Subpathway analysis, Regulatory networks, Causal inference

1 Introduction

Pathway analysis represents a fundamental class of computational methodologies for interpreting high-throughput data, allowing for the contextualization of molecular alterations within known biological processes [1, 2]. In this context, microRNAs (miRNAs) have emerged as post-transcriptional regulators of crucial importance, capable of orchestrating complex regulatory networks by binding to their target mRNAs to promote their degradation or inhibit their

Alessandro Laganà and Giovanni Nigita (eds.), *MicroRNA Target Identification*, Methods in Molecular Biology, vol. 3056, https://doi.org/10.1007/978-1-0716-5392-0_8,

translation [3]. The integration of miRNA information into pathway analysis has therefore promised to significantly increase the accuracy and depth of our understanding of disease mechanisms [1, 2].

The foundational methods for pathway analysis, historically classified into generations such as Over-Representation Analysis (ORA), Functional Class Scoring (FCS), and Pathway Topology (PT)-based methods, have laid the groundwork for this field [1, 2]. However, the last few years have witnessed a radical transformation. This evolution has not simply produced more accurate tools but has catalyzed a fundamental shift in the scientific questions that can be asked. Research has moved from the simple question, "Which pathways are enriched?" to more complex questions such as, "How do different regulatory layers, including miRNAs, interact to drive pathway activity in specific contexts, in certain cell types, or even in individual patients?".

Three main ideas have driven this transformation:

1. The proliferation of **multi-omics data** has made it essential to develop methods capable of integrating heterogeneous data layers (e.g., transcriptomics, methylomics, proteomics) far beyond the simple miRNA-mRNA pair to obtain a holistic view of biological processes [4, 5].
2. The maturation of **deep learning**, particularly Graph Neural Networks (GNNs), which offer entirely new approaches to modeling the complex non-linear relationships inherent in biological networks by learning directly from their structure [6, 7].
3. The advent of **single-cell sequencing** provides unprecedented resolution to study cellular heterogeneity and dynamic processes. This technology has required the development of innovative algorithms capable of handling the unique statistical properties of these data, such as sparsity and technical noise, and of inferring causal relationships from trajectory data [8–10].

The objective of this chapter is to provide an overview of these new paradigms. It will explore how foundational principles have been refined and how the new frontiers of multi-omics analysis, deep learning, and single-cell are reshaping the field.

2 Section 1: The Foundations of Pathway Analysis

While new paradigms emerge, the foundational concepts of pathway analysis continue to evolve. Modern methods have overcome the limitations of their predecessors by introducing more sophisticated statistical tests, integrating broader knowledge bases, and being encapsulated in robust and interoperable software platforms. This section explores how the basic principles of enrichment and topological analysis have been refined to offer more robust and biologically meaningful analyses.

2.1 A Concise Review of Foundational Principles

To understand the evolution of the field, it is useful to briefly revisit the three generations of methods that form the basis of pathway analysis [1, 2].

- **Over-Representation Analysis (ORA):** The first generation of methods, such as DIANA-miRPath and Onto-Express, is based on a simple hypothesis: a pathway is biologically relevant if it contains a statistically greater number of differentially expressed genes (DEGs) than would be expected by chance. These methods typically use statistical tests like the hypergeometric test, chi-squared test, or binomial test to calculate a *p*-value [1, 2]. Their main drawback is that they ignore the magnitude of expression change and only consider genes that pass an arbitrary significance threshold, effectively treating pathways as simple "gene bags."
- **Functional Class Scoring (FCS):** The second generation, which includes seminal methods like GSEA (Gene Set Enrichment Analysis), overcame some of ORA's limitations. FCS methods consider the expression of all genes in a pathway, not just the differentially expressed ones. They first calculate a single-gene statistic (e.g., t-test, signal-to-noise ratio) and then aggregate these statistics into a pathway-level score (e.g., Kolmogorov-Smirnov statistic). This approach allows for the detection of pathways where small but coordinated changes in expression have a significant biological effect [1, 2, 11]. However, FCS methods also treat pathways as sets of genes, ignoring the network topology, i.e., the specific interactions between genes.
- **Pathway Topology (PT)-based Methods:** The third generation of methods, such as SPIA and MITHrIL, introduced topological information into the analysis. These approaches model pathways as graphs, where nodes are genes/proteins and edges represent interactions (e.g., activation, inhibition). A pathway's score is calculated by considering not only the expression of genes but also their position and role within the network. For example, an upstream gene with many downstream connections may have a greater impact than an isolated gene [1, 2]. These methods represent a significant step forward, but their signal propagation rules are often based on predefined heuristics.

The inherent limitations of these three generations, such as sensitivity to thresholds, treating pathways as isolated entities, and using fixed rules for signal propagation, have driven the development of the more modern approaches discussed in the following sections.

2.2 Novel Approaches to Enrichment

The recent evolutions of enrichment analysis offer a suite of methods, rather than relying on a single test, recognizing that the optimal approach depends on the nature of the input data (e.g., binary gene list vs. continuous expression data) and the specific biological question, allowing researchers to customize the analysis based on the characteristics of their experiment.

An example of this evolution is the R package enrichMiR. This versatile toolkit implements several statistical tests to identify miRNAs whose targets are enriched in a gene set of interest or in a differential expression signature [12]. Its strength lies in offering tests optimized for different data types and hypotheses. enrichMiR distinguishes between tests based on "binary" signal (set membership) and "continuous" signal (a numerical score for each gene).

In **Binary Signal Tests**, the input is a list of genes of interest (e.g., significantly down-regulated genes) compared against a background. In this mode, two types of tests can be used: Fisher's exact test or weighted hypergeometric. The Fisher's exact test counts the number of miRNA binding sites on target transcripts. Although this technically violates the assumption of independence of counts (since all sites on a transcript are either in or out of the set), in practice, it proves excellent at identifying the most enriched miRNAs. The weighted hypergeometric test is a variant of Fisher's test that corrects for the length of UTRs (untranslated regions) using the Wallenius method, implemented via the goseq package. This correction mitigates biases because longer genes are more likely to contain binding sites by chance.

In **Continuous Signal Tests**, the entire spectrum of expression data, such as log-fold changes or *p*-values, is tested without the need for a threshold. Two types of analysis are supported in this mode: analytic Rank-based Enrichment Analysis (aREA) or a linear fit analysis. The aREA test, implemented in the viper package, is an analytical and faster version of GSEA. It can also incorporate the probability of set membership, using miRNA repression scores as prior probabilities if available. The linear fit analysis is an extension of the eBayes method from limma. It fits consecutive linear models by adding the most significant miRNAs one at a time. This approach is beneficial for identifying candidate miRNAs that are not redundant with the top-scoring miRNA, providing a more comprehensive view of the regulatory landscape.

The availability of multiple tests within a single framework like enrichMiR represents a maturation of the field, moving from monolithic solutions to a flexible, data-driven approach where the choice of statistical test becomes a critical and conscious step in the analysis process.

2.3 Integrated Database Platforms

In parallel with the development of new algorithms, databases, and web servers have evolved from simple data repositories to integrated and interactive analysis platforms.

miRPathDB 2.0 is a platform that represents a substantial update to its previous version. It has enormously expanded its knowledge base, now including data for tens of thousands of miRNAs, their targets, and associated pathways in Homo sapiens and Mus musculus [13]. Its new features include interactive visualizations like miRNA-pathway heatmaps, which intuitively show which pathways are regulated by a given set of miRNAs. One of the most significant innovations is the implementation of an **Integer Linear Program (ILP)**. This optimization approach allows one to identify the minimal set of miRNAs whose combined targetome covers a user-provided pathway or gene set.

miRTargetLink 2.0 is a tool designed as an "all-in-one" solution for the construction and analysis of miRNA regulatory networks in humans, mice, and rats [14]. Unlike a simple database, miRTargetLink 2.0 allows users to build networks starting from a single gene, a set of genes, a miRNA, or a set of miRNAs. Its strength lies in its integration with downstream enrichment tools like miEAA and GeneTrail, which enable a contextualized functional analysis of the network nodes. The interactive web interface, based on Cytoscape.js, facilitates the exploration and interpretation of these complex networks.

2.4 The Current Landscape of Bioconductor Packages for miRNA-Pathway Analysis

Bioconductor is a central ecosystem for the development and distribution of robust and interoperable bioinformatics tools. For miRNA-integrated pathway analysis, several R packages offer specialized functionalities.

- **miRNApath:** This package was specifically designed to address the complexity of "many-to-many" relationships between miRNAs and genes. It recognizes that one miRNA can target multiple genes and multiple miRNAs can target one gene. Its approach manages miRNA-gene and gene-pathway relationships separately, preserving the additive and combinatorial effects of these interactions, an aspect often overlooked by simpler methods [15].
- **MIRit:** An acronym for "miRNA integration tool," MIRit serves as a comprehensive framework that guides the researcher through the entire process of an integrated miRNA-mRNA analysis [16]. Its functionalities range from differential expression analysis (supporting data from RNA-Seq, microarrays, and other technologies) to expression level integration (via correlation for paired data or rotation tests for unpaired data) and functional and network enrichment analysis. Its "all-in-one" nature makes it a powerful tool for deciphering the complexity of miRNA biology.
- **topGO:** Although not specific to miRNAs, topGO remains a fundamental tool in the Bioconductor ecosystem for Gene Ontology (GO) term enrichment analysis. Its distinctive feature

is the ability to account for the topology of the GO graph, using algorithms like "eliminate" and "weight" to reduce dependencies between hierarchically related GO terms and decrease false positives [17].

The availability of these packages within a standardized environment like Bioconductor not only ensures high software quality but also promotes reproducibility and interoperability, allowing for the construction of complex and customized analysis pipelines.

3 Section 2: Multi-Omics Integration

The shift from two-layer analysis (miRNA-mRNA) to true multi-omics integration represents one of the most recent advances. The rationale behind this change is clear: biological processes are the result of a complex interplay between multiple molecular levels, including the genome, epigenome, transcriptome, and proteome. Analyzing a single omics layer provides a partial and often insufficient view for understanding pathway regulation [5]. Integrating miRNA and mRNA expression data with DNA methylation, copy number variation (CNV), or protein abundance data can reveal regulatory mechanisms that would otherwise remain invisible.

However, multi-omics integration presents unique computational challenges related to the heterogeneity, dimensionality, and different scales of the data. In response to these challenges, several algorithmic approaches have emerged. These approaches range from the statistical fusion of easily interpretable scores to the generation of abstract but powerful feature representations.

3.1 ActivePathways: Directional P-Value Merging

At the most interpretable end of the spectrum is ActivePathways, an R package that implements a pathway enrichment analysis by integrating multiple omics datasets through statistical fusion [18]. This approach is based on statistics familiar to biologists, such as *p*-values and fold-changes, making its results relatively easy to interpret in the context of a biological hypothesis. The main innovation of ActivePathways is its **Directional P-value Merging (DPM)** method. This method not only combines statistical significance (*p*-values) from different omics sources but also incorporates the expected direction of change (e.g., up- or down-regulation).

The method requires as input a matrix of *p*-values (scores), where rows are genes and columns are different omics datasets (e.g., mRNA expression, protein expression, methylation). Crucially, it also requires a corresponding matrix of directional values (typically log2 fold-changes) and a constraints vector that defines the expected directional relationship between the datasets. The constraints vector allows the user to specify biological hypotheses. For example, to integrate mRNA and protein expression, one would expect a positive correlation. The constraints vector would be [1, 1], rewarding genes where both levels increase or decrease concordantly. To integrate a miRNA and its target mRNA, one would expect an inverse relationship. The constraint could be (1, -1), which rewards cases where miRNA expression is high and mRNA expression is low, or vice versa.

The DPM method, an extension of Brown's method for combining dependent *p*-values, uses these constraints to calculate a combined *p*-value for each gene. Genes that satisfy the directional constraints receive a lower (more significant) combined *p*-value, while those that violate the constraints are penalized, obtaining a higher combined *p*-value. This merging process produces a single ranked list of genes that integrates evidence from all omics sources in a biologically informed way. The resulting gene list is then used for a classic pathway enrichment analysis. The final output is a list of significantly enriched pathways, with the added benefit of being able to trace which omics sources (and with what direction) contributed to the significance of each pathway.

ActivePathways is therefore ideal for hypothesis generation, where the clear biological interpretability of the results is necessary.

3.2 MOPA: Multi-Omics Pathway Analysis

MOPA (Multi-Omics Pathway Analysis) uses tensor decomposition to analyze multi-omics data, offering a trade-off between computational power and interpretability. Instead of producing a single list of enriched pathways for an entire dataset, MOPA calculates pathway activity scores for each sample [19].

MOPA is based on the decomposition of a multi-omics data tensor [25]. Multi-omics data (e.g., gene expression, miRNA expression, methylation) are organized into a three-dimensional tensor (genes × samples × omics types). MOPA uses non-negative PARAFAC tensor decomposition to break this tensor down into three lower-dimensional matrices: a **Sample (S)** matrix, a **Gene (G)** matrix, and an **Omics (O)** matrix. These matrices represent the latent "signatures" of samples, genes, and omics types in the data. Then, the **multi-omics Enrichment Score (mES)** is calculated for each pathway in each sample. It is conceptually similar to GSEA or GSVA, but instead of using expression data directly, it uses a combination of the latent S and G components. For each gene in a given sample, an aggregated multi-omics score is calculated. These scores are then used in a Kolmogorov-Smirnov-like random walk algorithm to determine if the genes of a particular pathway are enriched among the genes with the highest or lowest scores. Finally, an **Omics Contribution Rate (OCR)** is computed. Using the O matrix derived from the decomposition, the OCR quantifies the percentage contribution of each omics type to the mES score of a given pathway in a group of samples. For example, the analysis might reveal that in one cancer subtype, the "Cell Cycle Pathway" is driven 70% by gene expression and 30% by methylation. In contrast, in another subtype, it is driven primarily by miRNA expression.

MOPA represents a step towards personalized medicine by providing single-sample level scores. Although mES scores are more abstract than a *p*-value, the OCR offers a unique mechanism to "open the black box" and interpret which molecular layers are driving the pathway perturbation.

3.3 omicsGAN: Multi-Omics Data Integration by Generative Adversarial Network

omicsGAN is a novel deep-learning framework that does not seek to combine scores or decompose data but uses a Generative Adversarial Network (GAN) to generate entirely new synthetic data [4]. This synthetic data fuses information from two omics layers and their interaction network, with the primary goal of maximizing predictive performance in downstream machine learning tasks. It is a biologically motivated deep learning framework. The model is a GAN, composed of a Generator (**G**) and a Critic (**C**) (a discriminator in a variant known as a Wasserstein GAN, or wGAN, which improves training stability). The model takes as input two omics datasets, for example, mRNA expression (X) and miRNA expression (Y), and the normalized adjacency matrix of their interaction network ($\widetilde{S}$). It then iteratively updates them to find two new matrices that incorporate information from both omics data and their biological interactions. A classification model is then applied to the new matrices to predict the disease phenotype.

The core of omicsGAN is an iterative, cascaded process where each step, or "update," refines the omics datasets. Within each update block, two parallel wGANs are trained independently—one for each omics type. The process for generating a synthetic mRNA dataset in the k-th update serves as an example. First, the Generator, a three-layer fully connected neural network, takes the miRNA expression data from the previous update, $H_Y^{(k-1)}$, and the transposed interaction network, $\widetilde{S}^T$, to generate a new, intermediate mRNA dataset, $b_X^{(k)}$. This synthetic dataset explicitly uses the biological relationships in the network to translate information from the miRNA layer into the mRNA feature space. Second, the Critic's role is to distinguish between the real mRNA data, $H_X^{(k-1)}$, and the generated data, $b_X^{(k)}$, by assigning a higher score to real samples. Third, during training, the generator and critic are trained in an adversarial game where the generator tries to produce data so realistic that it "fools" the critic into assigning it a high score. A parallel process co-occurs to generate an updated miRNA dataset, $H_Y^{(k)}$, using the mRNA data, $H_X^{(k-1)}$, and the network as input.

The training of the wGANs is guided by specific objective functions. The intermediate data $b_X^{(k)}$ is generated by passing the miRNA data from the previous step and the network through the generator G:

$$b_X^{(k)} = G(H_Y^{(k-1)}, \widetilde{S}^T)$$

The critic, C, is trained by minimizing its loss function, L_C, which pushes the scores of real and synthetic data apart:

$$L_C = C\left(b_X^{(k)}\right) - C\left(H_X^{(k-1)}\right)$$

The generator, G, is trained by minimizing its loss function, L_G. This loss has two components: an adversarial term that encourages the generation of realistic data, and an L_2-norm regularization term that ensures the synthetic data preserves the characteristics of the original mRNA expression data (X).

$$L_G = -C\left(b_X^{(k)}\right) - \alpha \left|\left|b_X^{(k)} - X\right|\right|_2$$

Here, $\alpha$is a hyperparameter that balances the two terms of the loss function. The framework is not a single-pass system; the updated datasets from one block, $(H_Y^{(k)}, H_Y^{(k)})$ serve as the inputs for the next block, $(k+1)$. The initial inputs are the original datasets, where $H_X^{(0)} = X$ and $H_Y^{(0)} = Y$. This process is repeated for a predefined number of updates, K. A crucial step is the final selection of the synthetic data. Each of the K updated datasets is evaluated on a separate validation set, and the step that achieved the highest Area Under the Curve (AUC) score is selected as the optimal one.

The study demonstrates that the synthetic datasets generated by omicsGAN yield significantly better performance in both cancer outcome classification and patient survival prediction compared to the original omics data or a simple concatenation of the two. Most importantly, the authors proved the model's reliance on biologically meaningful relationships. When the true miRNA-mRNA interaction network was replaced with a randomized network of the same density, the predictive performance of the resulting synthetic data dropped significantly. This result confirms that omicsGAN is not merely blending two datasets; it is functionally leveraging the interaction network to guide the integration, resulting in synthetic data with stronger and more coherent biological signatures.

However, omicsGAN is a feature engineering tool, best suited for building high-performance predictive models where performance is prioritized over direct interpretability. The choice between Active-Pathways, MOPA, and omicsGAN thus depends on the final research goal: generating interpretable hypotheses, calculating sample-level scores, or building predictive classifiers.

4 Section 3: Deep Learning on Biological Networks

Another computational revolution in recent years has been the application of deep learning, and particularly Graph Neural Networks (GNNs), to the analysis of biological networks. Traditional PT methods rely on manually defined topological features (e.g., centrality, signal propagation rules) based on a priori biological hypotheses [1, 2, 20]. GNNs, in contrast, are designed to *learn* the relevant topological features directly from the data and the network structure in an end-to-end fashion, optimizing them for a specific task [7].

Pathways and molecular interaction networks are inherently graphs. Traditional methods, while considering topology, often simplify it into linear or fixed-rule-based models. GNNs overcome these limitations because they are deep learning models natively designed to operate on structured data like graphs. They are capable of capturing complex, non-linear, and context-dependent relationships between nodes (genes, miRNAs) through an iterative mechanism of aggregating information from each node's neighborhood. This aggregation allows for modeling the flow of information in a pathway in a much more flexible and data-driven way than previous approaches.

4.1 Fundamental Concepts: Graph Embeddings and Message Passing

To understand how GNNs work, two key concepts must be introduced.

Graph Embeddings: The central idea is to represent each node (e.g., a gene or miRNA) in a graph as a low-dimensional vector in a continuous space, called an "embedding." This vector captures the node's properties, particularly its relationships with other nodes in the network. A foundational method for creating these embeddings is DeepWalk. DeepWalk performs random walks on the graph starting from each node. The sequences of visited nodes are treated as "sentences," and a natural language processing model like word2vec is used to learn an embedding for each node. Nodes that appear in similar contexts (i.e., in the same random walks) will have similar vector embeddings [21]. These embeddings can then be used as features for downstream machine learning tasks.

Message Passing: The mechanism at the heart of most GNNs is "message passing." It is an iterative process that takes place in multiple layers. In each layer, for each node in the graph:

1. **Aggregation (or Message Passing):** The node collects the feature vectors (or embeddings) of its neighboring nodes.
2. **Update:** The node updates its own feature vector by combining its previous vector with the aggregated information from its neighbors, using a neural network.

After k layers of message passing, a node's embedding will contain information about its neighborhood up to k "hops" away, allowing the GNN to learn representations that encode the local network structure around each node. The update function, being a neural network, has weights that are learned during training to minimize a loss function, for example, the prediction error. In this sense, the GNN is not simply calculating a score based on topology but is learning a function *of* the topology that is most relevant to the biological outcome of interest.

4.2 GNN Architectures for Integrated Analysis with miRNAs

Several GNN architectures and related approaches have been proposed for biological analysis:

- **Graph Convolutional Networks (GCNs):** GCNs are one of the most popular GNN architectures and represent an analogy of image convolution applied to graphs. An example application is a model that uses a GCN to capture feature vectors from miRNA and disease similarity networks, preserving the network structure to improve the prediction of associations [22].
- **Convolutional Neural Networks (CNNs) on Network Features:** Although not an end-to-end GNN, the MDA-CNN model is an instructive example of how deep learning can be applied to network-derived features [6]. In this approach, interaction features are first extracted from a three-layer network (miRNA–Protein–Protein Interaction–Disease). These features are then fed into a standard Convolutional Neural Network (CNN) for the final prediction. It can be seen as an intermediate step towards GNNs, where feature extraction and prediction occur in a single learning process.

More recently, research on GNNs is rapidly evolving, adopting more advanced techniques [7], such as **Attention** or **Contrastive Learning**.

Attention allows the GNN to dynamically learn the importance of different neighboring nodes during aggregation, instead of treating them all equally. This is particularly useful in biological networks, where not all interactions have the same weight.

Contrastive Learning is a self-supervised learning technique that helps learn more robust embeddings. The model learns to make the embeddings of "similar" nodes (e.g., two genes in the same pathway) similar and the embeddings of "different" nodes dissimilar.

4.3 Applications: From Association Prediction to Pathway Interpretation

Currently, many applications of GNNs in this field focus on predicting miRNA-disease associations. However, the methodology is directly transferable and applicable to pathway analysis. For example:

- Gene and miRNA embeddings learned by a GNN can be used as input features for more traditional pathway scoring methods.
- A GNN could be trained end-to-end to directly predict the perturbation state of a pathway (e.g., active/inactive) starting from a multi-layer graph that integrates gene-gene, gene-miRNA, and miRNA-miRNA interactions, along with expression data as node attributes.

The shift to GNNs involves a trade-off. On one hand, these models have vastly superior expressive power and can discover new and non-obvious interaction patterns that rule-based models do not capture. On the other hand, they introduce the challenge of interpretability, often being considered "black box" models. Future research is focusing on developing methods to interpret the decisions of GNNs (e.g., using techniques like Shapley values, mentioned in a different context in [30]), a crucial step for translating their predictions into verifiable biological hypotheses.

5 Section 4: Single-Cell Resolution Analysis

Single-cell sequencing has opened a new frontier for the study of gene regulation. While "bulk" sequencing (on an aggregated tissue sample) provides an average signal from thousands or millions of cells, single-cell analysis allows for the analysis of the transcriptome of each cell [10]. This has revealed enormous heterogeneity within cell populations previously considered homogeneous and has made it possible to study dynamic biological processes, such as cell differentiation or tumor progression, with unprecedented resolution [9]. For miRNA-mediated pathway analysis, this means being able to investigate regulatory relationships directly in the context of the cell where they occur, overcoming the limitations of correlation analysis on aggregated data.

Single-cell level analysis is crucial for several reasons:

- **Cellular Heterogeneity:** It allows for the identification of rare cell populations and the study of pathway regulation specific to each cell type, information completely lost in bulk analysis [10].
- **Study of Dynamic Processes:** By ordering cells along a developmental or disease progression trajectory (a concept known as "pseudotime"), it is possible to transform static "snapshot" data into a pseudo-temporal sequence. This opens the door to inferring cause-and-effect relationships, a goal unattainable with static bulk data [8].
- **Direct Analysis of Regulation:** The development of techniques for co-profiling miRNAs and mRNAs from the same single cell ("co-assay") for the first time allows direct measurement of the effect of a miRNA on its target within the same cellular context, providing more direct evidence of regulation [23].

However, single-cell data analysis, particularly for miRNAs, presents unique challenges, such as high technical noise and sparsity. Indeed, scRNA-seq technologies have inherently higher technical noise than bulk RNA-seq. This can mask subtle biological signals, such as the variability that miRNAs exert on their target genes [24]. Furthermore, the data are extremely sparse. Many genes or miRNAs may have a count of zero in a cell, not because they are not expressed, but due to inefficiency in the capture and reverse transcription of RNA molecules. This phenomenon, known as "dropout," requires specific statistical methods to be handled correctly.

Finally, simultaneously profiling RNA molecules of very different sizes, like miRNAs (approx. 22 nucleotides) and mRNAs (thousands of nucleotides) from the same cell, is technically complex. Innovative approaches, such as "half-cell genomics," where the cell lysate is physically split into two fractions for separate analysis of small and large RNAs, are beginning to overcome this obstacle [23].

In response to these challenges and opportunities, new algorithms have been developed. An example is the **Pseudotime Causality (PTC)** algorithm, designed to infer causal miRNA-mRNA relationships from dynamic single-cell data [8]. This method represents an important advancement from **correlation** analysis to **causality** inference.

The algorithm is divided into three main phases.

1. **Phase 1: Pseudotime Analysis:** The first step is to transform the static dataset into a time series. Cells are ordered along a trajectory that represents a biological process (e.g., epithelial-mesenchymal transition, EMT). This ordering, or pseudotime, can be established using the expression of a known marker gene for the process (e.g., Vimentin for EMT) or through dedicated trajectory inference algorithms. The result is an expression matrix where the samples (cells) are temporally ordered.
2. **Phase 2: Gene Selection:** To reduce the search space, a two-step filter is applied. First, miRNAs and mRNAs with the highest expression variability across the trajectory (measured by Median Absolute Deviation, MAD) are selected, assuming they are the most relevant for the dynamic process. Second, for each chosen mRNA, a set of "plausible predictors" is defined using a target prediction database (e.g., TargetScan). This narrows the analysis to the most biologically likely miRNA-mRNA interactions.
3. **Phase 3: Causal Inference:** This is the core phase of the algorithm and is based on the principle of **causal invariance**. The idea is that if a set of miRNAs (X_S) is the true direct cause of an mRNA (Y), then the statistical relationship between them should remain stable (invariant) across different conditions or "environments."

   - **Linear Model:** The relationship between a target mRNA (Y) and its miRNA regulators (X_S) is modeled as a linear system: $Y_t = \mu + X_{t,S}\beta + \epsilon_t$
   - **Invariance Test:** The algorithm uses the different segments of the pseudotemporal trajectory as distinct "environments." It then tests whether the parameters of the linear model (specifically the regression coefficients beta and the noise variance σ^2) remain constant across these segments.

- **Identification of Causes:** If the model is invariant, the set of miRNAs S is accepted as a set of direct causes. If the **model** is *not* invariant, it means the statistical relationship changes along the process, suggesting that the set S contains spurious correlations or that a true cause is missing. The algorithm systematically tests subsets of the plausible predictors to find the largest set that satisfies the invariance condition.

The application of formal causal inference frameworks, made possible by the temporal dimension of single-cell data, marks a fundamental advance. It allows us to go beyond simply describing which molecules are correlated and to start building mechanistic and predictive models of gene regulation, describing how regulatory relationships change dynamically as a cell traverses a biological process. This is an essential step towards understanding and, in the future, controlling cell fate.

6 Section 5: Advanced Subpathway and Network Module Analysis

Subpathway analysis is based on the premise that biological perturbations rarely involve entire pathways uniformly. More often, specific local regions or functional modules are altered [1, 2]. Identifying these subpathways provides a more granular and biologically relevant view of disease mechanisms. Modern methods have evolved this concept, moving from the analysis of local structures within single pathways to the characterization of global network dynamics and higher-order organization.

6.1 The Rationale of Subpathway Analysis

As highlighted in the original chapter, identifying pathway regions that show significant alterations in response to a certain phenotypic condition offers greater detail about its biology. The high redundancy and cross-talk between pathways suggest that subpathways can respond more sensitively and effectively than entire pathways. Recent methods have integrated miRNAs into these analyses and introduced more advanced graph theory-based approaches to define and evaluate these functional modules.

6.2 Directed Random Walks (miDRW)

The miDRW (miRNA-mediated subpathway activities by directed random walk) method is an example of how subpathway analysis has shifted towards a global network perspective. Instead of searching for paths between nodes of interest within a single pathway, miDRW evaluates the importance of a gene in the context of the entire network of known pathways [25].

The miDRW methodology is a multi-step process that integrates network topology, differential expression, and miRNA-gene interactions.

1. **Network Construction:** The first step is the construction of a **Global Directed Pathway Network (GDPN)**. This network is created by merging all KEGG pathways into a single large graph. A "virtual node" that connects all other nodes is added to create a "closed" network, facilitating the random walk algorithm.
2. **Node Scoring:** The importance of each gene (node) in the GDPN is calculated using a **directed random walk with restart**. This algorithm simulates a "walker" moving through the network. The initial probability of each node is based on its statistical significance (e.g., t-score from differential expression analysis). The algorithm converges to a steady state (p_∞), where each gene's score reflects both its global topological importance (e.g., a gene influencing many other genes will have a high score) and its expression-level alteration.
3. **Subpathway Activity Calculation:** The activity of a subpathway mediated by a miRNA (miR_j) is defined as an aggregated score of its target genes. The formula for the activity, $a\left(miR_j\right)$, is a weighted sum that integrates multiple sources of information:

$$a\left(miR_j\right) = \frac{\sum_{i=1}^{n_j} p_\infty\left(g_j\right) \times \left[\operatorname{sgn}\left(t\left(miR_j\right) \times t\left(g_j\right)\right)\right] \times z\left(g_j\right)}{\sqrt{\sum_{i=1}^{n_j} \left(p_\infty(g_i)\right)^2}}$$

where $p_\infty\left(g_j\right)$ is the topological weight of the target gene g_j, t is the differential t-score of the miRNA or gene, $z\left(g_j\right)$ is the normalized expression level of the gene, and sgn is the sign function that applies the negative regulation constraint (an up-regulated miRNA is paired with down-regulated target genes, and vice versa).

This approach shifts the analysis from a local to a global view: the importance of a gene is not defined solely by its parent pathway but by its position and influence in the entire universe of known pathways.

6.3 Co-Activated Pathway Cluster Analysis (PanomiR)

The PanomiR framework represents a further step in abstraction, shifting the focus from subpathways to **clusters of entire pathways** that are coordinately regulated. This approach is designed to uncover the higher-order organization of cellular regulation, identifying "meta-pathways" or functional modules orchestrated by specific miRNAs [26].

PanomiR identifies groups of co-regulated pathways and the miRNAs that control them through a four-step workflow:

1. **Pathway Summarization:** Gene expression data are transformed into pathway activity profiles. For each sample, a score is calculated for each pathway, for example, as the mean of the

squared rank of the genes belonging to that pathway. This produces a pathway activity matrix (pathways × samples).

2. **Co-activation Identification:** To find groups of pathways that behave similarly, PanomiR uses a **Pathway Co-expression Network (PCxN)**. In this network, nodes are pathways, and edges are weighted by the correlation between their activity profiles. Clustering algorithms are applied to this network to identify groups of strongly co-activated or co-inhibited pathways.
3. **miRNA Prioritization:** The final step is to identify which miRNAs regulate these pathway clusters. This is done using a pre-calculated enrichment table that associates miRNAs with pathways (based, for example, on a Fisher's exact test between a miRNA's targets and a pathway's genes).
4. **Statistical Evaluation:** For each miRNA and each pathway cluster, an observed "targeting" score is calculated. This score is then compared to a null distribution, generated through repeated sampling, to obtain a *p*-value. This *p*-value indicates the probability that a miRNA targets a given pathway cluster so strongly by pure chance.

PanomiR allows for the discovery of regulatory principles at a higher systems level. For example, it might reveal that a single miRNA coordinates seemingly unrelated processes, such as the "cell cycle" and "lipid metabolism," because these processes are functionally coupled and co-activated in a specific disease condition. This provides a much more holistic understanding of miRNA function than was previously possible.

7 Conclusion and Future Perspectives

In the past few years, there has been profound innovation in the computational analysis of miRNA-integrated pathways. The field has evolved far beyond the simple identification of enriched pathways, embracing techniques that reflect the growing complexity of biological data and scientific questions. The synthesis of major trends reveals a coherent movement towards more integrated, dynamic, causal, and multilevel analysis.

Despite these remarkable advances, several challenges remain and outline future research directions, such as interpretability, patient-specific modeling, or large-scale data integration.

In conclusion, miRNA-integrated pathway analysis is a dynamic and rapidly expanding field. The algorithmic advances of recent years have provided researchers with powerful tools to unravel the complexity of gene regulation. The future challenges, though significant, promise to further push the boundaries of our ability to model, predict, and understand biological systems.

Conflict of Interest The authors have no conflicts of interest to declare that are relevant to the content of this chapter.

References

1. Khatri P, Sirota M, Butte AJ (2012) Ten years of pathway analysis: current approaches and outstanding challenges. PLOS Comput Biol 8:e1002375
2. Alaimo S, Micale G, La Ferlita A et al (2019) Computational methods to investigate the impact of miRNAs on pathways. Methods Mol Biol 1970:183–209
3. Bartel DP (2004) MicroRNAs: genomics, biogenesis, mechanism, and function. Cell 116:281–297
4. Ahmed KT, Sun J, Cheng S et al (2021) Multi-omics data integration by generative adversarial network. Bioinformatics 38:179–186
5. Hasin Y, Seldin M, Lusis A (2017) Multi-omics approaches to disease. Genome Biol 18:83
6. Peng J, Hui W, Li Q et al (2019) A learning-based framework for miRNA-disease association identification using neural networks. Bioinformatics 35:4364–4371
7. Zitnik M, Agrawal M, Leskovec J (2018) Modeling polypharmacy side effects with graph convolutional networks. Bioinformatics 34:i457–i466
8. Cifuentes-Bernal AM, Pham VV, Li X et al (2021) A pseudotemporal causality approach to identifying miRNA-mRNA interactions during biological processes. Bioinformatics 37:807–814
9. Trapnell C (2015) Defining cell types and states with single-cell genomics. Genome Res 25:1491–1498
10. Hwang B, Lee JH, Bang D (2018) Single-cell RNA sequencing technologies and bioinformatics pipelines. Exp Mol Med 50:1–14
11. Subramanian A, Tamayo P, Mootha VK et al (2005) Gene set enrichment analysis: a knowledge-based approach for interpreting genome-wide expression profiles. Proc Natl Acad Sci USA 102:15545–15550
12. Soutschek M, Germade T, Germain P-L et al (2022) enrichMiR predicts functionally relevant microRNAs based on target collections. Nucleic Acids Res 50:W280–W289
13. Kehl T, Kern F, Backes C et al (2020) miRPathDB 2.0: a novel release of the miRNA pathway dictionary database. Nucleic Acids Res 48:D142–D147
14. Kern F, Aparicio-Puerta E, Li Y et al (2021) miRTargetLink 2.0-interactive miRNA target gene and target pathway networks. Nucleic Acids Res 49:W409–W416
15. Cogswell JP, Ward J, Taylor IA et al (2008) Identification of miRNA changes in Alzheimer's disease brain and CSF yields putative biomarkers and insights into disease pathways. J Alzheimers Dis 14:27–41
16. Ronchi J, Foti M (2023) MIRit: an integrative R framework for the identification of impaired miRNA-mRNA regulatory networks in complex diseases. Bioinform Adv 3(1):vbad185. http://biorxiv.org/lookup/doi/10.1101/2023.11.24.568528
17. Adrian Alexa JR (2017) topGO. Bioconductor
18. Slobodyanyuk M, Bahcheli AT, Klein ZP et al (2024) Directional integration and pathway enrichment analysis for multi-omics data. Nat Commun 15: 5690
19. Jeon J, Han EY, Jung I (2023) MOPA: an integrative multi-omics pathway analysis method for measuring omics activity. PLOS One 18:e0278272
20. Alaimo S, Giugno R, Acunzo M et al (2016) Post-transcriptional knowledge in pathway analysis increases the accuracy of phenotypes classification. Oncotarget 7:54572–54582
21. Perozzi B, Al-Rfou R, Skiena S (2014) DeepWalk. In: Proceedings of the 20th ACM SIGKDD international conference on Knowledge discovery and data mining, ACM, New York, NY, USA
22. Han P, Yang P, Zhao P et al (2019) Gcn-mf. In: Proceedings of the 25th ACM SIGKDD international conference on knowledge discovery & data mining. ACM, New York, NY, USA
23. Faridani OR, Abdullayev I, Hagemann-Jensen M et al (2016) Single-cell sequencing of the

small-RNA transcriptome. Nat Biotechnol 34:1264–1266
24. Schmiedel JM, Klemm SL, Zheng Y, et al (2015) Gene expression MicroRNA control of protein expression noise. Science 348:128–132
25. Ning Z, Feng C, Song C et al (2019) Topologically inferring active miRNA-mediated subpathways toward precise cancer classification by directed random walk. Mol Oncol 13:2211–2226
26. Naderi Yeganeh, P. Teo, Y. Y. Karagkouni, D. Pita-Juárez, Y. Morgan, S. L. Slack, F. J. ... & Hide, W. A. (2023). PanomiR: a systems biology framework for analysis of multi-pathway targeting by miRNAs. Briefings in bioinformatics, 24(6), bbad418

Check for updates

Chapter 9

Computational Resources for Studying miRNA-lncRNA Interactions

Francesco Orilio, Rosario Nicola Brancaccio, Francesco Verona, Giorgio Stassi, Sebastiano Di Bella and Giovanni Nigita

Abstract

The discovery of microRNAs (miRNAs), small non-coding RNAs approximately 22 nucleotides in length, revolutionized our understanding of post-transcriptional gene regulation in the early part of the twenty-first century. Since then, non-coding RNAs (ncRNAs) have become a major focus of scientific research, revealing their critical roles in cellular processes. Advances in high-throughput sequencing technologies have brought long non-coding RNAs (lncRNAs)—transcripts longer than 200 nucleotides—into the spotlight. While their functions are still being unraveled, lncRNAs have been identified as important regulators within the RNA interference (RNAi) pathway. They contain multiple miRNA response elements (MREs), allowing them to compete with other RNAs for miRNA binding. This competition, known as the competing endogenous RNA (ceRNA) or "sponge RNA" mechanism, has provided new insights into the complexity of gene regulation. This chapter explores the biomedical significance of ceRNAs, highlighting their role in modulating gene expression and their potential implications in disease. The chapter also summarizes available tools, resources, and practical examples for studying lncRNA-miRNA interactions to aid researchers in navigating this sophisticated regulatory network. It aims to support further advancements in this rapidly evolving field by bridging theoretical knowledge with practical applications.

Keywords miRNA, lncRNA, Sponges, RNA–RNA interactions

1 Introduction

Over the past two decades, non-coding RNAs (ncRNAs) have emerged as a central focus in molecular biology, driving intense research due to their critical roles in gene regulation. Among the various ncRNA biotypes, microRNAs (miRNAs) and long non-coding RNAs (lncRNAs) have garnered significant attention due to their importance in cellular processes. This progress has been fueled by the rapid development of high-throughput sequencing (HTS) technologies, which, when paired with sophisticated computational tools,

Sebastiano Di Bella and Giovanni Nigita contributed equally to this work.

Alessandro Laganà and Giovanni Nigita (eds.), *MicroRNA Target Identification*, Methods in Molecular Biology, vol. 3056, https://doi.org/10.1007/978-1-0716-5392-0_9,

have enabled the identification and classification of regulatory ncRNAs from RNA sequencing data. These advancements have provided valuable insights into their structural and functional characteristics[1–3]. miRNAs are among the most extensively studied and well-understood regulatory ncRNAs. In recent years, the catalog of annotated miRNAs has expanded significantly, with miRBase now reporting a total of 1917 identified human miRNAs [4]. RNA interference (RNAi) is a fundamental biological process driven by miRNAs, which regulate gene expression at the post-transcriptional level. This regulation occurs when mature miRNAs—single-stranded RNAs 19–25 nucleotides long—are incorporated into the RNA-induced silencing complex (RISC). Within the RISC, the argonaute protein Ago2 plays a critical role by using the miRNA as a guide to bind target mRNAs through base complementarity. If the pairing is nearly perfect, Ago2's endonuclease activity cleaves the target transcript, leading to its degradation. In cases of partial complementarity, the RISC mediates translational repression, often through mechanisms involving transcript deadenylation and decay. As a result, the proteins encoded by the miRNA-targeted mRNAs are downregulated. These mature miRNAs are generated through the transcription and processing of precursor molecules, highlighting their essential role in cellular gene regulation.[5, 6]. Long non-coding RNAs (lncRNAs) represent a diverse class of transcripts typically longer than 200 nucleotides. Most lncRNAs are 5′-capped, polyadenylated at the 3′ end, and spliced. In mammals, the majority of lncRNAs are transcribed by RNA polymerase II, although RNA polymerase III produces a small subset. According to the GENCODE classification system, lncRNAs are primarily categorized by length, cellular localization, and functional roles, reflecting their complexity and versatility in regulating cellular processes [7]. In addition to their well-known roles, ncRNAs can also be generated from the transcription of pseudogenes, which frequently overlap with regions of protein-coding genes. Intriguingly, certain lncRNAs, traditionally classified as non-coding, have been found to harbor open reading frames (ORFs) within their sequences. These ORFs can, in some cases, be translated into peptides or proteins, blurring the conventional boundary between coding and non-coding RNAs. This discovery highlights the functional complexity and versatility of ncRNAs, revealing unexpected regulatory and biological activity layers. For example, the RNF217-AS1-encoded peptide has been implicated in the tumorigenesis of stomach cancer (SC) [8]. However, the majority of these peptides are predicted to be unstable and functionally inefficient [9, 10], highlighting the nuanced nature of ncRNA-derived peptides and their potential contributions to biological processes. Overall, more than 35,000 independently annotated lncRNA genes exist in the human genome [11]. So far, just ~200 lncRNA genes have been studied biologically and are listed in lncRNAdb (v2) [12], a specialized, manually curated database dedicated to annotating lncRNAs. While the functions and molecular mechanisms of miRNAs have been extensively explored, the functional understanding of lncRNAs remains in its early stages, with many details yet to be uncovered. LncRNAs exhibit tissue- and cell-specific expression patterns and play key roles in regulating the growth and development of embryonic and adult stem cells [13] and controlling cellular proliferation [14]. The primary mechanism through which lncRNAs perform their functions is acting as RNA scaffolds, facilitating the assembly of molecular complexes involved in gene transcription, DNA modification, and DNA repair. Additionally, lncRNAs can recruit RNA polymerase complexes to specific genomic regions, functioning as ribo-repressors or ribo-activators and inducing allosteric changes in enzymes and transcription factors [15]. lncRNAs play a multifaceted role in the RNA interference (RNAi) pathway, extending beyond their traditional functions. Some lncRNA genes host miRNA sequences within their structure and actively participate in miRNA maturation processes [16]. Additionally, lncRNAs can interfere with miRNA-mediated translation inhibition. This occurs through transcripts that contain multiple miRNA response elements (MREs), which are binding sites for miRNAs. These lncRNAs often share sequence homology with their protein-coding counterparts, enabling them to compete with coding transcripts for miRNA binding. This competitive interaction has led to classifying such lncRNAs as competing endogenous RNAs (ceRNAs) or "sponge RNAs." By acting as molecular decoys, lncRNAs can sequester miRNAs, preventing them from binding to their target mRNAs and thereby fine-tuning gene expression. This mechanism highlights the intricate regulatory networks orchestrated by lncRNAs within the RNAi pathway. These molecules enable precise, tissue-dependent regulation of gene expression, allowing for fine-tuned control over cellular processes [17].The disruption of the balance between lncRNAs and miRNAs can play a significant role in the development of various pathologies, contributing to their onset and progression [18]. Identifying novel miRNA–lncRNA interactions has emerged as a key area of research, driven by their potential regulatory significance. Many studies have relied on low-throughput expression methods to experimentally confirm specific miRNA–lncRNA pairs, advancing our knowledge of these intricate molecular interactions [19]. High-throughput sequencing technologies, such as CLIP-seq (cross-linking and immunoprecipitation followed by sequencing), offer a powerful alternative to traditional methods, identifying thousands of lncRNA–miRNA interactions in a single experiment. This approach provides a comprehensive view of regulatory networks, shedding light on the intricate relationships among mRNAs, miRNAs, and lncRNAs [20]. In contrast, conventional low-throughput techniques, while valuable, often yield only incremental insights and fail to capture the full complexity of these networks. As a result, CLIP-seq has become an essential tool for uncovering the dynamic and multifaceted nature of gene regulation. Complementing these advancements, several algorithms and prediction tools have been developed to analyze CLIP-seq data and identify new potential miRNA response elements (MREs) on lncRNAs. These computational resources further empower researchers by enabling them to explore and validate regulatory interactions, accelerating progress in this rapidly evolving field. Recent studies have revealed that lncRNAs can harbor multiple MREs, often clustered in specific regions of the transcript and distributed across relatively short spans. This spatial arrangement of MREs is crucial, as it thermodynamically influences the efficacy of competing endogenous RNA (ceRNA) activity [21]. While traditional prediction tools, originally designed to identify miRNA-binding sites on mRNAs [22], have been adapted to predict MREs in lncRNAs—considering factors such as complementarity, conservation, and thermodynamic stability—novel approaches are now emerging. These new methods align with the growing understanding of lncRNA complexity, offering more refined insights into the distribution and functional roles of MREs within non-coding transcripts. In this work, we provide a concise update on the biomedical relevance of competing endogenous RNAs (ceRNAs), emphasizing their role in gene regulation and disease. Additionally, we survey current web resources and bioinformatics tools that aid in the investigation of lncRNA–miRNA interactions. These tools enable researchers to predict, analyze, and validate interactions within ceRNA networks, offering valuable insights into their regulatory mechanisms. This chapter aims to report advancements in understanding the complex interplay between lncRNAs and miRNAs in health and disease by bridging theory and practice.

2 Brief Update on Recent lncRNA–miRNA Interactions Studies

Current research on ceRNAs (competing endogenous RNAs) is thriving, driven by a surge of studies that underline their critical role as master regulators of gene expression in normal biological processes and disease states. Among the diverse classes of ceRNAs, circRNAs (circular RNAs) have emerged as particularly important players. Unlike linear RNAs, circRNAs are characterized by a covalently closed loop structure, which confers exceptional stability and resistance to degradation. They are highly abundant in the brain and have been shown to function as efficient molecular sponges for miRNAs, sequestering them and preventing their interaction with target mRNAs. This unique mechanism allows circRNAs to act as key ceRNAs, modulating gene expression and influencing a wide range of cellular processes. Some ceRNAs, particularly circRNAs, have been closely linked to neurological pathologies such as epilepsy, where they play a central role in modulating neuronal excitability, inflammation, and cell survival. In epilepsy, circRNAs function as key ceRNAs by sequestering miRNAs and regulating the expression of target mRNAs, thereby influencing critical pathways involved in disease pathogenesis. For example, studies have demonstrated that circ_0003170 acts as a ceRNA by sponging miR-421, which in turn regulates the expression of CCL2, a chemokine implicated in neuroinflammation. This circ_0003170/miR-421/CCL2 axis has been shown to mitigate neuronal damage in epilepsy models, highlighting the therapeutic potential of targeting ceRNA interactions [23]. Similarly, in cardiovascular diseases (CVDs), ceRNAs play a crucial role in regulating gene expression and mediating crosstalk between pathways. For instance, the lncRNA CHRF sponges miR-489 to regulate Myd88, a gene involved in cardiac hypertrophy, while the circRNA HRCR sequesters miR-223 to modulate ARC expression, protecting against heart failure.

Additionally, the ceRNA pair PTEN-ZEB2 mediates crosstalk between the PI3K/Akt and TGF-β/Smad pathways, which is critical in cardiac remodeling. These interactions highlight the importance of ceRNA networks in coordinating cellular responses and their disruption in disease states [24]. In parallel, liver fibrosis also involves complex ceRNA regulatory networks. For example, the lncRNA ATB (Activated by TGF-β) sponges miR-200a to regulate β -catenin, promoting HSC activation and collagen production. This axis is particularly relevant in hepatitis C virus (HCV)-induced liver fibrosis, where lncRNA-ATB is upregulated, thereby increasing fibrotic activity [25]. Another lncRNA, PVT1 (Plasmacytoma Variant Translocation 1), interacts with miR-152 to influence the Hedgehog (HH) pathway, contributing to fibrosis progression. PVT1 acts as a ceRNA by competitively binding miR-152, which normally suppresses PTCH1 (Patched1), a negative regulator of the HH pathway. By sequestering miR-152, PVT1 promotes HH signaling and epithelial-mesenchymal transition (EMT), driving fibrosis [26]. The regulatory role of ceRNAs also extends to tumorigenesis. For example, the circular RNA circPVT1, derived from the PVT1 locus, plays a significant role in ERα-positive breast cancer by acting as a ceRNA. circPVT1 sponges miR-181a-2-3p, which normally targets and suppresses ESR1 (the gene encoding the estrogen receptor ERα). By sequestering miR-181a-2-3p, circPVT1 stabilizes ESR1 mRNA, leading to increased ERα expression and downstream estrogen-responsive gene expression, thereby promoting tumor growth and resistance to endocrine therapies such as tamoxifen. Additionally, circPVT1 interacts with the mitochondrial antiviral signaling protein (MAVS), disrupting the RIGI-MAVS complex and inhibiting the type I interferon (IFN) signaling pathway, which suppresses anti-tumor immunity. This dual role of circPVT1 in gene regulation and immune modulation highlights its potential as a therapeutic target in breast cancer [27]. On the other hand, the expression of long non-coding RNAs (lncRNAs) can also negatively impact disease progression. For instance, Claudin-4 (CLDN4), a protein crucial for the formation of tight junctions [28], is frequently overexpressed in gastric cancer. CLDN4 enhances the invasiveness and metastatic potential of gastric cancer cells, as well as promotes epithelial-mesenchymal transition (EMT), a key process in tumor progression [29, 30]

The levels of CLDN4 are downregulated by miR-596 and miR-3620-3p, two microRNAs that bind to CLDN4 and suppress its expression, thereby reducing proliferation and invasion rates. However, the expression of two lncRNAs, KRTAP5-AS1 and TUBB2A, can counteract this effect. These lncRNAs function as competing endogenous RNAs (ceRNAs), sequestering miR-596 and miR-3620-3p and preventing them from binding to CLDN4. As a result, CLDN4 expression is restored, indirectly promoting tumor cell invasiveness and metastasis. [31] In summary, we have highlighted the pivotal role of ceRNAs in gene regulation and disease progression. Now, we turn to the resources and computational tools essential for investigating lncRNA–miRNA interactions, which are key to understanding ceRNA-mediated regulatory networks.

3 Tools and Resources for Studying lncRNA–miRNA Regulatory Networks

Over the last few years, several online repositories have been developed that report experimentally validated and computationally predicted miRNA–lncRNA interactions. These resources have become invaluable for exploring the complex regulatory networks involving lncRNAs and miRNAs. Below, we provide a survey of the tools and resources currently available to researchers, along with brief tutorial examples to illustrate their application in discovering and elucidating lncRNA–miRNA interactions.

Recent evidence has greatly emphasized the importance of developing advanced tools to further promote and enhance research in this critical and emerging frontier of molecular biology. These tools are essential for unraveling the intricate mechanisms underlying lncRNA–miRNA interactions and their roles in gene regulation and disease.

3.1 Databases and Tools for lncRNA–miRNA Interactions: Verified and Predicted Data

StarBase or ENCORI

StarBase [32] is a comprehensive dataset that catalogs miRNA–target interactions, encompassing a wide range of RNA molecules, including mRNA, lncRNA, sncRNA, circRNA, and pseudogenes, as well as interactions involving ceRNA networks. Additionally, it documents interactions between proteins and various RNA types, such as lncRNA, ncRNA, and mRNA. These interactions are identified using high-throughput experimental techniques, including CLIP-Seq (Cross-linking Immunoprecipitation followed by sequencing) and Degradome-Seq data.

The second version of StarBase [20], released in 2014, integrated 111 CLIP-Seq datasets generated by methods such as PAR-CLIP, HITS-CLIP, iCLIP, and CLASH, all sourced from the *Gene Expression Omnibus (GEO)* database (available at www.ncbi.nlm.nih.gov/geo). At the time, this version provided the most extensive and detailed miRNA–lncRNA interaction networks available. StarBase is now accessible through the *ENCORI* platform, hosted at https://rnasysu.com/encori/.

ENCORI/starBase is a cutting-edge platform that integrates RNA interaction data using algorithms like rbsSeeker and rriScan, providing insights into regulatory networks involving miRNAs, lncRNAs, circRNAs, and mRNAs. It identifies RNA-binding preferences for hundreds of RBPs and uncovers millions of high-confidence RBP-RNA and RNA-RNA interactions. The platform integrates data from 23 species, including 2725 CLIP-seq and 100 Degradome-seq datasets, as well as 10,800 RNA-seq and 10,500 miRNA-seq samples from 32 cancer types. It also catalogs 1,800,000 mutations across 531 diseases and provides tools for pan-cancer analysis, including differential expression, survival analysis, and co-expression analysis.

NPInter

NPInter v5.0 [33] is a database cataloging interactions between non-coding RNAs (ncRNAs) and biomolecules like proteins, RNAs, and DNA. It integrates experimentally validated interactions from over 10,000 papers (2019–2022) and high-throughput data from technologies like CLIP-seq, ChIRP-seq, and GRID-seq, enabling large-scale identification of RNA-DNA, RNA-protein, and RNA-RNA interactions. The database includes a SARS-CoV-2 module for virus-host RNA interactions and a scNPInter module for cell-type-specific lncRNAs in tumors. Interactions are annotated with details such as interaction type, conservation scores, and functional tags, while molecules are linked to standard IDs from databases such as NONCODE and miRBase. NPInter v5.0 also features an RBP module for RNA-binding protein interactions, with annotations on localization and function. Accessible at http://bigdata.ibp.ac.cn/npinter5/, it offers tools for interaction network visualization and genomic binding site analysis. NPInter v5.0 stands out for its comprehensive coverage of RNA-DNA interactions and COVID-19-specific data, making it a key resource for ncRNA research in health and disease.

Diana-LncBase

DIANA-LncBase is a comprehensive resource for studying miRNA–lncRNA interactions, providing both experimentally validated and in silico-predicted data. The DIANA-LncBase v1 dataset [34], the first of its kind, included over 5000 experimentally validated miRNA–lncRNA interactions and more than 10 million in silico predictions. The experimentally validated interactions were identified using an in-house text-mining pipeline applied to HITS-CLIP and PAR-CLIP datasets. At the same time, the in silico predictions were generated using the DIANA-microT algorithm [35], leveraging the miRBase v18 annotation dataset [36] to identify miRNA recognition elements (MREs) on lncRNAs. The second version of DIANA-LncBase [37] expanded this collection, reporting over 70,000 experimentally validated interactions and in silico predictions based on an updated version of the miRBase annotation database [38]. DIANA-LncBase v 2.0 was structured around two main modules: one for experimentally validated interactions and another for predicted interactions. Its user-friendly web interface allowed advanced queries, enabling users to retrieve multiple miRNAs or lncRNAs in a single search. Results included lncRNA expression profiles and links to external databases, providing a comprehensive view of miRNA–lncRNA interactions. The database is also integrated with other DIANA tools to enhance analysis. In the experimentally validated module, users could expand results using tools like DIANA-miRPath [39] and DIANA-microT-CDS [35]. The predicted interactions module enabled the exploration of putative lncRNA MREs based on genomic loci, with filters for tissue-specific expression and interaction scores. The database expanded significantly with DIANA-LncBase v3.0 [40], the database expanded significantly, offering 500,000 entries (vs. 170,000 in v2.0), including 240,000 tissue- and cell-type-specific miRNA-lncRNA pairs. It introduced the microCLIP framework, a robust CLIP-Seq-guided algorithm, to analyze 236 AGO-CLIP-Seq libraries, cataloging 370,000 miRNA binding events with high confidence. The new version also integrated 67,966 miRNA binding sites with genetic variants from dbSNP [41], ClinVar [42], and COSMIC[43], providing insights into how variants might affect miRNA-lncRNA interactions.

Additionally, v3.0 introduced interactive visualizations for miRNA-lncRNA correlations and lncRNA expression profiles across various cell types and tissues. It also featured advanced filtering options that allowed users to refine searches by cell type, tissue, methodology, and more. A dedicated section for lncRNA expression profiles in nuclear and cytoplasmic compartments was added, with TPM values and Relative Concentration Index (RCI) to indicate subcellular localization trends. The interface was redesigned for improved usability, and integration with the UCSC Genome Browser [44] allowed users to visualize binding sites and variants interactively. The tool is freely usable at https://diana.e-ce.uth.gr/lncbasev3.

miRTarBase

MiRTarBase [45] is a dynamic and continuously updated database focused on experimentally validated miRNA-target interactions (Fig.1). The 7.0 version, released in 2018, cataloged 422,517 curated

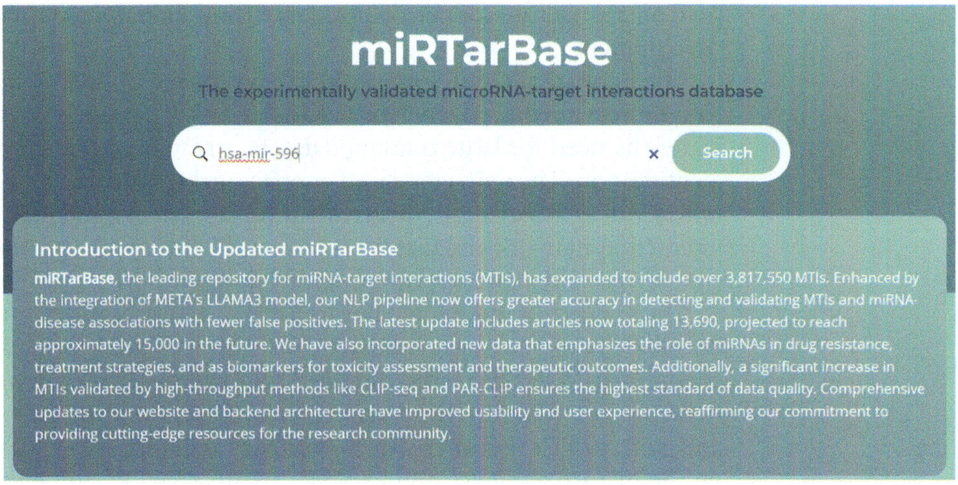

Fig. 1 miRTarBase input interface

interactions involving 4076 miRNAs and 23,054 targets, extracted from approximately 8500 scientific articles using text-mining systems and manually verified. Integrated into The Cancer Genome Atlas (TCGA) Research Network, it provides a richly annotated dataset for detailed exploration of miRNA-target interactions. Key resources include miRBase v21 [38] and the Human microRNA Disease Database (HMDD) [46] for disease-related miRNA data, NCBI RefSeq [47] and Entrez Gene [48] for target information, TCGA [49] and GEO [50] for expression profiles, and tools like TargetScan [51] and miRanda [52] for CLIP-seq target predictions. Functional annotations are provided by DAVID [53] and KEGG [54].

The 2022 update [55] expanded the database to over 2,200,449 validated interactions from 13,389 articles and CLIP-seq data, covering 37 species. New additions included data on extracellular vesicle miRNAs as biomarkers and tissue-specific miRNA expression profiles, offering insights into miRNA roles in specific contexts. Information on SNPs and disease-related variants (DRVs) that influence miRNA-target binding was also incorporated. Technologically, miRTarBase 2022 introduced a BioBERT-based text-mining system [56], combining natural language processing (NLP) with deep learning to improve interaction identification accuracy. The user interface was enhanced with advanced search capabilities and intuitive visualization tools, including regulatory network representations.

The 2025 version [57] further established miRTarBase as a critical resource, encompassing over 3.8 M validated interactions from 13,690 articles and covering 39 species. New features include data on oxidized miRs that affect target recognition under oxidative stress, and information on those that interact with miRNAs, focusing on drug resistance and therapeutic strategies. These updates are particularly relevant for personalized medicine and targeted therapy development.

Technologically, miRTarBase 2025 employs the LLAMA3 model [58] for text mining, paired with an advanced NLP pipeline that eliminates the need for large training datasets, streamlining interaction identification. The redesigned web interface includes new sections for miRNA-drug associations, drug resistance, biomarkers, and dynamic interaction network visualizations powered by the ECharts library (https://echarts.apache.org/en/index.html).

As an example, querying "hsa-miR-596" in miRTarBase returns the table of experimentally validated interactions (Fig. 2). By clicking on the first MIRT ID—the unique MiRTarBase record identifier (which groups all experimental evidence for a given miRNA-target-gene pair)—the microRNA summary panel (Fig. 3), details on the mature sequence and its SNP variants (Fig. 4), and links to prediction resources and expression profiles (Fig. 5) become visible. Selecting the other tabs associated with that record lets you explore additional layers of information, including miRNA, Target Gene, Evidence, CLIP-Seq, Expression, TCGA, Gene Set Enrichments, miRNA-Drug Associations, Network, and Error Report.

3.2 Advances in Computational Tools for Predicting and Analyzing lncRNA–miRNA Interactions

Recent advancements have focused on developing more sophisticated prediction methodologies and efficiently structured computational pipelines to better support researchers. Unlike the extensive array of tools available for predicting miRNA targets in mRNAs, there is now a shift toward more refined techniques for investigating MREs in lncRNAs that function as ceRNAs. In addition to the repositories and approaches that utilize traditional prediction tools, there is a

					Validation methods									
					Strong evidence			Less strong evidence						
ID	Species (miRNA)	Species (Target)	miRNA	Target	Reporter assay	Western blot	qPCR	Microarray	NGS	pSILAC	Other	CLIP-Seq	Sum	# of papers
MIRT007309	Homo sapiens	Homo sapiens	hsa-miR-596	LGALS3BP	✔	✔			✔			✔	4	3
MIRT099801	Homo sapiens	Homo sapiens	hsa-miR-596	SOX4					✔			✔	2	2
MIRT130525	Homo sapiens	Homo sapiens	hsa-miR-596	MCL1					✔			✔	2	6
MIRT443252	Homo sapiens	Homo sapiens	hsa-miR-596	C9orf170					✔			✔	2	2
MIRT447906	Homo sapiens	Homo sapiens	hsa-miR-596	CANX					✔			✔	2	2
MIRT460721	Homo sapiens	Homo sapiens	hsa-miR-596	FN3KRP					✔			✔	2	2
MIRT461846	Homo sapiens	Homo sapiens	hsa-miR-596	F2RL3					✔			✔	2	2
MIRT476904	Homo sapiens	Homo sapiens	hsa-miR-596	FBXL5					✔			✔	2	12
MIRT478254	Homo sapiens	Homo sapiens	hsa-miR-596	DDX3X					✔			✔	2	2

Fig. 2 Table from miRTarBase showing validated interactions of hsa-miR-596 with targets in homo sapiens, including LGALS3BP (strong evidence) and others like SOX4 and MCL1. Highlights miRNA-target interactions and validation levels

Fig. 3 Summary of hsa-mir-596 from miRTarBase, showing genomic coordinates (chr8: 1817231–1817307), synonyms, and functional data including target genes, CLIP-Seq, expression, and miRNA-drug associations. Highlights genomic context and regulatory roles

Fig. 4 Details of mature hsa-miR-596, including its sequence and experimental evidence (e.g., SAGE). Lists SNPs (e.g., rs771459873) from dbSNP, highlighting genetic variations within the miRNA sequence

growing need for computational environments that integrate these novel methodologies with rich online knowledge bases and existing approaches, including those previously described. Such integrated platforms enable researchers to more effectively evaluate lncRNA–miRNA interactions and further explore their roles in cellular networks and disease. Below, we briefly describe and provide a tutorial on three recent emerging resources to address this need.

GDCRNATools

This section presents a brief tutorial on ceRNA network analysis using GDCRNATools, with the TCGA-BRCA (Breast Cancer) dataset as an example. The tutorial provides a step-by-step guide to performing

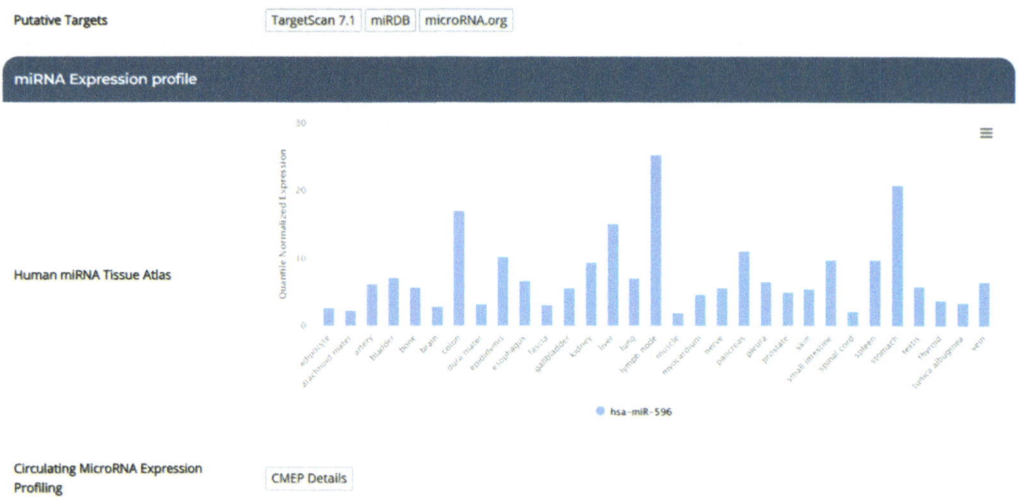

Fig. 5 Overview of hsa-miR-596 target prediction resources, including TargetScan7.1, miRDB, and microRNA.org, alongside its expression profiles from the Human miRNA Tissue Atlas and Circulating MicroRNA Expression (CMEP). Highlights tools and datasets for exploring miRNA targets and tissue-specific expression

key routine analyses, including gene expression, functional enrichment, and univariate survival analyses.

GDCRNATools[59] is an R/Bioconductor [60, 61] package designed to simplify the downloading, organization, and analysis of extensively annotated RNA data from the Genomic Data Commons (GDC) [62]. The GDC aggregates genomic and clinical data from various sources, including The Cancer Genome Atlas (TCGA) https://cancergenome.nih.gov/. Therapeutically Applicable Research To Generate Effective Treatments (TARGET) https://ocg.cancer.gov/programs/target, and Foundation Medicine (https://www.foundationmedicine.com/). With a focus on deciphering ceRNA regulatory networks in cancer, GDCRNATools offers a standardized, user-friendly pipeline that supports the integration of custom software and datasets.

The GDCRNATools package can be downloaded and installed following these lines:

```
#INSTALLATION
BiocManager::install("GDCRNATools")
#LOAD LIBRARY


library(GDCRNATools)
```

The GDCRNATools software package offers two methods for downloading data from the GDC: automatic and manual download. For automatic download, the package utilizes the Application Programming Interface (API) provided by the GDC package [41], allowing users to retrieve data of interest by specifying arguments such as "data.type" and "project.id." Alternatively, users can manually download genomics data using the GDC Data Transfer Tool, available on the GDC website (https://gdc.cancer.gov/access-data/gdc-data-transfer-tool). The GDC Data Transfer Tool's installation and usage instructions are provided at: https://docs.gdc.cancer.gov/Data_Transfer_Tool/Users_Guide/Getting_Started/.

To use the GDC Data Transfer Tool, users must first download the executable file from https://gdc.cancer.gov/access-data/gdc-data-transfer-tool, selecting the appropriate version for their operating system. After downloading, ensure the executable is placed in the desired directory, as this is where the data will be stored.

```
#Project of interest
project = 'TCGA-BRCA'
# Directories in which the data will be downloaded
rnadir=paste(project, 'RNAseq', sep='/')
mirdir=paste(project, 'miRNAs', sep='/')
WORKING DIRECTORY WITH CDC.exe file
setwd("path/to/file")
###################
## Data Download
###################
options(gdc.client = "path/to/file")
############
## RNASeq
############
gdcRNADownload(project.id='TCGA-BRCA', data.type=
'RNAseq',
write.manifest=FALSE, directory=rnadir)
############
############
## miRNAs
############
gdcRNADownload(project.id='TCGA-BRCA', data.type=
'miRNAs',
write.manifest = FALSE, directory=mirdir)
```

To filter the data, the functions gdcFilterDuplicate() are used to remove duplicate samples, and gdcFilterSampleType() is employed to select exclusively Primary Tumor (TCGA code: 01) and Solid Tissue Normal (TCGA code: 11) samples. These metadata, obtainable through the *gdcParseMetadata* function with the «project.id» and «data.type» arguments, or via a manually downloadable JSON file from the GDC portal, contain relevant clinical information such as patient ID, sample type, tumor stage, and age at diagnosis. Access to metadata is available on the Repository page of the GDC portal (https://portal.gdc.cancer.gov/repository), where the data cart (https://portal.gdc.cancer.gov/cart) can be viewed and the metadata download option selected. In subsequent analyses, such as differential expression studies, these data enable categorization of samples into specific groups based on parameters such as sex, days to death, and tumor grade, facilitating comparative investigations.

```
###########################
metaMatrix.RNA=gdcParseMetadata(project.id=
'TCGA-BRCA',
data.type='RNAseq',
write.meta=FALSE)
## Filtering Out Duplicated samples in RNA-Seq data
metaMatrix.RNA=gdcFilterDuplicate(metaMatrix.RNA)
## Filter out non-primary Tumor and non-solid
Tissue Normal samples
## in RNAseq metadata
metaMatrix.RNA=gdcFilterSampleType(metaMatrix.RNA)
############
## miRNAs
############
## Extracting miRNA data
metaMatrix.MIR=gdcParseMetadata(project.id=
'TCGA-BRCA',
data.type='miRNAs', write.meta=FALSE)
## Filtering Out Duplicated miRNA samples
metaMatrix.MIR=gdcFilterDuplicate(metaMatrix.MIR)
## Filter Out non-primary Tumor and non-solid
Tissue Normal samples
## in miRNA metadata
metaMatrix.MIR <- gdcFilterSampleType(metaMatrix.
MIR)
###########################
```

In the final step, the gdcRNAMerge() function is utilized to combine all miRNA isoform quantification data, including raw read counts for −5p and −3p miRNA strands, as well as HTS gene quantification data (read counts). These are merged into unified expression matrices. In these matrices, rows are represented by Ensembl IDs (for RNA data) or miRBase v21 identifiers (for miRNA data), while columns correspond to sample IDs.

```
###################
## Merging raw counts
###################
## For RNASeq data
###################
rnaCounts=gdcRNAMerge(metadata=metaMatrix.RNA,
path=rnadir,
data.type='RNAseq')
#################
## For miRNAs data
#################
mirCounts=gdcRNAMerge(metadata=metaMatrix.MIR,
path=mirdir,
data.type='miRNAs')
###########################
```

Following data acquisition and preprocessing (download, filtering, and quantification), normalization of raw counts is a critical initial step to ensure reliable comparative and enrichment analyses. The standard workflow involves applying the trimmed mean of M-values (TMM) normalization from edgeR [63, 64], followed by the voom transformation method in limma [65].These steps are streamlined via the gdcVoomNormalization() function within the GDCRNAtools package. By default (filter=TRUE), this function automatically filters out low-expression genes or miRNAs that fail to meet the threshold of counts per million (CPM) > 1 in at least 50% of samples, ensuring robust downstream results.

```
#################################
rnaExpr=gdcVoomNormalization(counts=rnaCounts,
filter=FALSE)
mirExpr=gdcVoomNormalization(counts=mirCounts,
filter=FALSE)
#################################
```

To identify differentially expressed genes (DEGs) or miRNAs between groups/conditions, GDCRNATools provides three methods via the *gdcDEAnalysis* function: limma [65], edgeR [63, 64], and DESeq2 [66].

```
####################################################
#########
DEGAll <- gdcDEAnalysis(counts=rnaCounts,
group=metaMatrix.RNA$sample_type,
comparison='PrimaryTumor-SolidTissueNormal',
method='limma')
####################################################
#########
```

The gdcDEReport() function allows users to generate separate reports for differentially expressed elements by transcript type (e.g., long non-coding genes, protein-coding genes, and miRNAs) using the "gene.type" parameter. Additionally, it enables filtering of genes/miRNAs that meet specific thresholds for fold change (*fc* parameter) and p-value (*pval* parameter).

```
################################
data(DEGAll)
deALL=gdcDEReport(deg=DEGAll, gene.type='all')
deLNC=gdcDEReport(deg=DEGAll, gene.type=
'long_non_coding')
dePC=gdcDEReport(deg=DEGAll, gene.type=
'protein_coding')
################################
```

Several criteria must be met to establish significant interactions between lncRNA and mRNA transcripts. First, a substantial number of miRNAs must be shared between the lncRNA and mRNA transcripts, and a parameter must be evaluated using a hypergeometric test. The analysis incorporates data from various miRNA–mRNA interaction databases, including StarBase v2.0 [20], miRcode [67], and mirTarBase v7.0 [45], as well as miRNA–lncRNA interaction resources like StarBase v2.0, miRcode, and spongeScan [68]. Users can also provide their datasets for miRNA–lncRNA and miRNA–mRNA interactions. Additionally, the expression levels of lncRNA and mRNA must show a positive correlation, as determined by Pearson correlation analysis. Lastly, the shared miRNAs between lncRNA and mRNA transcripts should regulate their expression coordinately. This coordination can be assessed using two methods: regulation similarity, which evaluates the similarity between the expression correlations of miRNAs with lncRNAs and mRNAs, and sensitivity correlation, which examines the role of shared miRNAs in mediating the correlation between the expression of a specific lncRNA–mRNA pair [69].

```
###################################################
################
##Example of ceRNAs network analysis by employing
internal resources
ceOutput=gdcCEAnalysis(lnc=rownames(deLNC),
pc=rownames(dePC),
lnc.targets='starBase',
pc.targets='starBase',
rna.expr=rnaExpr,mir.expr=mirExpr)
## ceRNAs network analysis using user-provided
datasets (an example
## dataset can be loaded from the library with
load)
data(lncTarget)
lncTarget
### Load miRNA-mRNA interactions
data(pcTarget)
ceOutput=gdcCEAnalysis(lnc=rownames(deLNC), pc=
rownames(dePC),
lnc.targets=lncTarget, pc.targets=pcTarget,
rna.expr=rnaExpr, mir.expr=mirExpr)
###################################################
################
```

Univariate survival analysis evaluates the survival potential of the factor of interest. This analysis uses the Cox Proportional-Hazards (CoxPH) model for continuous variables and the Kaplan–Meier (KM) method, which relies on a user-defined threshold to separate data into high- and low-expression groups (Fig. 6). The gdcSurvivalAnalysis( ) function accepts a vector of genes as input and outputs a dataframe containing the hazard ratio, 95% confidence intervals, and statistical significance for each transcript.

```
###################################################
################
```

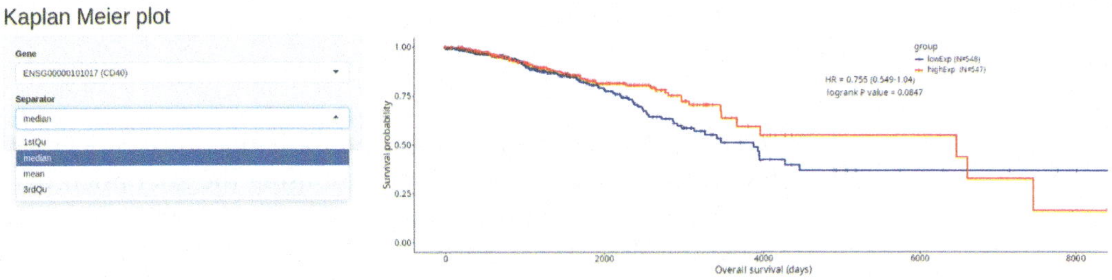

Fig. 6 KM plot using the shinyKMPlot function

```
## Check for missing genes
missing_genes <- setdiff(rownames(deALL),
rownames(rnaExpr))
deALL_filtered <- deALL[!rownames(deALL) %in%
missing_genes,]
rnaExpr_filtered <- rnaExpr[!rownames(rnaExpr) %in%
missing_genes,]
##################################################
################
##################################################
################
## CoxPHAnalysis ##
survOutput <- gdcSurvivalAnalysis(gene =
rownames(deALL_filtered),
method = 'coxph',
rna.expr = rnaExpr_filtered,
metadata = metaMatrix.RNA)
##################################################
################
##################################################
################
## KM analysis ##
survOutput=gdcSurvivalAnalysis(gene=
rownames(deALL_filtered),
method='KM',
rna.expr=rnaExpr_filtered,
metadata=metaMatrix.RNA,
sep='median')
##################################################
################
##################################################
################
## It is possible to use shinyKMPlot() function to
view the KMplot
## via a simple shiny app running on a local web
page.
##################################################
################
##################################################
################
## ShinyApp KM plot ##
```

```
shinyKMPlot(gene=rownames(deALL_filtered),
rna.expr=rnaExpr_filtered,
metadata=metaMatrix.RNA)
```

GDCRNATools enables users to perform functional enrichment analysis on a set of transcripts of interest using the gdcEnrichAnalysis() function. This function leverages annotation resources such as Gene Ontology (GO) (http://www.geneontology.org/) and the Kyoto Encyclopedia of Genes and Genomes (KEGG) (https://www.genome.jp/kegg/). The R packages clusterProfiler powers the GO and KEGG analyses [70] and DOSE [71]. By setting the argument simplify = TRUE, redundant GO terms can be filtered out. Graphical representations of the results can be generated using the gdcEnrichPlot() function, which provides options for creating bar plots or bubble plots to visualize functional enrichment analysis (see Figs. 7 and 8).

```
#######################################################
#################
## Function Enrichment Analysis. ##
enrichOutput=gdcEnrichAnalysis(gene=
rownames(deALL_filtered),
simplify=TRUE)
data(enrichOutput)
gdcEnrichPlot(enrichOutput, type='bar',
category='GO', num.terms = 10)
gdcEnrichPlot(enrichOutput, type='bubble',
category='GO',num.terms = 10)
#######################################################
#################
```

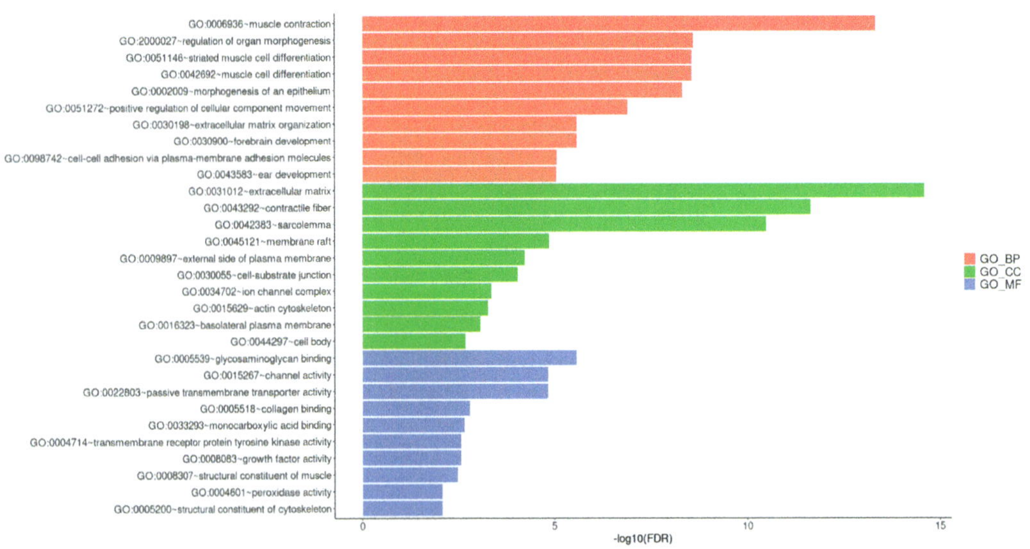

Fig. 7 Bar plot with the gdcEnrichPlot function

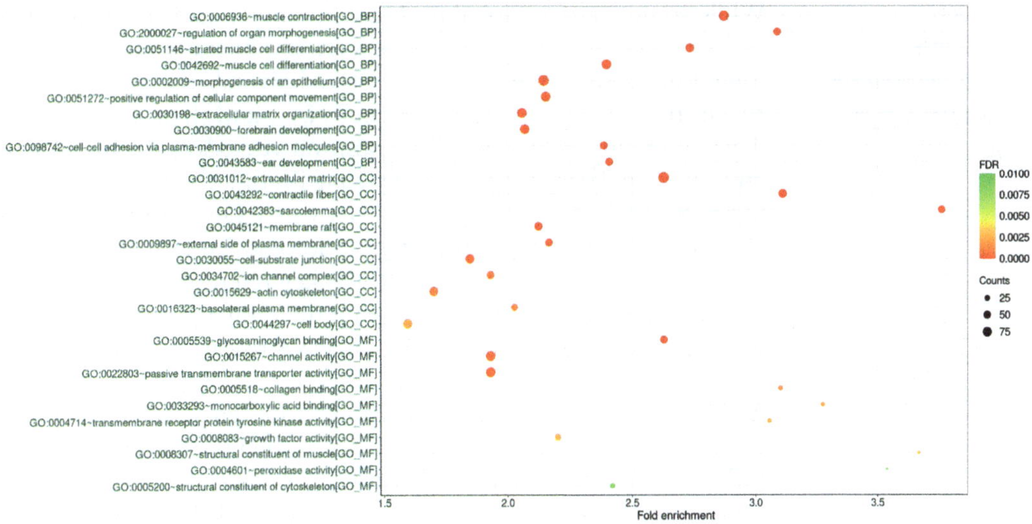

Fig. 8 Bubble plot with the gdcEnrichPlot function

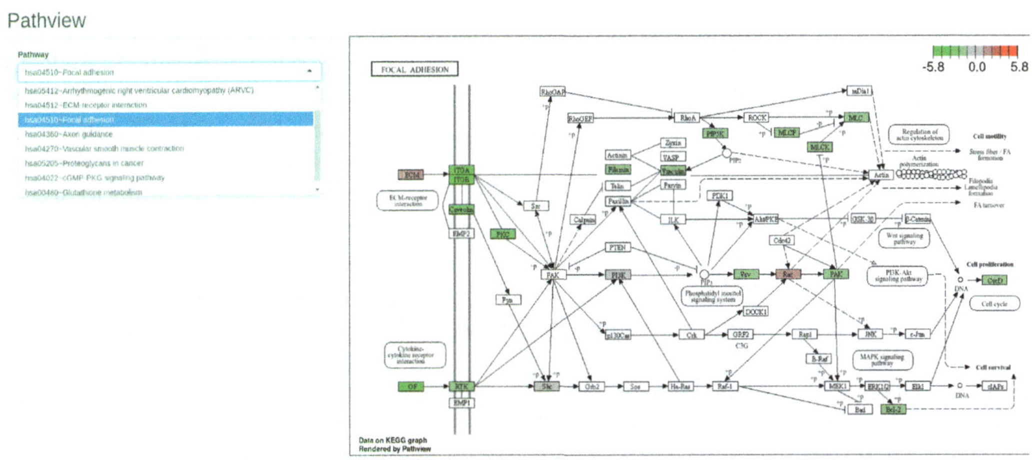

Fig. 9 Visual representation of pathway using shiny pathview

As for the univariate analysis, it is possible to use R Shiny to have a visible representation of the pathways (Fig. 9):

```
####################################################
################
library(pathview)
deg = deALL_filtered$logFC
names(deg) = rownames(deALL_filtered)
pathways=as.character(enrichOutput$Terms[enr-
ichOutput$Category=='KEGG'])
shinyPathview(deg,pathways= pathways,directory=
'pathview')
####################################################
#################
```

3.3 Tools of Historical Significance: miRNA-lncRNA Interaction Tools

While some computational tools are no longer actively maintained or have become inaccessible, others remain operational but are no longer updated. These tools have played pivotal roles in advancing the study of non-coding RNAs (ncRNAs) and their interactions. They offer unique functionalities, innovative analytical frameworks, and valuable datasets, significantly contributing to our understanding of miRNA-lncRNA regulatory networks. Despite limitations in maintenance or updates, their methodologies and outputs remain relevant, warranting their inclusion as key references. This section highlights these tools, acknowledging their impact on the field and the insights they have facilitated in ncRNA research.

miRcode

miRcode [67] is a comprehensive resource designed to identify putative and experimentally validated miRNA target sites across the entire GENCODE-annotated transcriptome (version 11), covering over 10,000 lncRNAs. Accessible via an intuitive web-based interface at http://www.mircode.org, miRcode enables researchers to explore miRNA–lncRNA interactions, with results available for download in tab-separated format. By leveraging seed complementarity and evolutionary conservation, the platform predicts miRNA target sites, providing valuable insights into miRNA regulatory roles in lncRNA networks.

Offering valuable insights into the regulatory roles of miRNAs in lncRNA networks. Although miRcode is still operational, it has not been updated for over 13 years. This lack of updates limits its ability to reflect the most recent advancements in ncRNA research and miRNA annotations. Nevertheless, its robust analytical framework and historical relevance continue to influence subsequent developments in computational prediction tools for ncRNA interactions.

Understanding RNA Editing in lncRNA–miRNA Networks: Contributions of LncEditing

One of the most intriguing aspects of lncRNA regulation involves post-transcriptional modifications, including RNA editing. Notably, adenosine-to-inosine (A-to-I) editing, driven by the adenosine deaminases acting on the RNA (ADAR) family, stands out as the most prevalent form of RNA editing in mammals [72]. Generally, A-to-I RNA editing targets double-stranded RNA (dsRNA), and depending on the size of the dsRNA region involved, this phenomenon can be categorized as site-selective (in shorter dsRNA regions) or promiscuous (in longer dsRNA regions) [73]. Once adenosine is deaminated to inosine, the latter is interpreted by the cellular splicing and translation machinery as guanosine (G), potentially influencing the biogenesis of the edited RNA molecule. Even a single base alteration can thus affect the function of both coding and non-coding transcripts.

With the rise of high-throughput sequencing technologies and advanced bioinformatics approaches, an increasing number of A-to-I editing events have been identified in lncRNAs [74]. However, their functional implications remain largely unexplored. LNCediting [75] was developed as a unique and comprehensive database to bridge this gap, cataloging approximately 200,000 A-to-I editing sites in lncRNAs across multiple species, including fly, rhesus monkey, mouse, and human. What set LNCediting apart was its ability to provide detailed insights into how these editing events influenced lncRNA secondary structures and miRNA–lncRNA interactions, a feature not replicated by other tools. Through its user-friendly web interface, LNCediting enabled researchers to examine key details of each editing site, including genomic location, repeat data, and the predicted impact of single editing events on lncRNA structure and function. Although LNCediting is no longer maintained, its dataset and analytical framework remain valuable references for studying RNA editing in lncRNAs. No other tool has fully replicated its ability to predict the functional consequences of RNA editing events in lncRNAs. The methodologies it employed continue to guide current research, shaping how scientists explore the regulatory functions of edited lncRNAs across various biological contexts.

Other Tools for RNA Editing

While LNCediting remains a unique resource for predicting the functional impact of A-to-I editing in lncRNAs, complementary tools like REDItools and REDIportal aid in RNA editing discovery and annotation:

- REDItools [76]: A Python suite for de novo detection of A-to-I editing events from RNA/DNA-Seq data. It employs advanced filters to distinguish true edits from artifacts (e.g., SNPs, sequencing errors) and supports large-scale analyses via HPC optimization. However, it focuses solely on event identification, lacking lncRNA-specific functional predictions.
- REDIportal [77]: A database cataloging ~16 million A-to-I events across 9642 human samples (GTEx project) and >9000 RNAseq datasets from TCGA, covering 31 cancer studies. It provides tissue-specific statistics, genomic annotations, and ncRNA annotations (e.g., miRNAs) via integration with RNAcentral. The new version includes AEI and REI indices for global and recoding-specific editing activity, dsRNA analysis with editing indices, and reliability scoring via the REDInet deep learning model. Integrated with Ensembl, UniProt, and PRIDE, it enables broader exploration of editing contexts but does not link editing events to structural/functional consequences in lncRNAs.
- **JACUSA2** [78]: A high-performance framework for RNA modification and A-to-I editing detection, compatible with Illumina and Nanopore sequencing data. It integrates multiple experimental conditions, detects complex RNA signatures (e.g., substitutions, insertions, deletions, reverse transcription

arrests), and models RNA modifications such as m6A and pseudouridine. While highly versatile, JACUSA2 primarily identifies and maps modification sites rather than predicting their structural or functional impact on lncRNAs.

Unlike LNCediting, which uniquely predicted how A-to-I editing alters lncRNA secondary structures and miRNA interactions, REDItools/REDIportal and Jacusa2 are **generalist tools** focused on editing site identification and cataloging. They address the "where" of RNA editing but not the "how" or "why" in lncRNA biology. The absence of tools replicating LNCediting's functional insights highlights a critical gap in lncRNA research, highlighting the need for resources integrating structural, interactional, and functional data.

Spongescan

Among the many approaches developed to predict miRNA targets, some tools have specifically aimed to identify competing endogenous RNAs (ceRNAs), a crucial aspect of understanding post-transcriptional regulation. One such tool was spongeScan, which departed from traditional prediction models by focusing on features that characterize miRNA sponges. Designed as a user-friendly web resource, spongeScan implemented a ceRNA-prediction algorithm capable of analyzing any user-specified sequence set and integrating candidate ceRNA and miRNA expression data.

The algorithm underlying spongeScan emphasized sequence patterns known to be enriched in miRNA sponges, optimizing target prediction through a specialized scanning method. Researchers could analyze potential miRNA binding sites by uploading a FASTA file containing lncRNA transcript sequences, either user-provided or retrieved from Ensembl [79] (2015 version). A GTF annotation file was also required to filter out non-lncRNA transcripts. Users could define the species of interest, specify relevant miRNAs for the analysis, integrate available expression data, and adjust key prediction parameters. Results were sent via email upon completion, offering flexibility in data retrieval.

spongeScan employed a sliding-window strategy to scan lncRNA sequences for k-mers of six, seven, or eight nucleotides, identifying potential miRNA seed matches and determining enriched sequence patterns. This approach maximized detection efficiency by capturing a wide range of k-mer distributions across transcripts. To refine predictions, the tool relied on three key parameters:

- LOD Score (Log-Odds Score)—Evaluated how frequently a k-mer matched an lncRNA sequence compared to all others, with higher values indicating greater specificity (default: 1).
- Dispersion Score—Measured the clustering pattern of matches for a given k-mer–lncRNA pair, assessing how evenly spaced the potential miRNA response elements (MREs) were (default: 10).

Table 1
Overview of each resource, including supported species (h = human, m = mouse, r = rat, f = fly), code availability (A = available, NA = not available), update status (Y = updated in the last 5 years, N = not updated), site coverage (5′UTR, CDS, 3′UTR), miRBase version, Relative Citation Ratio (RCR; [a]= not yet available), and references

Site coverageon miRNA											
Tool name	Organism	Type	Version or Date of last update	Status	URL	5′UTR	CDS	3′UTR	mirBase	Relativecitation	References
miRcode	h	Web Server [A]	Not updated since2012 [N]	Active	http://www.mircode.org/	V	V	V	–	15,43	[67]
LNCediting	h, m, r, f	Database [NA]	Inactive [N]	Inaccessible	NA	–	–	V	–	110,53	[75]
spongeScan	h, m, r	Web Server [NA]	Inactive[N]	Inaccessible	NA	–	–	V	–	110,53	[68]
StarBase (ENCORI)	h, m, r, f	Database/Web Server [A]	Last updated: 10/18/2024 [Y]	Active	https://rnasysu.com/encori/	V	V	V	22	2014 paper: [32]127.79	[32]
NPInter	h, m, r, f	Database [NA]	v 5.0 [Y]	Active	http://bigdata.ibp.ac.cn/npinter5/	V	V	V	22	*	[33]
DIANA-LncBase	h, m, r, f	Database/Web Server [A]	v 3.0 [Y]	Active	https://diana.e-ce.uth.gr/lncbasev3	V	V	V	22	13,28	[34]
miRTarBase 2025	h, m, r, f	Database [NA]	v 10.0 [Y]	Active	https://mirtarbase.cuhk.edu.cn/~miRTarBase/miRTarBase_2025	V	V	V	22,1	v 9.0 (2022) : 60.91	[45]
GDCRNA-Tools	h	Software (R package) [A]	v 1.26.0 [N]	Active	https://www.bioconductor.org/packages/release/bioc/html/GDCRNATools.html	–	–	–	–	110,53	[59]

(Continued)

Table 1
(Continued)

Site coverageon miRNA											
Tool name	**Organism**	**Type**	**Version or Date of last update**	**Status**	**URL**	**5′UTR**	**CDS**	**3′UTR**	**mirBase**	**Relativecitation**	**References**
Reditools	h	Python code [A]	v 3.0 [Y]	Active	https://github.com/BioinfoUNIBA/REDItools/releases/tag/REDItoolsFamily_V1.0	V	V	V	–	110,53	[76]
Rediportal	h	Web portal [A]	2024	Active	http://srv00.recas.ba.infn.it/atlas/	–	–	–	–	*	[77]
Jacusa2	h	Coded in multiple languages [A]	v 2.0 [y]	Active	https://github.com/dieterich-lab/JACUSA2	V	V	V	–	*	[78]

- K-mer Complexity Score—Filtered out k-mers that did not match canonical miRNA seed sequences and exhibited low complexity (default: 4).

Additionally, users could specify the minimum number of k-mer–lncRNA binding sites to be included in the analysis (default: 20) and whether to allow a single wobble in detected matches.

Final results were presented in a tabular format with up to 22 customizable columns, providing detailed statistics and allowing data export to Excel. A graphical visualization of the predicted binding-site distribution was also available, aiding interpretation of ceRNA interactions. Users could filter results by k-mers associated with canonical miRNA seeds or exclude them to explore non-canonical interactions.

One of the most distinctive aspects of spongeScan was its reliance solely on sequence-based prediction, without incorporating HITS- or PAR-CLIP data. This independence made it applicable across different organisms and datasets, broadening its utility in identifying ceRNAs capable of miRNA sequestration.

For convenience, all computational tools and databases cited in this study are summarised in Table 1, together with their main purpose, required input, and key references.

4 Conclusion

It is widely recognized that miRNAs and lncRNAs play pivotal roles in numerous physiological and pathological processes. A growing body of research has confirmed the significant association between miRNA–lncRNA interactions and the onset and progression of various diseases, particularly cancer. Consequently, identifying and exploring potential miRNA–lncRNA biomarkers are becoming integral to diagnostic methodologies and therapeutic strategies in the rapidly advancing field of personalized medicine. Leveraging existing computational tools and workflows remains an indispensable approach to unraveling the intricate regulatory networks involving these ncRNA molecules. Now, more than ever, there is a pressing need to develop innovative computational tools that build upon current methodologies and their discoveries, aiming to deepen our understanding of cellular physiology and disease through increasingly precise and effective in silico techniques.

Acknowledgments

This work was supported by the European Union's NextGenerationEU initiative under the Italian Ministry of University and Research as part of the PNRR – M4C2.-I1.3 PE00000019 "HEAL ITALIA," CUP B73C22001250006, awarded to Sebastiano Di Bella and Giorgio Stassi. Francesco Orilio was supported by the European Union under the Italian Ministry of University and Research as part of the Project: Health Extended Alliance for Innovative Therapies, Advanced Lab research, and Integrated Approaches of Precision Medicine: HealItalia Spoke 3 – Task 1.2 CUP B73C22001250006. Rosario Nicola Brancaccio was supported by the Italian Ministry of Health under the "Medicina rigenerativa, predittiva e personalizzata" program, project "GENESIS ATI – Studio degli eventi genetici alla base della carcinogenesi in aree ad alto tasso di inquinamento" (Project 0692, CUP B77G22000340005). Giovanni Nigita was supported in part by National Cancer Institute (National Institute of Health) grant (R21CA287180).

Conflict of Interest The authors have no conflicts of interest to declare that are relevant to the content of this chapter.

References

1. Kang W, Friedländer MR (2015) Computational Prediction of miRNA Genes from Small RNA Sequencing Data. Front Bioeng Biotechnol 3:7. https://doi.org/10.3389/fbioe.2015.00007
2. Veneziano D, Di Bella S, Nigita G, Laganà A, Ferro A, Croce CM (2016) Noncoding RNA: current deep sequencing data analysis approaches and challenges. Hum Mutat 37:1283–1298. https://doi.org/10.1002/humu.23066
3. Veneziano D, Nigita G, Ferro A (2015) Computational approaches for the analysis of ncRNA through deep sequencing techniques. Front Bioeng Biotechnol 3:77. https://doi.org/10.3389/fbioe.2015.00077
4. Kozomara A, Birgaoanu M, Griffiths-Jones S (2019) miRBase: from microRNA sequences to function. Nucleic Acids Res 47:D155–D162. https://doi.org/10.1093/nar/gky1141
5. Djuranovic S, Nahvi A, Green R (2012) miRNA-mediated gene silencing by translational repression followed by mRNA deadenylation and decay. Science 336:237–240. https://doi.org/10.1126/science.1215691
6. Leitão AL, Costa MC, Gabriel AF, Enguita FJ (2020) Interspecies communication in holobionts by non-coding RNA exchange. Int J Mol Sci 21:2333. https://doi.org/10.3390/ijms21072333
7. Black CM, Braden AA, Nasim S, Tripathi M, Xiao J, Khan MM (2024) The association between long non-coding RNAs and Alzheimer's disease. Brain Sci 14:818. https://doi.org/10.3390/brainsci14080818
8. Ma Q, Ma F, Zhang B, Zhang Y, Peng L, Li X (2024) The short peptide encoded by long non-coding RNA RNF217-AS1 inhibits stomach cancer tumorigenesis, macrophage recruitment, and pro-inflammatory responses. Amino Acids 56:45. https://doi.org/10.1007/s00726-024-03404-7
9. Housman G, Ulitsky I (2016) Methods for distinguishing between protein-coding and long noncoding RNAs and the elusive biological purpose of translation of long noncoding RNAs. Biochim Biophys Acta 1859:31–40. https://doi.org/10.1016/j.bbagrm.2015.07.017
10. Nelson BR, Makarewich CA, Anderson DM, Winders BR, Troupes CD, Wu F, Reese AL, McAnally JR, Chen X, Kavalali ET, Cannon SC, Houser SR, Bassel-Duby R, Olson EN (2016) A peptide encoded by a transcript annotated as long noncoding RNA enhances SERCA activity in muscle. Science 351:271–275. https://doi.org/10.1126/science.aad4076
11. Mudge JM, Carbonell-Sala S, Diekhans M, Martinez JG, Hunt T, Jungreis I, Loveland JE, Arnan C, Barnes I, Bennett R, Berry A, Bignell A, Cerdán-Vélez D, Cochran K, Cortés LT, Davidson C, Donaldson S, Dursun C, Fatima R, Hardy M, Hebbar P, Hollis Z, James BT, Jiang Y, Johnson R, Kaur G, Kay M, Mangan RJ, Maquedano M, Gómez LM, Mathlouthi N, Merritt R, Ni P, Palumbo E, Perteghella T, Pozo F, Raj S, Sisu C, Steed E, Sumathipala D, Suner M-M, Uszczynska-Ratajczak B, Wass E, Yang YT, Zhang D, Finn RD, Gerstein M, Guigó R, Hubbard TJP, Kellis M, Kundaje

A, Paten B, Tress ML, Birney E, Martin FJ, Frankish A (2025) GENCODE 2025: reference gene annotation for human and mouse. Nucleic Acids Res 53:D966–D975. https://doi.org/10.1093/nar/gkae1078
12. Quek XC, Thomson DW, Maag JLV, Bartonicek N, Signal B, Clark MB, Gloss BS, Dinger ME (2015) lncRNAdb v2.0: expanding the reference database for functional long noncoding RNAs. Nucleic Acids Res 43:D168–D173. https://doi.org/10.1093/nar/gku988
13. Zhu L, Chen R, Jiang C, Xie Q, Zhao W, Gao X, Huang H (2021) Mechanism underlying long non-coding RNA ILF3-AS1-mediated inhibition of cervical cancer cell proliferation, invasion and migration, and promotion of apoptosis. Mol Med Rep 24:554. https://doi.org/10.3892/mmr.2021.12193
14. Zeng L, Lyu X, Yuan J, Wang W, Zhao N, Liu B, Sun R, Meng X, Yang S (2020) Long non-coding RNA LINC01116 is overexpressed in lung adenocarcinoma and promotes tumor proliferation and metastasis. Am J Transl Res 12:4302–4313
15. Laurent G, Wahlestedt C, Kapranov P (2015) The landscape of long noncoding RNA classification. Trends Genet TIG 31:239–251. https://doi.org/10.1016/j.tig.2015.03.007
16. Dey BK, Pfeifer K, Dutta A (2014) The H19 long noncoding RNA gives rise to microRNAs miR-675-3p and miR-675-5p to promote skeletal muscle differentiation and regeneration. Genes Dev 28:491–501. https://doi.org/10.1101/gad.234419.113
17. Ounzain S, Micheletti R, Beckmann T, Schroen B, Alexanian M, Pezzuto I, Crippa S, Nemir M, Sarre A, Johnson R, Dauvillier J, Burdet F, Ibberson M, Guigó R, Xenarios I, Heymans S, Pedrazzini T (2015) Genome-wide profiling of the cardiac transcriptome after myocardial infarction identifies novel heart-specific long non-coding RNAs. Eur Heart J 36:353–368. https://doi.org/10.1093/eurheartj/ehu180
18. Xue J, Huang S, Chen J, Chen Y-Z, Shao Z-M, Wu J, Chi Y (2021) Overexpression of long non-coding RNA Linc01315 predicts poor prognosis in breast cancer. Front Oncol 11:562378. https://doi.org/10.3389/fonc.2021.562378
19. Fan M, Li X, Jiang W, Huang Y, Li J, Wang Z (2013) A long non-coding RNA, PTCSC3, as a tumor suppressor and a target of miRNAs in thyroid cancer cells. Exp Ther Med 5:1143–1146. https://doi.org/10.3892/etm.2013.933
20. Li J-H, Liu S, Zhou H, Qu L-H, Yang J-H (2014) starBase v2.0: decoding miRNA-ceRNA, miRNA-ncRNA and protein–RNA interaction networks from large-scale CLIP-Seq data. Nucleic Acids Res 42:D92–D97. https://doi.org/10.1093/nar/gkt1248
21. Sellars E, Gabra M, Salmena L (2020) The complex landscape of PTEN mRNA regulation. Cold Spring Harb Perspect Med 10:a036236. https://doi.org/10.1101/cshperspect.a036236
22. Roberts JT, Borchert GM (2017) Computational prediction of MicroRNA target genes, target prediction databases, and web resources. Methods Mol Biol Clifton NJ 1617:109–122. https://doi.org/10.1007/978-1-4939-7046-9_8
23. Chen F, Zheng H, Zhang W, Kang J, Liu Q, Pu J, Yang L (2021) circ_0003170 aggravates human hippocampal neuron injuries by regulating the miR-421/CCL2 axis in cells models of epilepsy. Gen Physiol Biophys 40:115–126. https://doi.org/10.4149/gpb_2020045
24. Song C, Zhang J, Liu Y, Hu Y, Feng C, Shi P, Zhang Y, Wang L, Xie Y, Zhang M, Zhao X, Cao Y, Li C, Sun H (2022) Characterization and validation of ceRNA-mediated pathway-pathway crosstalk networks across eight major cardiovascular diseases. Front Cell Dev Biol 10:762129. https://doi.org/10.3389/fcell.2022.762129
25. Fu N, Zhao S-X, Kong L-B, Du J-H, Ren W-G, Han F, Zhang Q-S, Li W-C, Cui P, Wang R-Q, Zhang Y-G, Nan Y-M (2017) LncRNA-ATB/microRNA-200a/β-catenin regulatory axis involved in the progression of HCV-related hepatic fibrosis. Gene 618:1–7. https://doi.org/10.1016/j.gene.2017.03.008
26. Zheng J, Yu F, Dong P, Wu L, Zhang Y, Hu Y, Zheng L (2016) Long non-coding RNA PVT1 activates hepatic stellate cells through competitively binding microRNA-152. Oncotarget 7:62886–62897. https://doi.org/10.18632/oncotarget.11709
27. Yi J, Wang L, Hu G-S, Zhang Y-Y, Du J, Ding J-C, Ji X, Shen H-F, Huang H-H, Ye F, Liu W (2023) CircPVT1 promotes ER-positive breast tumorigenesis and drug resistance by targeting ESR1 and MAVS. EMBO J 42:e112408. https://doi.org/10.15252/embj.2022112408
28. Lal-Nag M, Morin PJ (2009) The claudins. Genome Biol 10: 235. https://doi.org/10.1186/gb-2009-10-8-235
29. Jung H, Jun KH, Jung JH, Chin HM, Park WB (2011) The expression of claudin-1, claudin-2, claudin-3, and claudin-4 in gastric cancer tissue. J Surg Res 167:e185–191. https://doi.org/10.1016/j.jss.2010.02.010
30. Lin X, Shang X, Manorek G, Howell SB (2013) Regulation of the epithelial-mesenchymal

transition by Claudin-3 and Claudin-4. PlOS One 8:e67496. https://doi.org/10.1371/journal.pone.0067496
31. Song Y-X, Sun J-X, Zhao J-H, Yang Y-C, Shi J-X, Wu Z-H, Chen X-W, Gao P, Miao Z-F, Wang Z-N (2017) Non-coding RNAs participate in the regulatory network of CLDN4 via ceRNA mediated miRNA evasion. Nat Commun 8: 289. https://doi.org/10.1038/s41467-017-00304-1
32. Yang J-H, Li J-H, Shao P, Zhou H, Chen Y-Q, Qu L-H (2011) starBase: a database for exploring microRNA-mRNA interaction maps from Argonaute CLIP-Seq and Degradome-Seq data. Nucleic Acids Res 39:D202–209. https://doi.org/10.1093/nar/gkq1056
33. Zheng Y, Luo H, Teng X, Hao X, Yan X, Tang Y, Zhang W, Wang Y, Zhang P, Li Y, Zhao Y, Chen R, He S (2023) NPInter v5.0: ncRNA interaction database in a new era. Nucleic Acids Res 51:D232–D239. https://doi.org/10.1093/nar/gkac1002
34. Paraskevopoulou MD, Georgakilas G, Kostoulas N, Reczko M, Maragkakis M, Dalamagas TM, Hatzigeorgiou AG (2013) DIANA-LncBase: experimentally verified and computationally predicted microRNA targets on long non-coding RNAs. Nucleic Acids Res 41:D239–245. https://doi.org/10.1093/nar/gks1246
35. Paraskevopoulou MD, Georgakilas G, Kostoulas N, Vlachos IS, Vergoulis T, Reczko M, Filippidis C, Dalamagas T, Hatzigeorgiou AG (2013) DIANA-microT web server v5.0: service integration into miRNA functional analysis workflows. Nucleic Acids Res 41:W169–173. https://doi.org/10.1093/nar/gkt393
36. Griffiths-Jones S, Grocock RJ, van Dongen S, Bateman A, Enright AJ (2006) miRBase: microRNA sequences, targets and gene nomenclature. Nucleic Acids Res 34:D140–144. https://doi.org/10.1093/nar/gkj112
37. Paraskevopoulou MD, Vlachos IS, Karagkouni D, Georgakilas G, Kanellos I, Vergoulis T, Zagganas K, Tsanakas P, Floros E, Dalamagas T, Hatzigeorgiou AG (2016) DIANA-LncBase v2: indexing microRNA targets on non-coding transcripts. Nucleic Acids Res 44:D231–238. https://doi.org/10.1093/nar/gkv1270
38. Kozomara A, Griffiths-Jones S (2014) miRBase: annotating high confidence microRNAs using deep sequencing data. Nucleic Acids Res 42:D68–73. https://doi.org/10.1093/nar/gkt1181
39. Vlachos IS, Kostoulas N, Vergoulis T, Georgakilas G, Reczko M, Maragkakis M, Paraskevopoulou MD, Prionidis K, Dalamagas T, Hatzigeorgiou AG (2012) DIANA miRPath v.2.0: investigating the combinatorial effect of microRNAs in pathways. Nucleic Acids Res 40:W498–504. https://doi.org/10.1093/nar/gks494
40. Karagkouni D, Paraskevopoulou MD, Tastsoglou S, Skoufos G, Karavangeli A, Pierros V, Zacharopoulou E, Hatzigeorgiou AG (2020) DIANA-LncBase v3: indexing experimentally supported miRNA targets on non-coding transcripts. Nucleic Acids Res 48:D101–D110. https://doi.org/10.1093/nar/gkz1036
41. Sherry ST, Ward M-H, Kholodov M, Baker J, Phan L, Smigielski EM, Sirotkin K (2001) dbSNP: the NCBI database of genetic variation. Nucleic Acids Res 29:308–311
42. Landrum MJ, Lee JM, Benson M, Brown GR, Chao C, Chitipiralla S, Gu B, Hart J, Hoffman D, Jang W, Karapetyan K, Katz K, Liu C, Maddipatla Z, Malheiro A, McDaniel K, Ovetsky M, Riley G, Zhou G, Holmes JB, Kattman BL, Maglott DR (2018) ClinVar: improving access to variant interpretations and supporting evidence. Nucleic Acids Res 46:D1062–D1067. https://doi.org/10.1093/nar/gkx1153
43. Forbes S, Clements J, Dawson E, Bamford S, Webb T, Dogan A, Flanagan A, Teague J, Wooster R, Futreal PA, Stratton MR (2006) COSMIC 2005. Br J Cancer 94:318–322. https://doi.org/10.1038/sj.bjc.6602928
44. Nassar LR, Barber GP, Benet-Pagès A, Casper J, Clawson H, Diekhans M, Fischer C, Gonzalez JN, Hinrichs AS, Lee BT, Lee CM, Muthuraman P, Nguy B, Pereira T, Nejad P, Perez G, Raney BJ, Schmelter D, Speir ML, Wick BD, Zweig AS, Haussler D, Kuhn RM, Haeussler M, Kent WJ (2023) The UCSC genome browser database: 2023 update. Nucleic Acids Res 51:D1188–D1195. https://doi.org/10.1093/nar/gkac1072
45. Chou C-H, Shrestha S, Yang C-D, Chang N-W, Lin Y-L, Liao K-W, Huang W-C, Sun T-H, Tu S-J, Lee W-H, Chiew M-Y, Tai C-S, Wei T-Y, Tsai T-R, Huang H-T, Wang C-Y, Wu H-Y, Ho S-Y, Chen P-R, Chuang C-H, Hsieh P-J, Wu Y-S, Chen W-L, Li M-J, Wu Y-C, Huang X-Y, Ng FL, Buddhakosai W, Huang P-C, Lan K-C, Huang C-Y, Weng S-L, Cheng Y-N, Liang C, Hsu W-L, Huang H-D (2018) miRTarBase update 2018: a resource for experimentally validated microRNA-target interactions. Nucleic Acids Res 46:D296–D302. https://doi.org/10.1093/nar/gkx1067
46. Li Y, Qiu C, Tu J, Geng B, Yang J, Jiang T, Cui Q (2014) HMDD v2.0: a database for experimentally supported human microRNA and disease associations. Nucleic Acids Res 42:D1070–1074. https://doi.org/10.1093/nar/gkt1023
47. O'Leary NA, Wright MW, Brister JR, Ciufo S, Haddad D, McVeigh R, Rajput B, Robbertse B, Smith-White B, Ako-Adjei D, Astashyn A,

Badretdin A, Bao Y, Blinkova O, Brover V, Chetvernin V, Choi J, Cox E, Ermolaeva O, Farrell CM, Goldfarb T, Gupta T, Haft D, Hatcher E, Hlavina W, Joardar VS, Kodali VK, Li W, Maglott D, Masterson P, McGarvey KM, Murphy MR, O'Neill K, Pujar S, Rangwala SH, Rausch D, Riddick LD, Schoch C, Shkeda A, Storz SS, Sun H, Thibaud-Nissen F, Tolstoy I, Tully RE, Vatsan AR, Wallin C, Webb D, Wu W, Landrum MJ, Kimchi A, Tatusova T, DiCuccio M, Kitts P, Murphy TD, Pruitt KD (2016) Reference sequence (RefSeq) database at NCBI: current status, taxonomic expansion, and functional annotation. Nucleic Acids Res 44:D733–745. https://doi.org/10.1093/nar/gkv1189

48. Maglott D, Ostell J, Pruitt KD, Tatusova T (2005) Entrez gene: gene-centered information at NCBI. Nucleic Acids Res 33:D54–58. https://doi.org/10.1093/nar/gki031
49. Tomczak K, Czerwińska P, Wiznerowicz M (2015) The cancer genome atlas (TCGA): an immeasurable source of knowledge. Contemp Oncol Poznan Pol 19:A68–77. https://doi.org/10.5114/wo.2014.47136
50. Barrett T, Wilhite SE, Ledoux P, Evangelista C, Kim IF, Tomashevsky M, Marshall KA, Phillippy KH, Sherman PM, Holko M, Yefanov A, Lee H, Zhang N, Robertson CL, Serova N, Davis S, Soboleva A (2013) NCBI GEO: archive for functional genomics data sets–update. Nucleic Acids Res 41:D991–D995. https://doi.org/10.1093/nar/gks1193
51. Agarwal V, Bell GW, Nam J-W, Bartel DP (2015) Predicting effective microRNA target sites in mammalian mRNAs. eLife 4:e05005. https://doi.org/10.7554/eLife.05005
52. Betel D, Koppal A, Agius P, Sander C, Leslie C (2010) Comprehensive modeling of microRNA targets predicts functional non-conserved and non-canonical sites. Genome Biol 11: R90. https://doi.org/10.1186/gb-2010-11-8-r90
53. Huang DW, Sherman BT, Lempicki RA (2009) Systematic and integrative analysis of large gene lists using DAVID bioinformatics resources. Nat Protoc 4:44–57. https://doi.org/10.1038/nprot.2008.211
54. Kanehisa M, Sato Y, Kawashima M, Furumichi M, Tanabe M (2016) KEGG as a reference resource for gene and protein annotation. Nucleic Acids Res 44:D457–462. https://doi.org/10.1093/nar/gkv1070
55. Huang H-Y, Lin Y-C-D, Cui S, Huang Y, Tang Y, Xu J, Bao J, Li Y, Wen J, Zuo H, Wang W, Li J, Ni J, Ruan Y, Li L, Chen Y, Xie Y, Zhu Z, Cai X, Chen X, Yao L, Chen Y, Luo Y, LuXu S, Luo M, Chiu C-M, Ma K, Zhu L, Cheng G-J, Bai C, Chiang Y-C, Wang L, Wei F, Lee T-Y, Huang H-D (2022) miRTarBase update 2022: an informative resource for experimentally validated miRNA-target interactions. Nucleic Acids Res 50:D222–D230. https://doi.org/10.1093/nar/gkab1079
56. Lee J, Yoon W, Kim S, Kim D, Kim S, So CH, Kang J (2020) BioBERT: a pre-trained biomedical language representation model for biomedical text mining. Bioinforma Oxf Engl 36:1234–1240. https://doi.org/10.1093/bioinformatics/btz682
57. Cui S, Yu S, Huang H-Y, Lin Y-C-D, Huang Y, Zhang B, Xiao J, Zuo H, Wang J, Li Z, Li G, Ma J, Chen B, Zhang H, Fu J, Wang L, Huang H-D (2025) miRTarBase 2025: updates to the collection of experimentally validated microRNA-target interactions. Nucleic Acids Res 53:D147–D156. https://doi.org/10.1093/nar/gkae1072
58. Grattafiori A, Dubey A, Jauhri A, Pandey A, Kadian A, Al-Dahle A, Letman A, Mathur A, Schelten A, Vaughan A, Yang A, Fan A, Goyal A, Hartshorn A, Yang A, Mitra A, Sravankumar A, Korenev A, Hinsvark A, Rao A, Zhang A, Rodriguez A, Gregerson A, Spataru A, Roziere B, Biron B, Tang B, Chern B, Caucheteux C, Nayak C, Bi C, Marra C, McConnell C, Keller C, Touret C, Wu C, Wong C, Ferrer CC, Nikolaidis C, Allonsius D, Song D, Pintz D, Livshits D, Wyatt D, Esiobu D, Choudhary D, Mahajan D, Garcia-Olano D, Perino D, Hupkes D, Lakomkin E, AlBadawy E, Lobanova E, Dinan E, Smith EM, Radenovic F, Guzmán F, Zhang F, Synnaeve G, Lee G, Anderson GL, Thattai G, Nail G, Mialon G, Pang G, Cucurell G, Nguyen H, Korevaar H, Xu H, Touvron H, Zarov I, Ibarra IA, Kloumann I, Misra I, Evtimov I, Zhang J, Copet J, Lee J, Geffert J, Vranes J, Park J, Mahadeokar J, Shah J, Linde J van der, Billock J, Hong J, Lee J, Fu J, Chi J, Huang J, Liu J, Wang J, Yu J, Bitton J, Spisak J, Park J, Rocca J, Johnstun J, Saxe J, Jia J, Alwala KV, Prasad K, Upasani K, Plawiak K, Li K, Heafield K, Stone K, El-Arini K, Iyer K, Malik K, Chiu K, Bhalla K, Lakhotia K, Rantala-Yeary L, Maaten L van der, Chen L, Tan L, Jenkins L, Martin L, Madaan L, Malo L, Blecher L, Landzaat L, Oliveira L de, Muzzi M, Pasupuleti M, Singh M, Paluri M, Kardas M, Tsimpoukelli M, Oldham M, Rita M, Pavlova M, Kambadur M, Lewis M, Si M, Singh MK, Hassan M, Goyal N, Torabi N, Bashlykov N, Bogoychev N, Chatterji N, Zhang N, Duchenne O, Çelebi O, Alrassy P, Zhang P, Li P, Vasic P, Weng P, Bhargava P, Dubal P, Krishnan P, Koura PS, Xu P, He Q, Dong Q, Srinivasan R, Ganapathy R, Calderer R, Cabral RS, Stojnic R, Raileanu R, Maheswari R, Girdhar R, Patel R, Sauvestre R, Polidoro R, Sumbaly R, Taylor R, Silva R, Hou R, Wang R, Hosseini S, Chennabasappa

S, Singh S, Bell S, Kim SS, Edunov S, Nie S, Narang S, Raparthy S, Shen S, Wan S, Bhosale S, Zhang S, Vandenhende S, Batra S, Whitman S, Sootla S, Collot S, Gururangan S, Borodinsky S, Herman T, Fowler T, Sheasha T, Georgiou T, Scialom T, Speckbacher T, Mihaylov T, Xiao T, Karn U, Goswami V, Gupta V, Ramanathan V, Kerkez V, Gonguet V, Do V, Vogeti V, Albiero V, Petrovic V, Chu W, Xiong W, Fu W, Meers W, Martinet X, Wang X, Wang X, Tan XE, Xia X, Xie X, Jia X, Wang X, Goldschlag Y, Gaur Y, Babaei Y, Wen Y, Song Y, Zhang Y, Li Y, Mao Y, Coudert ZD, Yan Z, Chen Z, Papakipos Z, Singh A, Srivastava A, Jain A, Kelsey A, Shajnfeld A, Gangidi A, Victoria A, Goldstand A, Menon A, Sharma A, Boesenberg A, Baevski A, Feinstein A, Kallet A, Sangani A, Teo A, Yunus A, Lupu A, Alvarado A, Caples A, Gu A, Ho A, Poulton A, Ryan A, Ramchandani A, Dong A, Franco A, Goyal A, Saraf A, Chowdhury A, Gabriel A, Bharambe A, Eisenman A, Yazdan A, James B, Maurer B, Leonhardi B, Huang B, Loyd B, Paola BD, Paranjape B, Liu B, Wu B, Ni B, Hancock B, Wasti B, Spence B, Stojkovic B, Gamido B, Montalvo B, Parker C, Burton C, Mejia C, Liu C, Wang C, Kim C, Zhou C, Hu C, Chu C-H, Cai C, Tindal C, Feichtenhofer C, Gao C, Civin D, Beaty D, Kreymer D, Li D, Adkins D, Xu D, Testuggine D, David D, Parikh D, Liskovich D, Foss D, Wang D, Le D, Holland D, Dowling E, Jamil E, Montgomery E, Presani E, Hahn E, Wood E, Le E-T, Brinkman E, Arcaute E, Dunbar E, Smothers E, Sun F, Kreuk F, Tian F, Kokkinos F, Ozgenel F, Caggioni F, Kanayet F, Seide F, Florez GM, Schwarz G, Badeer G, Swee G, Halpern G, Herman G, Sizov G, Guangyi, Zhang, Lakshminarayanan G, Inan H, Shojanazeri H, Zou H, Wang H, Zha H, Habeeb H, Rudolph H, Suk H, Aspegren H, Goldman H, Zhan H, Damlaj I, Molybog I, Tufanov I, Leontiadis I, Veliche I-E, Gat I, Weissman J, Geboski J, Kohli J, Lam J, Asher J, Gaya J-B, Marcus J, Tang J, Chan J, Zhen J, Reizenstein J, Teboul J, Zhong J, Jin J, Yang J, Cummings J, Carvill J, Shepard J, McPhie J, Torres J, Ginsburg J, Wang J, Wu K, U KH, Saxena K, Khandelwal K, Zand K, Matosich K, Veeraraghavan K, Michelena K, Li K, Jagadeesh K, Huang K, Chawla K, Huang K, Chen L, Garg L, A L, Silva L, Bell L, Zhang L, Guo L, Yu L, Moshkovich L, Wehrstedt L, Khabsa M, Avalani M, Bhatt M, Mankus M, Hasson M, Lennie M, Reso M, Groshev M, Naumov M, Lathi M, Keneally M, Liu M, Seltzer ML, Valko M, Restrepo M, Patel M, Vyatskov M, Samvelyan M, Clark M, Macey M, Wang M, Hermoso MJ, Metanat M, Rastegari M, Bansal M, Santhanam N, Parks N, White N, Bawa N, Singhal N, Egebo N, Usunier N, Mehta N, Laptev NP, Dong N, Cheng N, Chernoguz O, Hart O, Salpekar O, Kalinli O, Kent P, Parekh P, Saab P, Balaji P, Rittner P, Bontrager P, Roux P, Dollar P, Zvyagina P, Ratanchandani P, Yuvraj P, Liang Q, Alao R, Rodriguez R, Ayub R, Murthy R, Nayani R, Mitra R, Parthasarathy R, Li R, Hogan R, Battey R, Wang R, Howes R, Rinott R, Mehta S, Siby S, Bondu SJ, Datta S, Chugh S, Hunt S, Dhillon S, Sidorov S, Pan S, Mahajan S, Verma S, Yamamoto S, Ramaswamy S, Lindsay S, Lindsay S, Feng S, Lin S, Zha SC, Patil S, Shankar S, Zhang S, Zhang S, Wang S, Agarwal S, Sajuyigbe S, Chintala S, Max S, Chen S, Kehoe S, Satterfield S, Govindaprasad S, Gupta S, Deng S, Cho S, Virk S, Subramanian S, Choudhury S, Goldman S, Remez T, Glaser T, Best T, Koehler T, Robinson T, Li T, Zhang T, Matthews T, Chou T, Shaked T, Vontimitta V, Ajayi V, Montanez V, Mohan V, Kumar VS, Mangla V, Ionescu V, Poenaru V, Mihailescu VT, Ivanov V, Li W, Wang W, Jiang W, Bouaziz W, Constable W, Tang X, Wu X, Wang X, Wu X, Gao X, Kleinman Y, Chen Y, Hu Y, Jia Y, Qi Y, Li Y, Zhang Y, Zhang Y, Adi Y, Nam Y, Yu, Wang, Zhao Y, Hao Y, Qian Y, Li Y, He Y, Rait Z, DeVito Z, Rosnbrick Z, Wen Z, Yang Z, Zhao Z, Ma Z (2024) The Llama 3 herd of models. arXiv preprint arXiv:2407.21783

59. Li R, Qu H, Wang S, Wei J, Zhang L, Ma R, Lu J, Zhu J, Zhong W-D, Jia Z (2018) GDCRNATools: an R/Bioconductor package for integrative analysis of lncRNA, miRNA and mRNA data in GDC. Bioinforma Oxf Engl 34:2515–2517. https://doi.org/10.1093/bioinformatics/bty124
60. R: A language and environment for statistical computing. Semantic Scholar. https://www.semanticscholar.org/paper/R%3A-A-language-and-environment-for-statistical-Team/659408b243cec55de8d0a3bc51b81173007aa89b. Accessed 22 Jan 2025
61. Gentleman RC, Carey VJ, Bates DM, Bolstad B, Dettling M, Dudoit S, Ellis B, Gautier L, Ge Y, Gentry J, Hornik K, Hothorn T, Huber W, Iacus S, Irizarry R, Leisch F, Li C, Maechler M, Rossini AJ, Sawitzki G, Smith C, Smyth G, Tierney L, Yang JYH, Zhang J (2004) Bioconductor: open software development for computational biology and bioinformatics. Genome Biol 5: R80. https://doi.org/10.1186/gb-2004-5-10-r80
62. Grossman RL, Heath AP, Ferretti V, Varmus HE, Lowy DR, Kibbe WA, Staudt LM (2016) Toward a shared vision for cancer genomic data. N Engl J Med 375:1109–1112. https://doi.org/10.1056/NEJMp1607591

63. Robinson MD, McCarthy DJ, Smyth GK (2010) edgeR: a Bioconductor package for differential expression analysis of digital gene expression data. Bioinforma Oxf Engl 26:139–140. https://doi.org/10.1093/bioinformatics/btp616
64. McCarthy DJ, Chen Y, Smyth GK (2012) Differential expression analysis of multifactor RNA-seq experiments with respect to biological variation. Nucleic Acids Res 40:4288–4297. https://doi.org/10.1093/nar/gks042
65. Ritchie ME, Phipson B, Wu D, Hu Y, Law CW, Shi W, Smyth GK (2015) limma powers differential expression analyses for RNA-sequencing and microarray studies. Nucleic Acids Res 43:e47. https://doi.org/10.1093/nar/gkv007
66. Love MI, Huber W, Anders S (2014) Moderated estimation of fold change and dispersion for RNA-seq data with DESeq2. Genome Biol 15: 550. https://doi.org/10.1186/s13059-014-0550-8
67. miRcode: a map of putative microRNA target sites in the long non-coding transcriptome. https://pubmed.ncbi.nlm.nih.gov/22718787/. Accessed 27 Jan 2025
68. Furió-Tarí P, Tarazona S, Gabaldón T, Enright AJ, Conesa A (2016) spongeScan: a web for detecting microRNA binding elements in lncRNA sequences. Nucleic Acids Res 44: W176–180. https://doi.org/10.1093/nar/gkw443
69. Paci P, Colombo T, Farina L (2014) Computational analysis identifies a sponge interaction network between long non-coding RNAs and messenger RNAs in human breast cancer. BMC Syst Biol 8: 83. https://doi.org/10.1186/1752-0509-8-83
70. Yu G, Wang L-G, Han Y, He Q-Y (2012) clusterProfiler: an R package for comparing biological themes among gene clusters. Omics J Integr Biol 16:284–287. https://doi.org/10.1089/omi.2011.0118
71. Yu G, Wang L-G, Yan G-R, He Q-Y (2015) DOSE: an R/Bioconductor package for disease ontology semantic and enrichment analysis. Bioinforma Oxf Engl 31:608–609. https://doi.org/10.1093/bioinformatics/btu684
72. Nishikura K (2016) A-to-I editing of coding and non-coding RNAs by ADARs. Nat Rev Mol Cell Biol 17(2):83–96 https://doi.org/10.1038/nrm.2015.4
73. Wahlstedt H, Ohman M (2011) Site-selective versus promiscuous A-to-I editing. WIREs RNA 2(6):761–771. https://doi.org/10.1002/wrna.89
74. Nigita G, Veneziano D, Ferro A (2015) A-to-I RNA editing: current knowledge sources and computational approaches with special emphasis on non-coding RNA molecules. Front Bioeng Biotechnol 3. https://doi.org/10.3389/fbioe.2015.00037
75. Gong J, Liu C, Liu W, Xiang Y, Diao L, Guo A-Y, Han L (2017) LNCediting: a database for functional effects of RNA editing in lncRNAs. Nucleic Acids Res 45(D1):D79–D84. https://doi.org/10.1093/nar/gkw835
76. Picardi E, Pesole G (2013) REDItools: high-throughput RNA editing detection made easy. Bioinformatics 29(14):1813–1814. https://doi.org/10.1093/bioinformatics/btt287
77. D'Addabbo P, Cohen-Fultheim R, Twersky I, Fonzino A, Silvestris DA, Prakash A, Mazzacuva PL, Vizcaino JA, Green A, Sweeney B, Yates A, Lussi Y, Luo J, Martin MJ, Eisenberg E, Levanon EY, Pesole G, Picardi E (2025) REDIportal: toward an integrated view of the A-to-I editing. Nucleic Acids Res 53(D1):D233–D242. https://doi.org/10.1093/nar/gkae1083
78. Piechotta M, Naarmann-de Vries IS, Wang Q, Altmüller J, Dieterich C (2022) RNA modification mapping with JACUSA2. Genome Biology 23(1). https://doi.org/10.1186/s13059-022-02676-0
79. Cunningham F, Amode MR, Barrell D, Beal K, Billis K, Brent S, Carvalho-Silva D, Clapham P, Coates G, Fitzgerald S, Gil L, Girón CG, Gordon L, Hourlier T, Hunt SE, Janacek SH, Johnson N, Juettemann T, Kähäri AK, Keenan S, Martin FJ, Maurel T, McLaren W, Murphy DN, Nag R, Overduin B, Parker A, Patricio M, Perry E, Pignatelli M, Riat HS, Sheppard D, Taylor K, Thormann A, Vullo A, Wilder SP, Zadissa A, Aken BL, Birney E, Harrow J, Kinsella R, Muffato M, Ruffier M, Searle SM, Spudich G, Trevanion SJ, Yates A, Zerbino DR, Flicek P (2015) Ensembl 2015. Nucleic Acids Res 43(D1):D662–D669. https://doi.org/10.1093/nar/gku1010

Check for updates

Chapter 10

Construction and Analysis of circRNA–miRNA–mRNA Regulatory Networks

Simone Avesani, Luciano Cascione and Rosalba Giugno

Abstract

Circular RNAs (circRNAs) represent an emerging class of noncoding RNAs characterized by their covalently closed-loop structure and generated through the back-splicing event. This structural configuration provides circRNAs with enhanced stability and unique functional properties. Functionally, circRNAs act as protein scaffolds, enhance protein activity, can be translated, and frequently serve as competing endogenous RNAs (ceRNAs), particularly by sponging microRNAs (miRNAs) and modulating downstream gene expression. Since these regulatory capacities have significant implications in cancer, neurodegenerative, and cardiovascular diseases, in response, numerous computational approaches have been developed to reconstruct circRNA–miRNA–mRNA networks, ranging from specific circRNA-oriented applications to sequencing-based tools, each with distinct strengths and limitations. This chapter provides an overview of these strategies, highlighting their application in bioinformatics research and their potential to explore novel regulatory mechanisms mediated by circRNAs.

Keywords circRNA, circRNA–miRNA–mRNA networks, Network reconstruction, Regulative analysis

1 Introduction

Circular RNAs (circRNAs) represent an increasingly important class of noncoding RNAs that are distinguished from linear RNAs by their distinctive closed-loop structures. This circular formation occurs during the RNA maturation process through a noncanonical splicing event known as back-splicing. In this process, a downstream 5′ splice site is joined to an upstream 3′ splice site, resulting in a circular RNA molecule that lacks the free ends found in linear RNAs. This structural characteristic provides the circRNAs with functional advantages and differentiates them from other RNA types, particularly in terms of stability and function [1, 2]. First discovered in the genomes of pathogens, circRNAs were initially believed to be rare and functionally irrelevant. For many years, they were regarded as "junk" RNA, a perception that changed dramatically with the improvement of RNA

Alessandro Laganà and Giovanni Nigita (eds.), *MicroRNA Target Identification*, Methods in Molecular Biology, vol. 3056, https://doi.org/10.1007/978-1-0716-5392-0_10,

sequencing technologies that allowed researchers to more accurately detect and quantify circRNAs across many organisms and tissues. This shift in perspective has led to the discovery of thousands of circRNAs in different cell types and has opened the door to a deeper understanding of their potential biological relevance. Recent studies have investigated the diverse potential functions of circRNAs [1], uncovering their roles beyond conventional RNA mechanisms. These functions include acting as protein scaffolds, enhancing protein activities, encoding proteins, and serving as competing endogenous RNAs (ceRNAs) in relation to other RNA molecules, such as microRNAs (miRNAs). One of the most studied functions of circRNAs is their ability to "sponge" miRNAs, binding them and preventing their interaction with target messenger RNAs (mRNAs). This interaction initiates a regulatory cascade that can significantly affect gene expression. The capacity of circRNAs to modulate miRNA activity has gained growing attention in biomedical research, particularly due to its implications in the development and progression of diseases such as cancer, neurodegenerative disorders, and cardiovascular diseases [3–6]. This growing attention has also been reflected in the field of bioinformatics, leading to the development of numerous computational methods and tools aimed at exploring circRNAs' regulatory effects and functional implications. The functional analysis of circRNA regulatory effects primarily investigates their interactions with microRNAs (miRNAs) and the downstream genes modulated by these miRNAs, modeling the regulatory cascade process as complex regulatory networks involving circRNAs, miRNAs, and corresponding genes. These networks encode cascade regulatory effects that can ultimately influence diverse cellular processes and phenotypic outcomes. This chapter presents an overview of the main computational strategies used to investigate the regulatory roles of circRNAs through the reconstruction of circRNA–miRNA–mRNA interaction networks. It focuses on two main categories of approaches: specific circRNA-oriented and sequencing data-oriented tools, including CircInteractome [7], CircNet 2.0 [8], CircNetVis [9], CircMiMi [10], CircScan [11], and EasyCircR [12]. In the following sections, each tool is described in depth, and its functionalities are illustrated through simple case studies, demonstrating its usability and applicability in bioinformatics research.

2 Materials

The web applications have been used with Google Chrome. R (version 4.4.0) packages have been tested on a machine with Apple M3 Pro, RAM 36 GB LPDDR5 with macOS 14.4.

3 Methods

Computational approaches designed to investigate circRNA–miRNA–mRNA regulatory networks can be classified into two major categories: specific circRNA-oriented tools and sequencing data-oriented tools. The first category focuses on one or more circRNAs of interest and typically constructs regulatory networks by integrating known molecular interactions retrieved from existing databases. These tools allow users to explore the regulatory potential of specific circRNAs without requiring sequencing data, making them more accessible and easier to use. This group of tools includes CircInteractome [7], CircNet 2.0 [8], CircNetVis [9], and CircMiMi [10], all web applications that are easy to use and require no specific computational skills. In contrast, sequencing data-oriented tools reconstruct regulatory networks directly from transcriptomic datasets through data mining and filtering procedures, retaining only the molecules that are experimentally detected in the analyzed sequencing data. This approach generally provides more accurate and biologically relevant analysis but requires complex and resource-intensive sequencing data, which may not always be available. This group of tools includes the CircScan [11] and EasyCircR [12] R packages. Each approach presents distinct strengths and limitations: specific circRNA-oriented tools are more user-friendly and computationally lightweight, while sequencing data-oriented tools offer higher accuracy and specificity at the cost of greater experimental and computational demands. To evaluate the performance of different tools for circRNA–miRNA–mRNA regulatory network reconstruction in real-case scenarios, distinct case studies were designed, reflecting the different input requirements of circRNA-oriented and sequencing data-oriented tools. For the first category, namely the circRNA-oriented tools, the *hsa_circ_0001946* circRNA was selected as the case study, as this molecule has been extensively explored in the literature [13, 14]. Instead, for the sequencing-based tools, due to the difficulty of finding suitable raw sequencing data for circRNAs reconstruction and comprehensive multilevel transcriptomic datasets, we used, as case studies, the datasets described in the original publications [11, 12]. Specifically, CircScan was tested using the colorectal cancer dataset [15] and EasyCircR using a dataset of triple-negative breast cancer (TNBC) slow-cycling and rapid-cycling cells [16].

3.1 CircInteractome

CircInteractome [7] is a database developed to facilitate the exploration of circRNA interactions and their potential regulatory functions at the gene expression level. The platform mainly relies on TargetScan [17, 18] for predicting molecular interactions, particularly miRNA-circRNA associations, and integrates multiple analysis modules to support circRNA research. Users can query circRNAs by name, genomic position, or by retrieving the best-matching transcript, and can obtain both genomic and mature circRNA sequences. In addition, CircInteractome provides tools to identify RNA-binding proteins (RBPs) predicted to interact with specific circRNAs and to detect miRNAs that may target circRNAs. The resource also incorporates utilities for experimental design, such as generating divergent primers for circRNA validation and designing siRNAs targeting circRNAs of interest. Although the tool does not allow the full reconstruction of circRNA–miRNA–mRNA regulatory networks, its features make CircInteractome a versatile resource for investigating the biological roles of circRNAs and for supporting both computational analyses and experimental studies.

CircInteractome Web Search

CircInteractome web application (https://circinteractome.irp.nia.nih.gov/index.html) provides several dedicated modules to perform analyses through an intuitive window-based interface. In the first window, "Circular RNA," users can enter either a circRNA ID or a gene symbol of interest to visualize the RBP binding sites associated with the circRNA, together with basic annotation details. The second window, "RBP on CircRNA," enables the search for specific associations between an RBP and a given circRNA. In the "miRNA Target Sites" window, users can input a circRNA ID to explore all predicted miRNA interactions for that circRNA, or alternatively provide a miRNA ID to check for potential associations with circRNAs. Two additional modules support experimental design: one for generating divergent primers and another for designing siRNAs specific to a given circRNA. Finally, a Help window provides detailed guidelines and step-by-step instructions on how to query and effectively use the CircInteractome database.

In our case, we queried the database searching for the *hsa_circ_0001946* circRNA in the "Circular RNA" window, to retrieve the list of RNA-binding proteins. Figure 1 shows the output window, including general information about the circRNA and several result tables. These tables allow users to explore the full list of RNA-binding proteins (RBPs) along with the corresponding number of predicted binding sites interacting with the circRNA. In this example, AGO2 exhibits the highest number of binding sites. Additional tables report the RBPs binding specifically to the back-splice junction (BSJ) or to the flanking regions of the circRNA. Moving to the "miRNA Target Sites" section, we queried the *hsa_circ_0001946* circRNA to explore the miRNAs predicted to be sponged by this molecule. Once the query is submitted, the output window displays a summary table listing all miRNAs predicted to bind the circRNA (Fig. 2). Scrolling further down, a more detailed table is presented, containing information about each miRNA–circRNA interaction, including base-pairing details, site types, and various alignment scores. Although the CircInteractome database does not provide direct visualization of the complete circRNA–miRNA–mRNA regulatory network, it enables

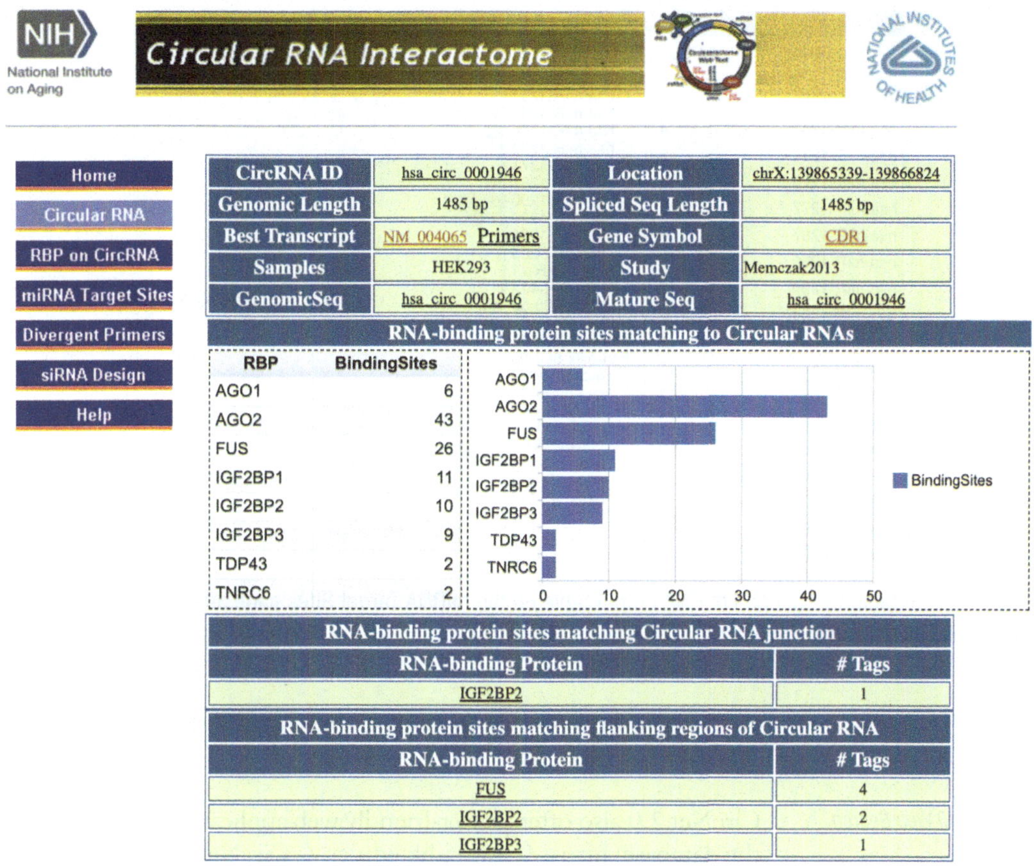

Fig. 1 CircInteractome output window of the Circular RNA Search page after querying the *hsa_circ_0001946* molecule

users to retrieve detailed data on individual interactions that can be downloaded and used for further downstream analyses.

3.2 CircNet 2.0

CircNet 2.0 [8] is the updated version of the original CircNet [19], first published in 2016. It contains a comprehensive catalog of circRNAs detected from raw RNA-seq data of six cancer types obtained from The Cancer Genome Atlas (TCGA). In addition, CircNet 2.0 integrates circRNA data from several external resources, including over 1000 RNA-seq samples stored in circAtlas [20] and circRNAs derived from 1000 human tumor tissues available in MiOncoCirc [21]. circRNA identification and collection were carried out using a set of well-established bioinformatics tools: CIRI2 [22], CIRC-explorer2 [23], find_circ [24], and DCC [25] for circRNA reconstruction, and miRanda [26], PITA [27], TargetScan [17, 18], and miRTarBase [28] for the prediction of circRNA–miRNA–mRNA interaction networks. Altogether, the database integrates information from more than 2732 samples representing 37 types of human

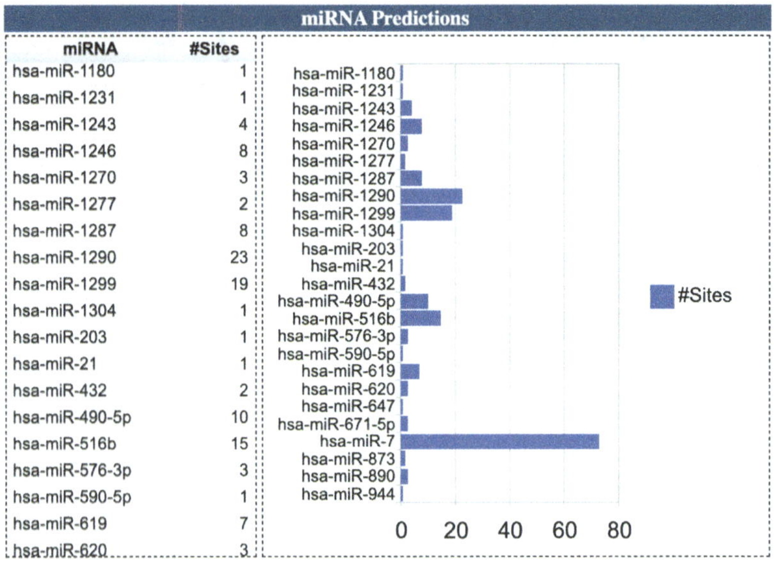

Fig. 2 CircInteractome sponged miRNA output table on the miRNA Target Sites page after querying the *hsa_circ_0001946* molecule

cancers, making it one of the most comprehensive circRNA resources available in cancer research.

CircNet 2.0 Web Search

CircNet 2.0 also offers a user-friendly web application to easily query the database: https://awi.cuhk.edu.cn/~CircNet/php/index.php. Through its top navigation bar, the application provides several functionalities that facilitate the exploration of circRNA data. The "Data" page displays the complete list of circRNAs stored in the database, which can be filtered by chromosome, and provides key details for each circRNA, including strand, circRNA type, host gene, and the algorithm used for reconstruction. The "Download" page allows users to retrieve the entire dataset, including circRNA–miRNA interaction lists, circRNA information for specific cancer types, and the comprehensive list of circRNA read counts per sample. The "Statistics and Tutorial" pages provide, respectively, an overview of the data sources and detailed instructions for performing searches within the database.

The CircNet 2.0 search page (Fig. 3) further expands these capabilities by providing multiple strategies for retrieving circRNA information, supporting flexible and targeted data exploration. Users can perform a basic search by entering a specific circRNA, miRNA ID, or gene symbol; a BLAST search by submitting a FASTA sequence; or a cancer type search by selecting a specific cancer type and specifying the number of circRNAs to explore. Additional options include querying by CircBase ID, circAtlas ID, or host gene (Ensembl format). Once a circRNA is queried, the platform displays a series

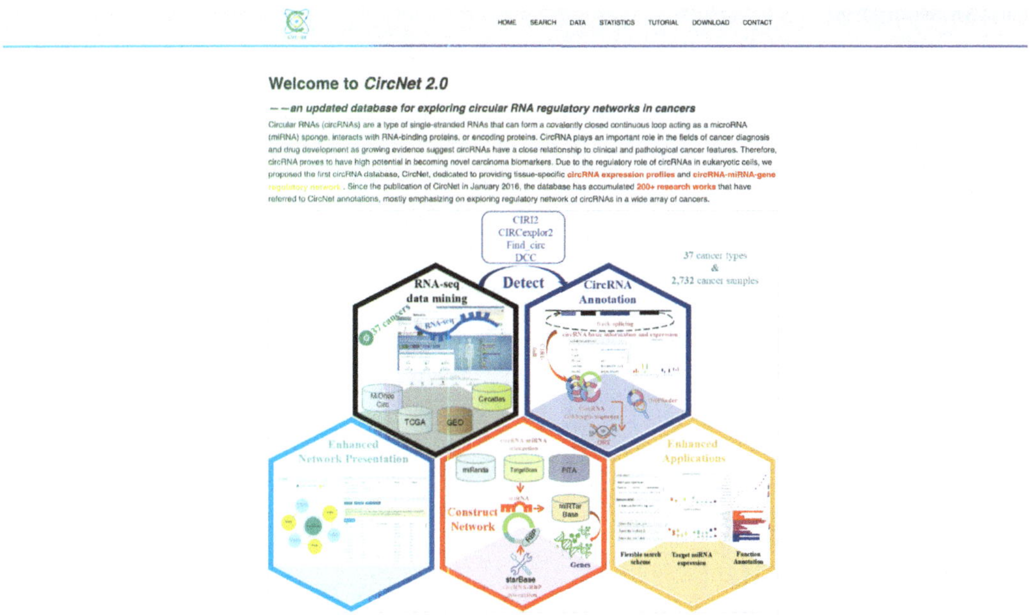

Fig. 3 CircNet 2.0 web application homepage

Target miRNAs

CircID	miRNA	PITA	miRanda	targetScan
chrX:140783175|140784659	hsa-miR-125b-2-3p	0	1	1
chrX:140783175|140784659	hsa-miR-7-5p	0	1	1
chrX:140783175|140784659	hsa-miR-139-3p	0	1	1

Fig. 4 Table displaying the miRNAs targeted by hsa_circ_0001946 as identified by CircNet 2.0

of panels containing comprehensive information, including core attributes (type, host gene, identification algorithm, and nucleotide sequence), expression profiles across cancer tissues, and the inferred circRNA–miRNA–mRNA regulatory network with sponged miRNAs, downstream target genes, and functional enrichment results generated through GO [29], KEGG [30], and DisGeNET [31]. For our case study, we searched for hsa_circ_0001946 using its CircBase ID. The first panel displays its genomic coordinates (chrX: 140783175-140784659) and sequence, while the second presents the corresponding regulatory network and the enrichment results (Fig. 4). Additional panels provide complementary information, including miRNA expression profiles across cancer types, open reading frames, and RNA-binding proteins with related statistics such as upstream, downstream, and exon site counts (Fig. 5).

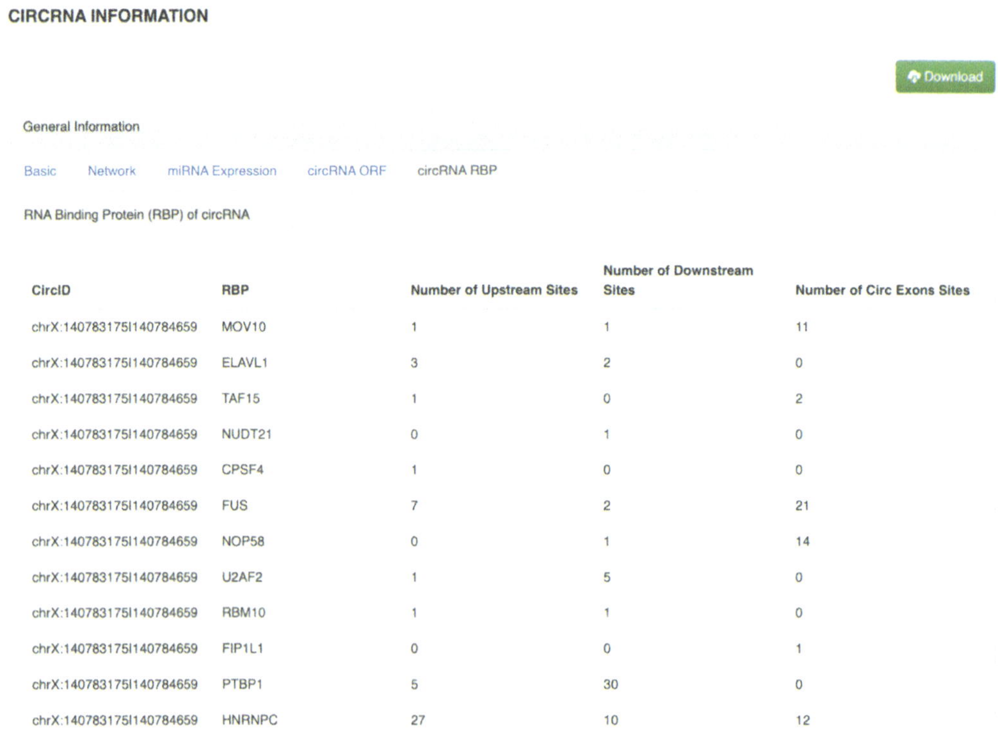

CircID	RBP	Number of Upstream Sites	Number of Downstream Sites	Number of Circ Exons Sites
chrX:140783175I140784659	MOV10	1	1	11
chrX:140783175I140784659	ELAVL1	3	2	0
chrX:140783175I140784659	TAF15	1	0	2
chrX:140783175I140784659	NUDT21	0	1	0
chrX:140783175I140784659	CPSF4	1	0	0
chrX:140783175I140784659	FUS	7	2	21
chrX:140783175I140784659	NOP58	0	1	14
chrX:140783175I140784659	U2AF2	1	5	0
chrX:140783175I140784659	RBM10	1	1	0
chrX:140783175I140784659	FIP1L1	0	0	1
chrX:140783175I140784659	PTBP1	5	30	0
chrX:140783175I140784659	HNRNPC	27	10	12

Fig. 5 Table showing the RNA-binding proteins of hsa_circ_0001946 identified by CircNet 2.0

3.3 CircNetVis

CircNetVis [9] is an interactive web application specifically developed for the visualization and exploration of circRNA regulatory networks. The platform enables users to investigate multiple types of interactions, including circRNA–miRNA, miRNA–mRNA, and circRNA–RBP (RNA-binding protein) associations, thereby providing a comprehensive view of the potential regulatory landscape mediated by circRNAs. To ensure reliability, CircNetVis integrates information from several established resources, such as circBase [32], miRBase [33–38], CircInteractome [7], TargetScan [17, 18], miRanda [26], RNAhybrid [39], and Circall-simulator [40]. By combining these heterogeneous data sources, the tool facilitates the construction of rich and multilayered circRNA interaction networks. CircNetVis supports network visualization, interactive querying, and filtering features, and it is freely accessible through the following link: https://www.meb.ki.se/shiny/truvu/CircNetVis/

CircNetVis Web Search

CircNetVis provides a five-panel menu interface. The "Getting Started" window allows users to select the reference species; currently, only two species are available: human and mouse. Once the reference species is selected, the user can move to the "circRNA–miRNA–mRNA Interaction" window, where they can insert the ID of the circRNA of interest or its nucleotide sequence and execute the analysis to visualize the circRNA–miRNA–mRNA interaction network. On the left menu, the user can select specific thresholds to filter out nonrelevant interactions retrieved from the different sources integrated inside CircNetVis, including removing circRNA–miRNA interactions based on the type of binding seed or the significance of the interaction and removing miRNA–mRNA interactions setting a maximum number of target genes or based on a weighted-context score. In our case, we input *hsa_circ_0001946* circRNA, setting the default parameters except for the maximum number of gene interactions, which was set to 100. Figure 6 shows the full circRNA–miRNA–mRNA regulatory network for the selected parameters.

In the "circRNA–RBP Interaction" window, users can explore the circRNA–RBP interaction network and refine the visualization by applying several filtering options. These include restricting the output to only the top-ranked RBP binding sites or selecting a specific number or subset of proteins of interest. In our case study, Fig. 7 shows the circRNA–RBP network setting the minimum number of

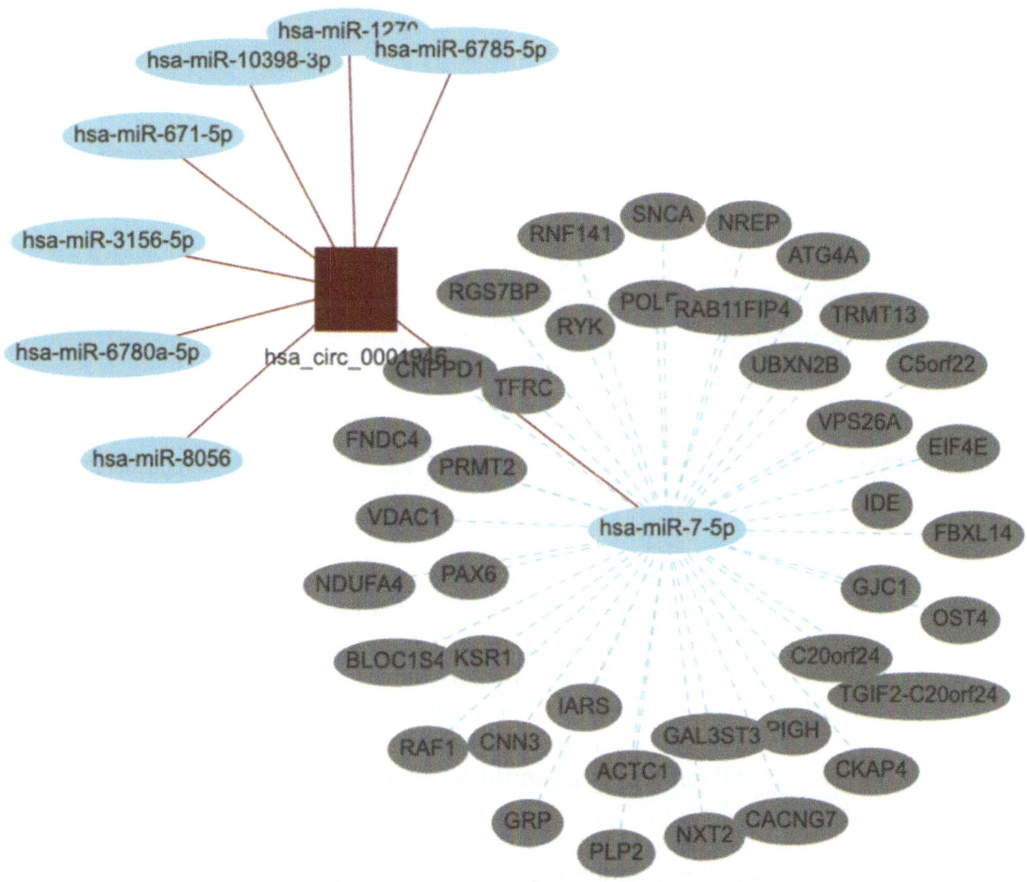

Fig. 6 CircNetVis circRNA–miRNA–mRNA regulatory network after querying *hsa_circ_0001946* molecule

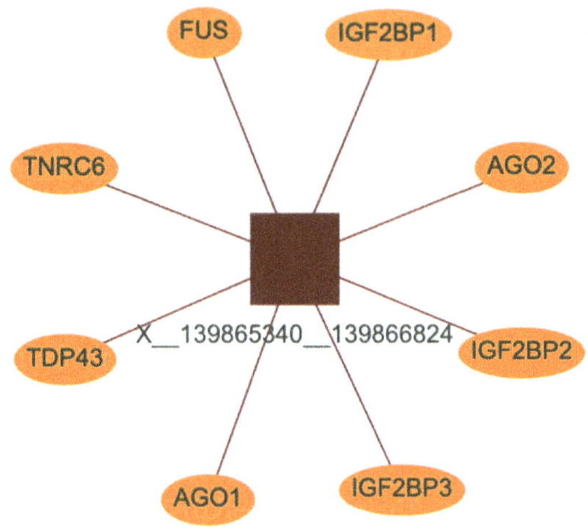

Fig. 7 CircNetVis circRNA–RBP network after querying *hsa_circ_0001946* molecule

Fig. 8 CircNetVis GSEA enrichment results for mRNAs involved in the hsa_circ_0001946 regulatory network

binding sites as 1. To conclude the analysis, the user can move to the Pathway analysis and run Gene Set Enrichment Analysis [41, 42] on the set of genes targeted by the circRNA.

Figure 8 shows the enrichment result of GSEA on the mRNAs involved in *hsa_circ_0001946* regulatory network. The last window of the CircNetVis web application is the "Help" page, where details and guidelines about the tool are presented.

3.4 CircMiMi

CircMiMi [10] is a Python-based software designed to identify circRNA–miRNA–mRNA interactions across 18 different species, based on circRNA coordinates provided by the user. Starting from back-splice junction (BSJ) coordinates, CircMiMi automatically retrieves annotation data, verifies whether BSJs overlap annotated exon boundaries, and filters out false positives coming from ambiguous alignments or multiple mapping. Exonic flanking sequences are concatenated and aligned against the reference genome with two different parameter sets to ensure mapping reliability; only candidates passing both checks are retained. CircMiMi then predicts exonic circRNA sequences according to user-specified genome annotations and identifies potential miRNA binding sites with miRanda [26], applying stringent thresholds and considering BSJ-spanning sites. For human and mouse species, predictions are further supported by CLIP-seq data from ENCORI [43]. To extend the regulatory landscape, miRNA–mRNA interactions are integrated from miRDB [44], miRTarBase [28], and ENCORI, limited to species with Ensembl annotations. By combining circRNA–miRNA and miRNA–mRNA associations, CircMiMi builds circRNA–miRNA–mRNA regulatory networks. A hypergeometric test with FDR correction is used to determine whether circRNA–mRNA pairs are significantly co-regulated by shared miRNAs (FDR < 0.05). Significant interactions are retained and can be exported for visualization, while a user-friendly web interface is also available to facilitate usage without requiring local installation.

CircMiMi Web App

The CircMiMi web application is accessible at: https://circmimi.genomics.sinica.edu.tw/. Figure 9 shows its homepage. In the "Run" section, users can select the species of interest, enter the circRNA coordinates, and submit the query. The platform then generates a compressed folder for download, containing multiple .tsv files with all predicted circRNA–miRNA-mRNA interactions and summary statistics related to the queried circRNA. To evaluate the tool on our case study, we provided the coordinates of *hsa_circ_0001946*; however, the query returned no predicted interactions, resulting in empty output files.

3.5 CircScan

CircScan [11], unlike the previously presented tools, belongs to the class of sequencing data-oriented approaches, as it infers circRNA–miRNA–mRNA regulatory networks directly from transcriptomic data collected under two distinct experimental conditions (e.g., treated vs. untreated). The tool requires multilevel transcriptomic expression data as input, including circRNA, mRNA, and miRNA from human samples, preprocessed and organized in CSV files. Network inference is carried out through two main steps: (i) construction of a multi-criteria scoring matrix to assess interaction reliability and the functional relevance of circRNA-mediated sponging, and (ii) calculation of a sponging efficiency score for each circRNA–miRNA–mRNA subnetwork. Two biological prerequisites are applied to ensure meaningful interaction detection: miRNAs

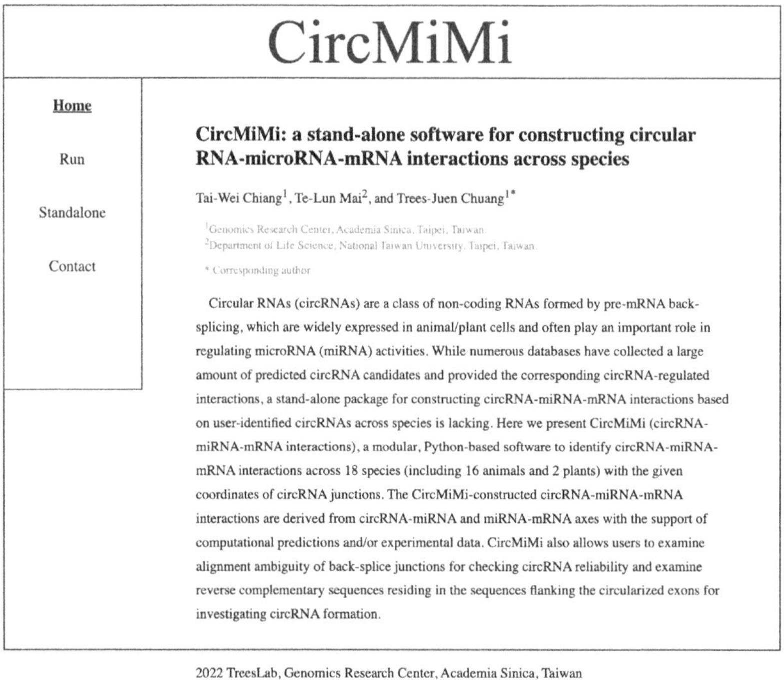

Fig. 9 CircMiMi web application homepage

must be sufficiently expressed in all conditions, but not necessarily differentially expressed, while circRNAs and mRNAs must be co-expressed and differentially expressed in the same direction. The tool integrates several interaction reliability scores, evaluating factors such as binding strength, number of binding sites, sequence length, and expression levels. A multi-criterion ranking method (TOPSIS) combines these features into two final metrics: a sponge score for ranking circRNA–miRNA interactions and a global score for ranking miRNA–mRNA interactions. CircScan also provides a user-friendly Shiny application, which includes an interactive visualization panel built with the visNetwork R package [45], enabling use by non-programmers. Additionally, it supports gene enrichment analysis via the Enrichr tool [46–48] to enhance the biological interpretability of the targeted genes.

CircScan Regulative Network Reconstruction

CircScan tool provides a Shiny application that can be used to run the analysis. To open the Shiny app, the user should install some required packages and import CircScan:

```
$install.packages(c("BiocManager", "circlize",
"data.table", "dplyr", "DT", "enrichR", "plyr",
"shiny", "shinycssloaders", "shinydashboard",
"shinyjs", "stringr", "tidyr", "topsis",
"visNetwork", "vroom"))
```

```
$BiocManager::install("fgsea")
$shiny::runApp('/cirscan_Rshiny-main/')
```

Once the Shiny application is running (Fig. 10), using the left panel, the user can easily upload the different necessary data in CSV file format, including mRNA, circRNA, miRNA data, and the sample annotation file. Moreover, the user can set interaction filtering criteria, adjust score thresholds, and run the analysis.

As a case study, we selected the colorectal cancer data provided by the authors [15] of the tool. We uploaded the four .csv files in the Shiny app: three files storing RNA expression scores for 20 samples (10 tumor and 10 healthy) of mRNAs, miRNAs, and circRNAs, respectively, and one file storing sample information. We then ran the analysis, and CircScan reconstructed circRNA–miRNA–mRNA regulatory networks for the various differentially expressed circRNAs provided as input.

3.6 *EasyCircR*

EasyCircR [12] is a sequencing data-oriented R package designed to infer circRNA–miRNA–mRNA regulatory networks. Unlike other sequencing data-oriented tools, EasyCircR reconstructs full-length circRNAs directly from FASTQ files across multiple sample groups, identifies condition-specific target genes from the same raw data, and predicts their interacting miRNAs. Since circRNAs function as competing endogenous RNAs at the post-transcriptional level, EasyCircR improves biological relevance by keeping only post-transcriptionally regulated genes, identified through exon–intron split analysis (EISA) [49], which distinguishes transcriptional from post-transcriptional regulation based on exonic and intronic expression changes. The

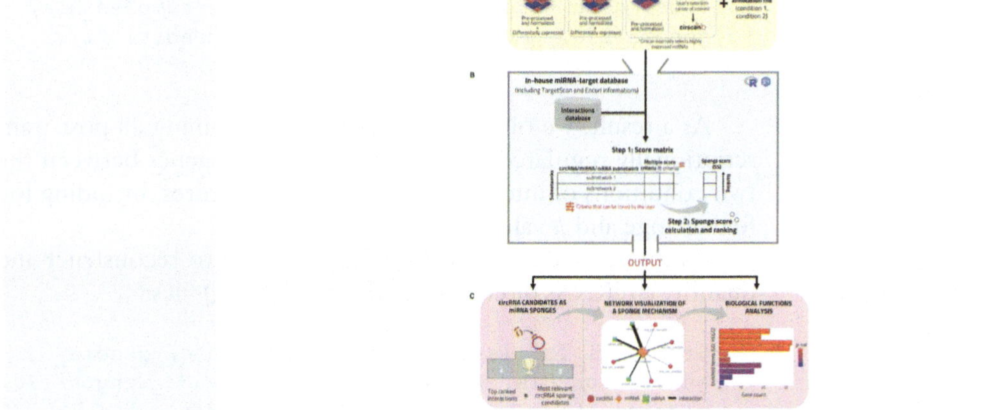

Fig. 10 CircScan shiny application homepage

package performs a comprehensive multistep analysis starting directly from raw RNA-seq data. First, it reconstructs full-length circRNAs using CIRI-full [50], a highly accurate algorithm that integrates back-splice junctions and reverse overlaps for robust circRNA detection. It then identifies differentially expressed circRNAs between experimental conditions using Limma-voom [51, 52] statistical methods and links them to their host genes through the circBank [53] database. Next, EasyCircR predicts miRNA binding sites on circRNAs using TargetScan [17, 18], generating potential circRNA–miRNA interaction maps. To capture downstream regulatory effects, it executes EISA to identify genes likely regulated by circRNAs at the post-transcriptional level. Finally, EasyCircR integrates all results, using the multiMiR tool [54], to build a circRNA–miRNA–mRNA regulatory network, which can be interactively explored through an R Shiny web application. Through it, users can browse, filter, and export interaction data, visualize circRNA structures, and examine specific regulatory relationships in detail.

EasyCircR Regulative Network Reconstruction

To demonstrate a simple application of EasyCircR, we analyzed the triple-negative breast cancer (TNBC) dataset (GSE267759) described in the EasyCircR article for the sake of simplicity. We applied the pipeline to RNA-seq data collected from three slow-cycling and three rapid-cycling TNBC samples, with the goal of reconstructing circRNA–miRNA–mRNA regulatory networks.

As a first step, a sample annotation file is created, linking each FASTQ file to its corresponding sample name. Next, post-transcriptionally regulated genes are identified using a dedicated function that runs EISA.

```
$samples_file <- "sample.txt"
$genome_file <- "/Data/hg38.fa"
$genome_annotation_file <- "/Data/hg38.gtf"
$condition <
- factor(c("DMSO","DMSO","DMSO","PQR","PQR","PQR"))
$cellLine <
- factor(c("12h","8h","4h","12h","8h","4h"))
$post_gene <- get_postregulated_gene(samples_file,
genome_file, genome_annotation_file, condition,
aligner="Rbowtie", force=FALSE, n_core=10)
```

As a result, we obtained a data frame containing all post-transcriptionally regulated differentially expressed genes between the two conditions of interest, along with several scores, including log fold change and *p*-value.

As the second step, CIRI-full is executed to reconstruct and quantify circRNAs directly from the raw FASTQ files:

```
$run_ciri_full(samples_file, genome_file, genome_
annotation_file, force=FALSE, trim_reads_length=130,
n_core= 10)
```

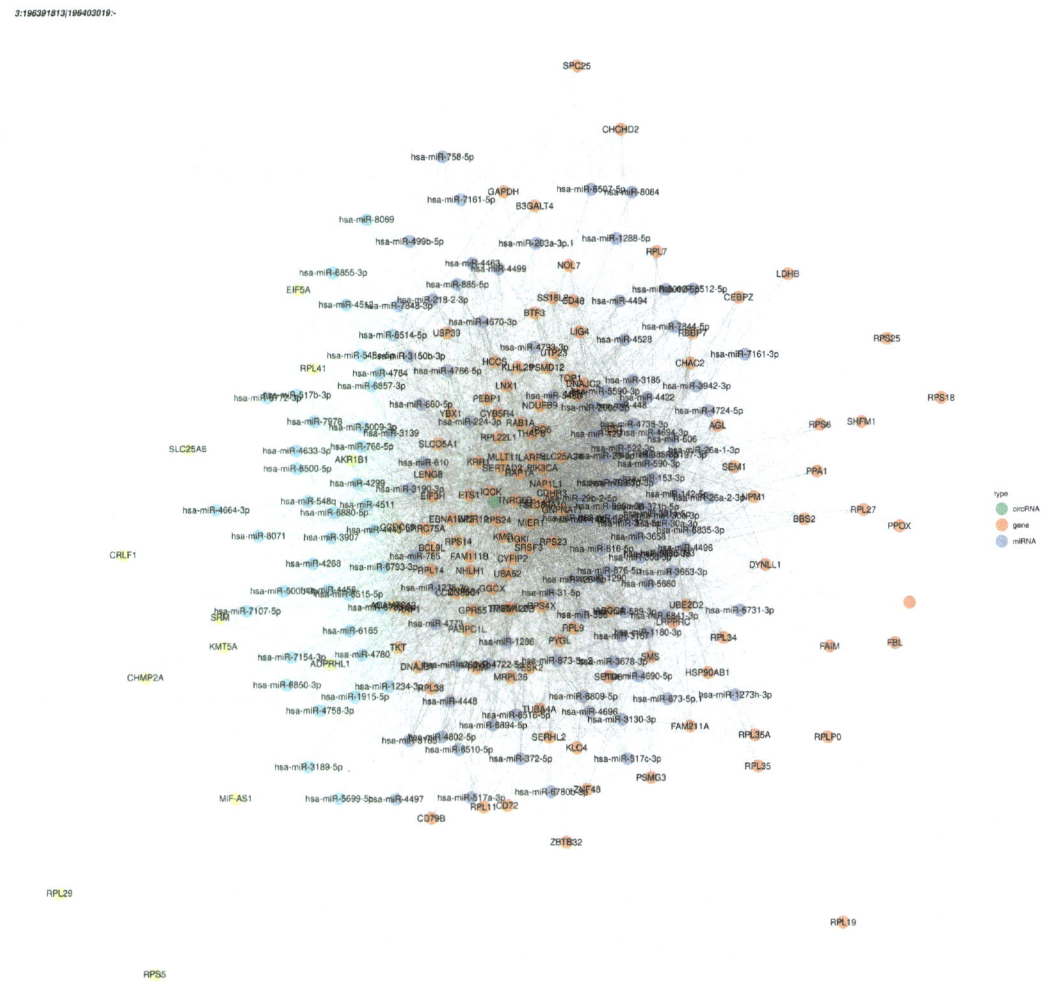

Fig. 11 Regulatory network of circRNA chr3_196391813_196403019 reconstructed by EasyCircR

```
$circ <- read_ciri_output(samples_file)
$circ_df <- circ$circ_df
$circ_mtx <- circ$circ_mtx
```

The function also trims all reads to a uniform length, a necessary pre-processing step for CIRI-full. The *read_ciri_output()* function is then used to post-process the CIRI-full output, generating both a circRNA quantification matrix and a dataframe containing detailed information about the reconstructed circRNAs:

```
$design <- model.matrix(~0+condition+cellLine)
$contr <- limma::makeContrasts("PQRvsDMSO"
= conditionPQR - conditionDMSO, levels =
colnames(design))
$circ_de <- de_circrna(circ_mtx, condition, design,
contr, lfc=0, p.value=0.05
voomWithQualityWeights=FALSE, min_num_samples=2)
```

```
$circ_df_de <- circ_df[circ_df$bsj_id %in%
rownames(circ_de),]
```

At the end, only the differentially expressed circRNAs are retained for the subsequent analysis steps. The fourth step involves inferring circRNA–miRNA interactions using the TargetScan tool:

```
$circ_mirna <- get_mirna_binding(circ_df_de, force=
F)
```

To finalize the regulatory network construction, the fifth step involves inferring miRNA–mRNA associations through the multiMiR package associations:

```
$gene_mirna_circ_all <- connect_circ_gene(circ_
mirna, post_gene, only_significant_genes=TRUE,
force=T, tabletype="all")
```

At the end, a Shiny application can be launched to visualize and further explore the results.

```
$launch_shiny(shiny_host = "127.0.0.1", shiny_port
= 7775)
```

Figure 11 shows an example of a circRNA–miRNA–mRNA network generated from one of the reconstructed circRNAs: chr3_196391813_196403019.

Conflict of Interest

The authors have no conflicts of interest to declare that are relevant to the content of this chapter.

References

1. Bolha L, Ravnik-Glavač M, Glavač D (2017) Circular RNAs: biogenesis, function, and a role as possible cancer biomarkers. Int J Genom 2017(1):6218353
2. Feng Z, Meng S, Zhou H et al (2019) Functions and potential applications of circular RNAs in cancer stem cells. Front Oncol 9:500
3. Qu S, Liu Z, Yang X et al. (2018) The emerging functions and roles of circular RNAs in cancer. Cancer Lett 414:301–309
4. Xiao F, He Z, Wang S et al (2024) Regulatory mechanism of circular RNAs in neurodegenerative diseases. CNS Neurosci Ther 30(4):e14499
5. Wu DP, Zhao YD, Yan QQ et al (2023) Circular RNAs: emerging players in brai n aging and neurodegenerative diseases. J pathol 259(1):1–9
6. Fan X, Weng X, Zhao Y et al (2017) Circular RNAs in cardiovascular disease: an overview. BioMed Res Int 2017(1):5135781
7. Dudekula DB, Panda AC, Grammatikakis I et al (2016) CircInteractome: a web tool for exploring circular RNAs and their interacting proteins and microRNAs. RNA Biol 13(1):34–42
8. Chen Y, Yao L, Tang Y et al (2022) CircNet 2.0: an updated database for exploring circular RNA regulatory networks in cancers. Nucleic Acids Res 50(D1):D93–D101
9. Nguyen TH, Nguyen HN, Vu TN (2024) CircNetVis: an interactive web application for visualizing interaction networks of circular RNAs. BMC Bioinform 25(1):31
10. Chiang TW, Mai TL, Chuang TJ (2022) CircMiMi: a stand-alone software for constructing circular RNA-microRNA-mRNA interactions across species. BMC Bioinform 23(1):164
11. Fraboulet RM, Si Ahmed Y, Aubry M et al (2024) Cirscan: a shiny application to identify differentially active sponge mechanisms and

visualize circRNA–miRNA–mRNA networks. BMC Bioinform 25(1):53
12. Aparo A, Avesani S, Parmigiani L et al (2025) EasyCircR: detection and reconstruction of circular RNAs post-transcriptional regulatory interaction networks. Comput Biol Med 188:109846
13. Huang MS, Liu JY, Xia XB et al (2019) Hsa_circ_0001946 inhibits lung cancer progression and mediates cisplatin sensitivity in non-small cell lung cancer via the nucleotide excision repair signaling pathway. Front Oncol 9:508
14. Huang S, Zeng Z, Sun Y et al (2021) Association study of hsa_circ_0001946, hsa-miR-7-5p and PARP1 in coronary atherosclerotic heart disease. Int J Cardiol 328:1–7
15. Chen Z, Ren R, Wan D et al (2019) Hsa_circ_101555 functions as a competing endogenous RNA of miR-597-5p to promote colorectal cancer progression. Oncogene 38(32):6017–6034
16. Dong Y, Bai J, Zhou J (2024) Developing a dormancy-associated ECM signature in TNBC that is linked to immunosuppressive tumor microenvironment and selective sensitivity to MAPK inhibitors. Heliyon 10:11
17. Agarwal V, Bell GW, Nam JW, Bartel DP (2015) Predicting effective microRNA target sites in mammalian mRNAs. elife 4:e05005
18. McGeary SE, Lin KS et al (2019) The biochemical basis of microRNA targeting efficacy. Science 366(6472):eaav1741
19. Liu YC, Li JR, Sun CH et al (2016) CircNet: a database of circular RNAs derived from transcriptome sequencing data. Nucleic Acids Res 44(D1):D209–D215
20. Wu W, Ji P, Zhao F (2020) CircAtlas: an integrated resource of one million highly accurate circular RNAs from 1070 vertebrate transcriptomes. Genome Biol 21(1):101
21. Zhao S, Li S, Liu W et al (2020) Circular RNA signature in lung adenocarcinoma: a MiOncoCirc database-based study and literature review. Front Oncol 10:523342
22. Gao Y, Zhang J, Zhao F (2018) Circular RNA identification based on multiple seed matching. Br Bioinform 19(5):803–810
23. Zhang XO, Dong R et al (2016) Diverse alternative back-splicing and alternative splicing landscape of circular RNAs. Genome Res 26(9):1277–1287
24. Memczak S, Jens M, Elefsinioti A et al (2013) Circular RNAs are a large class of animal RNAs with regulatory potency. Nature 495(7441):333–338
25. Cheng J, Metge F, Dieterich C (2016) Specific identification and quantification of circular RNAs from sequencing data. Bioinformatics 32(7):1094–1096
26. Enright A, John B, Gaul U et al (2003) MicroRNA targets in Drosophila. Genome Biol 4(11):P8
27. Kertesz M, Iovino N, Unnerstall U et al (2007) The role of site accessibility in microRNA target recognition. Nature Genet 39(10):1278–1284
28. Hsu SD, Lin FM, Wu WY et al (2011) miRTarBase: a database curates experimentally validated microRNA–target interactions. Nucleic Acids Res 39(suppl_1):D163–D169
29. Ashburner M, Ball CA, Blake JA et al (2000) Gene ontology: tool for the unification of biology. Nature Genet 25(1):25–29
30. Kanehisa M, Goto S (2000) KEGG: kyoto encyclopedia of genes and genomes. Nucleic Acids Res 28(1):27–30
31. Piñero J, Bravo À, Queralt-Rosinach N et al (2016) DisGeNET: a comprehensive platform integrating information on human disease-associated genes and variants. Nucleic Acids Res 45(D1):D833–D839
32. Glažar P, Papavasileiou P, Rajewsky N (2014) circBase: a database for circular RNAs. RNA 20(11):1666–1670
33. Kozomara A, Birgaoanu M, Griffiths-Jones S (2019) miRBase: from microRNA sequences to function. Nucleic Acids Res 47(D1):D155–D162
34. Kozomara A, Griffiths-Jones S (2014) miRBase: annotating high confidence microRNAs using deep sequencing data. Nucleic Acids Res 42(D1):D68–D73
35. Kozomara A, Griffiths-Jones S (2010) miRBase: integrating microRNA annotation and deep-sequencing data. Nucleic Acids Res 39(suppl_1):D152–D157
36. Griffiths-Jones S, Saini HK, Van Dongen S, Enright AJ (2007) miRBase: tools for microRNA genomics. Nucleic Acids Res 36(suppl_1):D154–D158
37. Griffiths-Jones S, Grocock RJ, Van Dongen S et al (2006) miRBase: microRNA sequences, targets and gene nomenclature. Nucleic Acids Res 34(suppl_1):D140–D144
38. Griffiths-Jones S (2004) The microRNA registry. Nucleic Acids Res 32(suppl_1):D109–D111
39. Krüger J, Rehmsmeier M (2006) RNAhybrid: microRNA target prediction easy, fast and flexible. Nucleic Acids Res 34(suppl_2):W451–W454
40. Nguyen DT, Trac QT, Nguyen TH et al (2021) Circall: fast and accurate methodology for discovery of circular RNAs from paired-end RNA-sequencing data. BMC Bioinform 22(1):495
41. Subramanian A, Tamayo P, Mootha VK et al (2005) Gene set enrichment analysis: a knowledge-based approach for interpreting genome-wide expression profiles. Proc Natl Acad Sci 102(43):15545–15550
42. Mootha VK, Lindgren CM, Eriksson KF et al (2003) PGC-1α-responsive genes involved in

oxidative phosphorylation are coordinately downregulated in human diabetes. Nature Genet 34(3):267–273
43. Li JH, Liu S, Zhou H et al (2014) starBase v2. 0: decoding miRNA-ceRNA, miRNA-ncRNA and protein–RNA interaction networks from large-scale CLIP-Seq data. Nucleic Acids Res 42(D1):D92–D97
44. Chen Y, Wang X (2020) miRDB: an online database for prediction of functional microRNA targets. Nucleic Acids Res 48(D1):D127–D131
45. Almende B V, Thieurmel B, Robert T (2019). Package 'visnetwork'. Network visualization using 'vis. js' Library. Version, 2(9).
46. Chen EY, Tan CM, Kou Y et al (2013) Enrichr: interactive and collaborative HTML5 gene list enrichment analysis tool. BMC Bioinform 14(1):128
47. Kuleshov MV, Jones MR, Rouillard AD et al (2016) Enrichr: a comprehensive gene set enrichment analysis web server 2016 update. Nucleic Acids Res 44(W1):W90–W97
48. Xie Z, Bailey A, Kuleshov MV et al (2021) Gene set knowledge discovery with Enrichr. Curr Protoc 1(3):e90
49. Gaidatzis D, Burger L, Florescu M, Stadler MB (2015) Analysis of intronic and exonic reads in RNA-seq data characterizes transcriptional and post-transcriptional regulation. Nat Biotechnol 33(7):722–729
50. Zheng Y, Ji P, Chen S, Hou L, Zhao F (2019) Reconstruction of full-length circular RNAs enables isoform-level quantification. Genome Med 11(1):2
51. Law CW, Chen Y, Shi W, Smyth GK (2014) voom: precision weights unlock linear model analysis tools for RNA-seq read counts. Genome Biol 15(2):R29
52. Ritchie ME, Phipson B, Wu DI et al (2015) limma powers differential expression analyses for RNA-sequencing and microarray studies. Nucleic Acids Res 43(7):e47–e47
53. Liu M, Wang Q, Shen J et al (2019) Circbank: a comprehensive database for circRNA with standard nomenclature. RNA Biol 16(7):899–905
54. Ru Y, Kechris KJ, Tabakoff B et al.(2014) The multiMiR R package and database: integration of microRNA–target interactions along with their disease and drug associations. Nucleic Acids Res 42(17):e133–e133

Check for updates

Chapter 11

Identification of Disease–miRNA Networks Across Different Human Complex Diseases Using SWIM and MIENTURNET Tools

Federica Conte, Giulia Fiscon, Davide Capozzi, Federica D'Annunzio and Paola Paci

Abstract

miRNAs are small endogenous noncoding RNAs widely known for their pivotal role as key post-transcriptional gene regulators. They exert their effects through binding to specific regions of transcripts with imperfect sequence complementarity. miRNAs are deeply involved in cell growth, differentiation, development, and apoptosis and have a critical role in numerous diseases. In this context, the prediction of miRNA targets stands as a crucial step toward the development of new treatments and innovative therapeutic approaches. However, the prediction of miRNA-binding sites and the prioritization of miRNA–target interactions are often based exclusively on sequence complementarity and present notable challenges. In this context, network-based approaches have emerged as powerful tools to capture complex relationships among miRNAs and their targets and to provide relevant insights into their regulation. In this chapter, we present the application of SWIM to a wide range of complex diseases to identify a group of "switch miRNAs" characterized by distinctive topological properties and a major involvement in key phenotypic transitions. Furthermore, we highlight the use of MIENTURNET, an innovative web-based tool designed to perform statistical and network-based analyses, enabling the identification and prioritization of miRNA–target interactions.

Keywords MicroRNA, Network biology, Human diseases, Switch genes

1 Introduction

Although coding regions in the human genome account for only 1.1% of the total length [1], at least 93% of it is transcribed in different cells [2]. On the other hand, noncoding regions in eukaryotes are mostly known to regulate gene expression [3, 4], but recent studies have shown that their role is far from being fully uncovered [5, 6]. The biological significance of such regions is emphasized by the positive correlation of the amount of non-protein-coding DNA with biological complexity [7–9]. Transcription of noncoding

Alessandro Laganà and Giovanni Nigita (eds.), *MicroRNA Target Identification*, Methods in Molecular Biology, vol. 3056, https://doi.org/10.1007/978-1-0716-5392-0_11,

regions produces a variety of different types of noncoding RNAs (ncRNAs) that are involved in sophisticated regulatory processes such as RNA interference, co-suppression, transgene silencing, imprinting, and methylation [10]. Species of ncRNAs are classified into housekeeping ncRNAs, which mainly include rRNAs and tRNAs, and regulatory ncRNAs such as microRNAs (miRNAs) and long noncoding RNAs (lncRNAs) [11]. In particular, miRNAs are short single-stranded transcripts (21–25 nt) that negatively regulate the expression of specific genes (target genes). This repression mechanism is post-transcriptional: miRNAs bind to target mRNAs and, based on sequence-specific base pairing, can either repress translation or degrade target mRNAs [12]. This silencing process was proven to affect several biological processes such as cell division and death [13], intracellular signaling [14], and cell migration [15]. As a consequence of their ubiquitous role, dysregulation of miRNA expression is associated with a range of diverse pathologies, which include cardiovascular diseases [16, 17], neurodegenerative diseases [18, 19], and cancer [20–22]. Interestingly, both miRNAs and lncRNAs were shown to regulate key genes in oncogenic pathways [23], thus the importance of such molecules cannot be overstated, far from the "junk DNA" that was once thought. Notably, miRNAs are target-specific, thus they are considered as promising candidates for both diagnostic and therapeutic purposes [24, 25]. For these reasons, the identification of miRNA targets is pivotal for understanding the molecular mechanisms underlying complex diseases. In this context, network-based approaches have emerged as powerful tools for analyzing complex datasets originating from a wide range of sources. Their effectiveness lies in their ability to capture intricate relationships and patterns that traditional methods often overlook. Specifically, network medicine provides relevant insights into the integrative nature of human pathogenesis by considering the dense interconnections among disease genes rather than merely considering those genes in isolation [26]. In this chapter, we leveraged SWIM [27, 28] to identify a small group of miRNAs (referred to as switch miRNAs) putatively associated with a given disease/phenotype of interest. SWIM [27, 28] is a sophisticated network-based algorithm designed to detect a group of genes, characterized by unique topological properties, linked to intriguing patterns of molecular co-abundance and known to play crucial roles in phenotypic transitions. Next, to identify the targets of the detected switch miRNAs, we employed MIENTURNET [29], a network-based web tool designed to prioritize miRNA–target interactions by integrating computational predictions and experimentally validated data. In the last years, MIENTURNET has gained notable attention, as shown by the steadily increasing number of citations it has received, due to its user-friendly interface and its accurate miRNA-target predictions. The primary goal of the proposed analysis is to provide an enhanced version of the framework outlined in the earlier chapter [30], introducing two key innovations: the application of SWIM to a broad spectrum of complex diseases and the employment of MIENTURNET to identify and study miRNA–target interactions. This chapter is organized as follows: in Sect. 2, we summarize the datasets and their retrieval, followed by a brief explanation of the SWIM and MIENTURNET algorithms. In Sect. 2, we show the application of SWIM and MIENTURNET to a large panel of human complex diseases. In this chapter, we demonstrate how SWIM is able to precisely identify key switch miRNAs in datasets of different diseases. Using this network-based approach, we investigate the roles of miRNAs that act as "switch genes" in different types of complex diseases. As an updated version of the earlier chapter, we extend the analysis by evaluating the miRNA–target interactions using MIENTURNET, which precisely identified several targets associated with the switch miRNAs. This study summarizes the importance of network-based tools to uncover interesting insights into fundamental interactions that regulate key cellular processes across different diseases.

2 Materials

2.1 The Cancer Genome Atlas (TCGA) and Gene Expression Omnibus (GEO)

A total of nine datasets of different human complex diseases were analyzed. These datasets include mature miRNA expression data, sequenced by high-throughput technology, across at least 18 samples in each disease. In particular, five datasets were retrieved from The Cancer Genome Atlas (TCGA) [31] and include miRNA-seq data from breast invasive carcinoma (BRCA), lung adenocarcinoma (LUAD), lung squamous cell carcinoma (LUSC), prostate adenocarcinoma (PRAD),and thyroid carcinoma (THCA). The remaining four datasets were downloaded from the Gene Expression Omnibus database (GEO) [32] and include miRNA-seq data for COVID (GSE230092), diabetes (GSE262414), ataxia telangiectasia (GSE266440), and Friedreich's ataxia (GSE105052). These datasets were selected due to the availability of data on mature miRNAs. A total of 656 samples were analyzed; specifically, the number of samples and the number of total miRNAs available for each disease are reported in Table 1. The TCGA data matrices were directly downloaded in a normalized format, while raw count matrices of the other diseases were downloaded from the GEO database and normalized using the DESeq2 R package [33].

2.2 SWIM

SWIM (SWItchMiner) [27] is a well-established network-based methodology that considers differentially expressed genes within the co-expression network framework to predict important genes affected by a disease of interest and combines this information with a structured network of correlated patterns. Considering the topological properties of the nodes and assessing their functional roles according

Table 1
SWIM parameters and results for each dataset

	BRCA	LUAD	PRAD	LUSC	THCA	COVID	AT	FA	Diabetes
# miRNAs	2221	2197	2080	2205	2190	1216	2656	897	1719
# cases	103	18	52	45	59	40	10	25	22
# controls	103	18	52	45	59	21	6	17	11
FC threshold	1	1	1	1	1.5	1	1	1	1
FDR threshold	0.05	0.05	0.05	0.05	0.05	0.05	0.1	0.1	1
correlation threshold	0.42	0.69	0.44	0.45	0.49	0.46	0.64	0.63	0.47
prc threshold	0.75	0.88	0.80	0.90	0.85	0.85	0.75	0.92	0.97
FDR correlation threshold	0.05	0.05	0.05	0.05	0.05	0.05	0.05	0.05	0.05
# clusters	3	3	3	3	4	3	3	3	4
# DE miRNAs	621 (up: 364, down: 257)	234 (up: 162, down: 72)	371 (up: 295, down: 76)	626 (up: 459, down: 167)	188 (up: 38, down: 150)	347 (up: 139, down: 208)	47 (up: 12, down: 35)	168 (up: 80, down: 88)	–
# switch miRNAs	97 (up: 96, down: 1)	20 (up: 20, down: 0)	11 (up: 10, down: 1)	52 (up: 2, down: 50)	27 (up: 20, down: 7)	44 (up: 44, down: 0)	9 (up: 9, down: 0)	7 (up: 0, down: 7)	45

The table includes the total number of mature miRNAs present in each data matrix, the number of samples for each condition, the utilized thresholds for the adjusted p-value (FDR), the fold -change (FC), and the chosen percentile (prc) for the correlation distribution. In addition, the number of clusters given as input to the k-means algorithm and the number of obtained differentially expressed miRNAs and switch miRNAs are reported *BRCA* breast invasive carcinoma, *LUAD* lung adenocarcinoma, *LUSC* lung squamous cell carcinoma, *PRAD* prostate adenocarcinoma, *THCA* thyroid carcinoma, *FA* Friedreich's ataxia, *AT* ataxia telangiectasia to their ability to convey information within and between modules in the network, SWIM identifies a small pool of genes (known as *switch genes*) that are associated with intriguing patterns of molecular co-abundance and play a crucial role in the observed phenotype (transitions). Switch genes are defined as a special set of genes with peculiar topological features within gene co-expression network, i.e., (i) they show coherent patterns of correlation, suggesting they may be co-regulated or functionally related; (ii) they form localized connected subnetworks/modules; (iii) they are not local hubs within their module, but they act as connectors able to convey information among modules of the correlation network. SWIM has been widely and successfully applied in several contexts, spanning from plant [34] to human complex diseases [27], along with experimental validation of the candidate biomarker predicted by studying stem-like cells of glioblastoma multiforme [35]. SWIM was originally developed in MATLAB [27], and it is also now available in an open-source version written in R, named SWIMmeR [28].

2.3 MIENTURNET

MIENTURNET (MicroRNA ENrichment TURned NETwork) [29] is a new and easy-to-use web tool that we leveraged to assess the significance of an over-representation of miRNA–target interactions in the list of switch genes obtained by running SWIM. Indeed, MIENTURNET is devised for receiving as input a list of mature miRNAs and inferring possible evidence (computational or experimental) of their regulation of target genes. In particular, we will consider both the predicted miRNA–target interactions retrieved from TargetScan [36] and the experimentally validated miRNA–target interactions retrieved from miRTarBase [37], filtering for those obtained using strong experimental methods (e.g., Luciferase assay, Western blot). Functional enrichment analysis was performed on the target genes of a group of selected miRNAs, taking advantage of the functional enrichment analysis feature provided by MIENTURNET [29]. An adjusted p-value (FDR) equal to or lower than 0.05 was considered statistically significant.

2.4 Cytoscape

Networks representing disease–miRNA associations and miRNA–target interactions were constructed using Cytoscape [38], an open-source software project for integrating biomolecular interaction networks with high-throughput expression data and other molecular states into a unified conceptual framework.

3 Methods

3.1 SWIM Application on Human Diseases Datasets

To investigate common patterns in different complex human diseases and to unveil critical biomarkers, SWIM was applied to mature miRNA datasets of nine diseases. As a first step, SWIM performs a

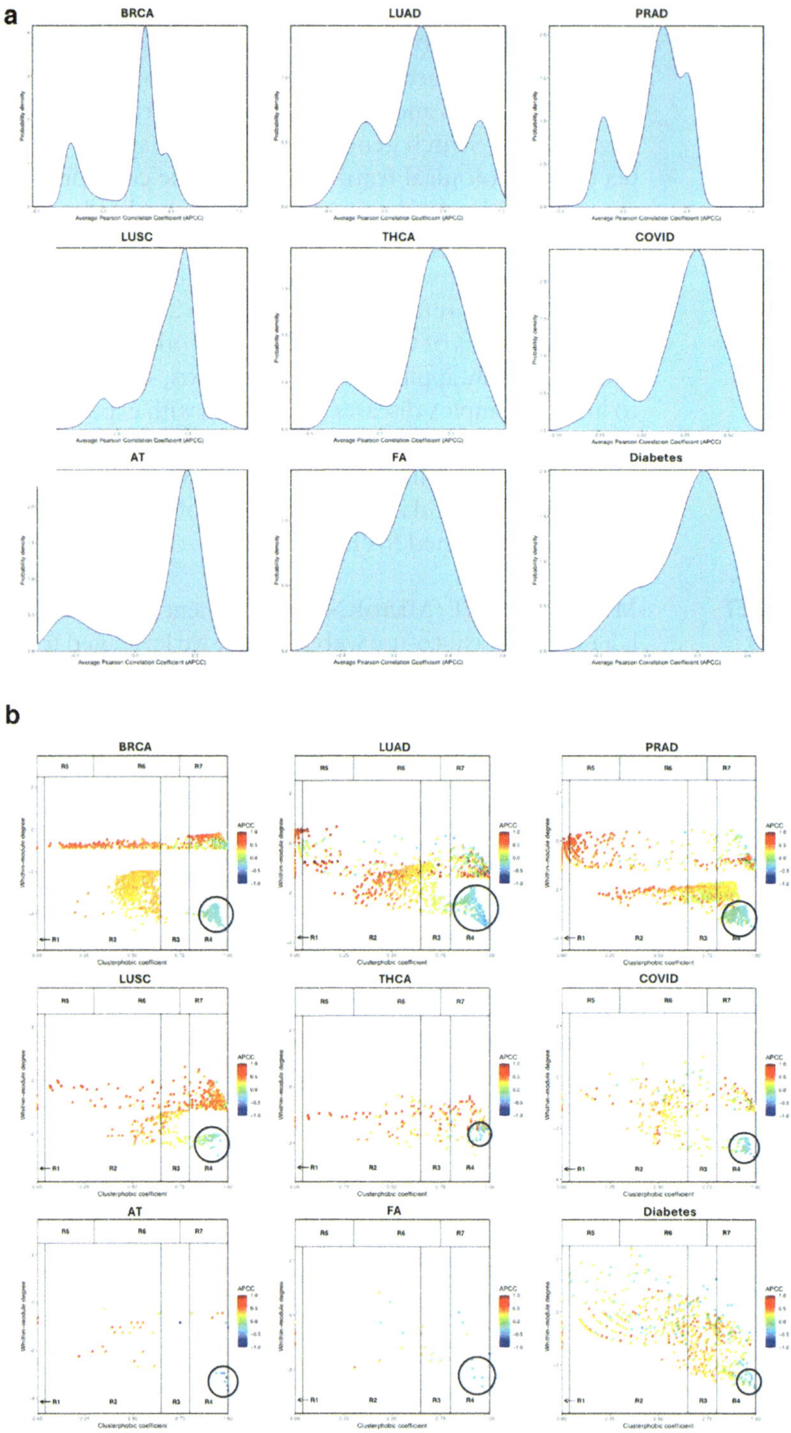
a
BRCA
LUAD
PRAD
LUSC
THCA
COVID
AT
FA
Diabetes
b
BRCA
LUAD
PRAD
LUSC
THCA
COVID
AT
FA
Diabetes

Fig. 1 Integrated network analysis identified switch genes. (**a**) Probability distribution of APCC for hubs (i.e., nodes with degree greater than 5) identified in the correlation network built from miRNA expression datasets. In all datasets, the APCC distribution shows a peak on the left side, representing hub genes with negative APCC (fight-club hubs). (**b**) Heat cartography map where dots correspond to nodes in the correlation networks. The nodes are distributed across seven regions (R1–R7) according to their clusterphobic coefficient (*x*-axis), which measures the propensity of a node to share edges with nodes in different modules, and to their within-module degree (*y*-axis). Black circles highlight switch genes in each plot. Dot color renders the APCC value in a red (positive APCC) to blue (negative APCC) scale. BRCA breast invasive carcinoma, LUAD lung adenocarcinoma, LUSC lung squamous cell carcinoma, PRAD prostate adenocarcinoma, THCA thyroid carcinoma, FA Friedreich's ataxia, AT ataxia telangiectasia

differential expression analysis to retain the subgroup of genes that vary in a statistically significant way between the two conditions of interest. Using these genes, SWIM builds a gene correlation network and identifies network modules using the k-means algorithm. Subsequently, SWIM identifies network hubs and classifies them as party, date, or fight-club hubs based on the strength and the direction of the Average Pearson Correlation Coefficient (APCC) between each hub and its nearest neighbors. Hub nodes are then distributed across seven regions (R1–R7) in a heat cartography map according to their clusterphobic coefficient (a measure of the propensity of a node to communicate with nodes in different modules) and to their within-module degree. Lastly, SWIM identifies switch genes, which are a subgroup of the fight-club hubs, located in the R4 region of the heat cartography map (Fig. 1). All parameters utilized in the SWIM analyses for the different diseases are reported in Table 1. Note that, for the diabetes dataset, no differentially expressed genes were identified and, therefore, the analysis was performed using all available miRNAs. Following the SWIM workflow, a group of switch miRNAs was identified across all studied diseases. The distribution of the APCC showed either a bimodal or trimodal distribution for all datasets, with visible peaks corresponding to the three types of hubs (Fig. 1a). A heat cartography map was produced for each dataset, highlighting in the R4 region the switch miRNAs identified by SWIM (Fig. 1b). Overall, 249 unique switch miRNAs were identified; notably, the number of switch genes associated with each disease is reported in Table 1.

3.2 Disease–miRNA Networks

Given the fundamental role of miRNAs as powerful and dynamic genetic regulators [39], we focused our analysis on miRNAs acting as switch genes in different complex diseases. This study led to the identification of both disease-specific miRNAs and miRNAs common to different diseases. To further investigate switch miRNAs, we constructed a network with nodes divided into two distinct sets representing, respectively, the 9 diseases and the 249 switch miRNAs identified by SWIM (Fig. 2). The edges represent the association

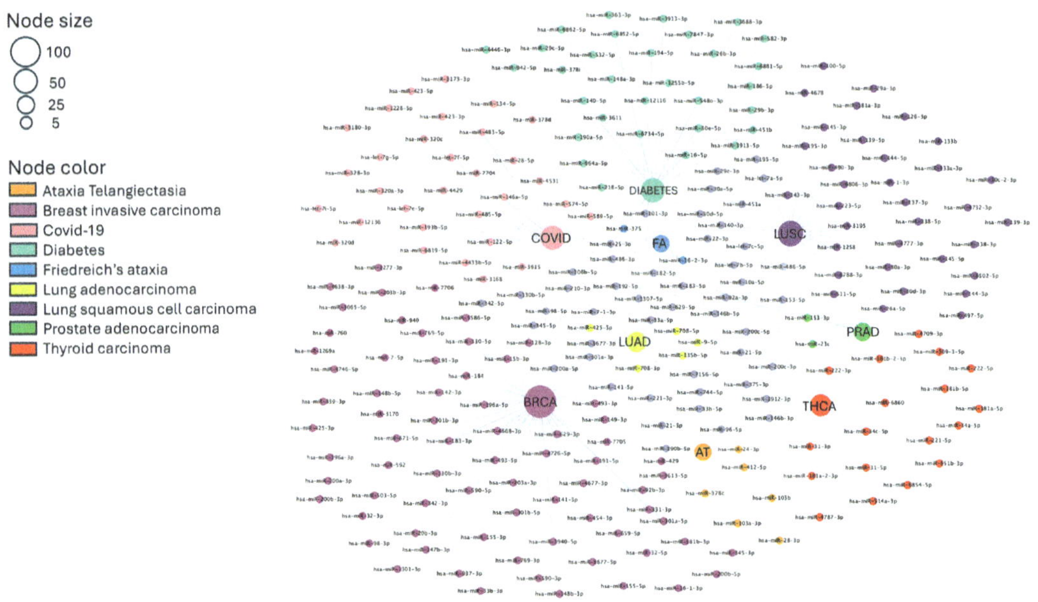

Fig. 2 miRNA-diseasome. The network is made up of two sets of nodes representing the diseases and the switch miRNAs identified by SWIM, respectively. The larger nodes correspond to the analyzed diseases, while the smaller ones correspond to all miRNAs acting as switch genes. Specifically, the dimension of disease nodes is proportional to their degree (i.e., number of associated miRNAs). A link is placed between a disease and a miRNA if the miRNA has been identified by SWIM as a switch gene for that particular disease. Different colors are associated with different diseases. Disease-specific miRNAs are colored based on the disease they are associated with. miRNAs shared by more than one disease are colored in grey. BRCA breast invasive carcinoma, LUAD lung adenocarcinoma, LUSC lung squamous cell carcinoma, PRAD prostate adenocarcinoma, THCA thyroid carcinoma, FA Friedreich's ataxia, AT ataxia telangiectasia

between switch miRNAs and the diseases. This representation reflects the human diseasome network very closely [40], and for this reason, we named it "miRNA-diseasome." From the constructed network, it is clear that some switch miRNAs are shared among a range of different diseases, reflecting their role in regulating various cellular processes across disparate disease contexts. This evidence is pivotal for understanding the underlying basis of common deregulated patterns in different complex diseases. Furthermore, from this network representation (Fig. 2), diseases linked to a high number of switch miRNAs and switch miRNAs linked specifically to one disease are highlighted, leading potentially to the identification of novel disease-specific therapeutic targets. Subsequently, we computed the Hamming distance to group diseases based on the number of common switch miRNAs associated with them. As shown in Fig. 3a, b, this analysis pinpointed several switch miRNAs that were disease-specific, but also a reasonable number of switch miRNAs shared among different diseases. For instance, COVID and LUSC were found to have five switch miRNAs in common, while diabetes shares seven switch miRNAs with LUSC and six with BRCA. Moreover, many

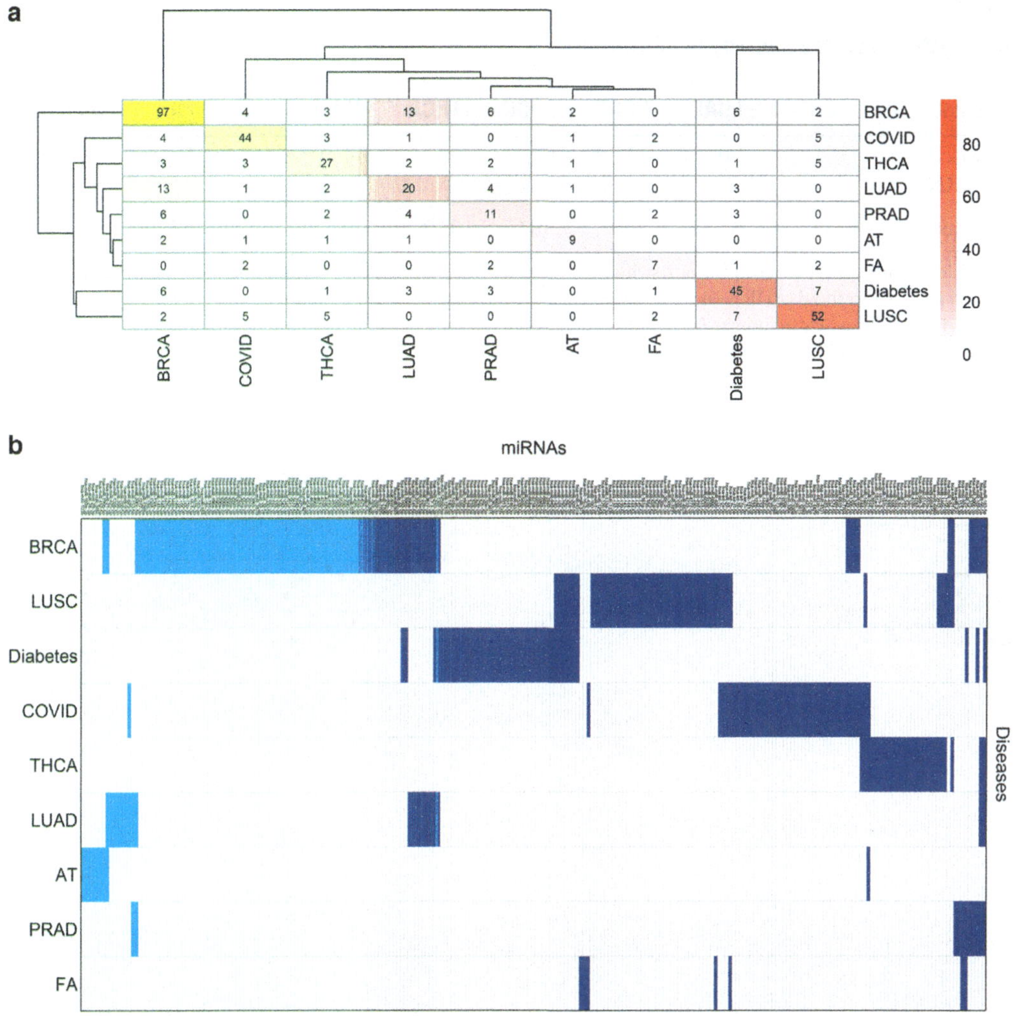

Fig. 3 Comparative analysis of switch miRNAs across different diseases.(**a**) Comparative analysis of switch miRNAs across a panel of nine complex diseases (five human cancer types and four other diseases). The distance matrix, computed based on the Hamming distance of binary-encoded (1 = present; 0 = absent) annotation of miRNAs across diseases, is rendered as a symmetrical heatmap where decreasing distance is displayed on a white-to-red scale. The dendrograms on columns and rows of the distance matrix indicate disease clustering based on Hamming distance. (**b**) The distribution of miRNAs acting as switch genes across a panel of nine complex diseases (five human cancer types and four other diseases) is represented in a heatmap where the blue color indicates that a given miRNA listed in the columns is a switch gene for the disease listed in the rows. BRCA breast invasive carcinoma, LUAD lung adenocarcinoma, LUSC lung squamous cell carcinoma, PRAD prostate adenocarcinoma, THCA thyroid carcinoma, FA Friedreich's ataxia, AT ataxia telangiectasia

miRNAs were also found to be shared among different cancer types, underscoring their potential role as key regulators in common oncogenic pathways.

Table 2
MIENTURNET results according to miRTarBase

Disease	BRCA	LUAD	PRAD	LUSC	THCA	COVID	AT	FA	Diabetes
# switch miRNAs	97	20	11	52	27	44	9	7	45
# targets	53	8	299	384	310	607	7	553	585

The table includes the number of input switch miRNAs for each disease and the number of identified targets. The adjusted *p*-value threshold (FDR) was set to 0.05, while the threshold for the minimum number of interactions was set to 1. *BRCA* breast invasive carcinoma, *LUAD* lung adenocarcinoma, *LUSC* lung squamous cell carcinoma, PRAD prostate adenocarcinoma, *THCA* thyroid carcinoma, *FA* Friedreich's ataxia, *AT* ataxia telangiectasia

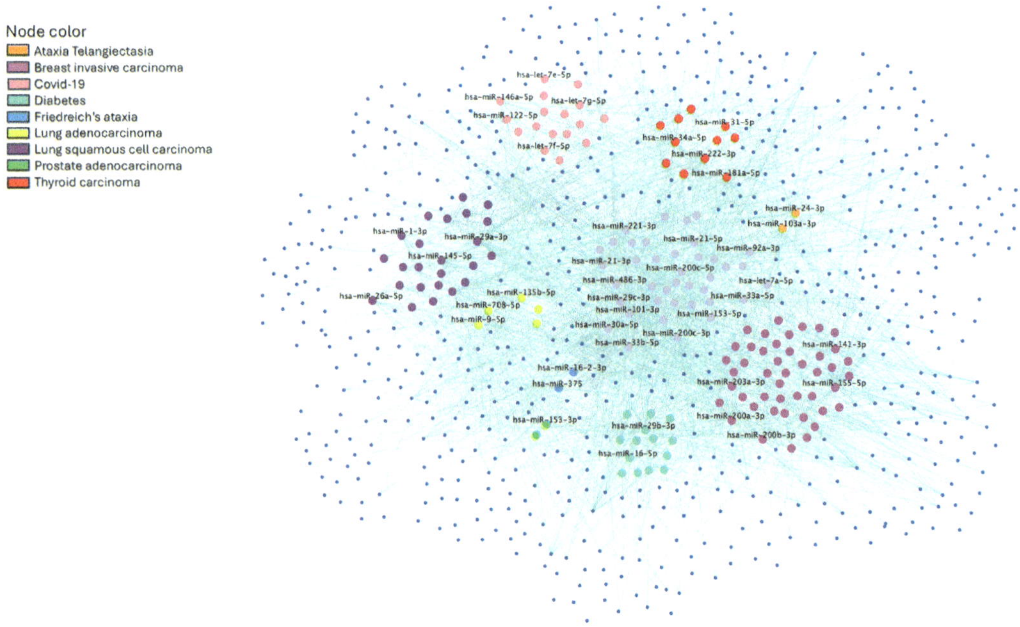

Fig. 4 miRNA–target network. Nodes colored in blue represent targets identified by MIENTURNET; all other nodes represent switch miRNAs of all datasets previously identified with SWIM. miRNAs found to be univocally associated with one disease were colored following the same color code utilized in Fig. 2;miRNAs nodes associated with several diseases were colored in grey. The miRNAs with the highest degree in each disease were selected and labeled in the constructed network

3.3 MIENTURNET Application on Human Diseases Datasets

In order to investigate the roles of miRNAs identified as switch genes by SWIM, we employed MIENTURNET [29] to predict their targets and uncover cellular processes in which they act as key regulators. We began our analysis by providing as input the list of the predicted switch miRNAs related to each disease (249 switch genes) and setting the adjusted *p*-value (FDR) threshold to 0.05 for the miRNA–target enrichment analysis. Across all diseases, a total of 641 genes were identified as statistically significant targets of 161

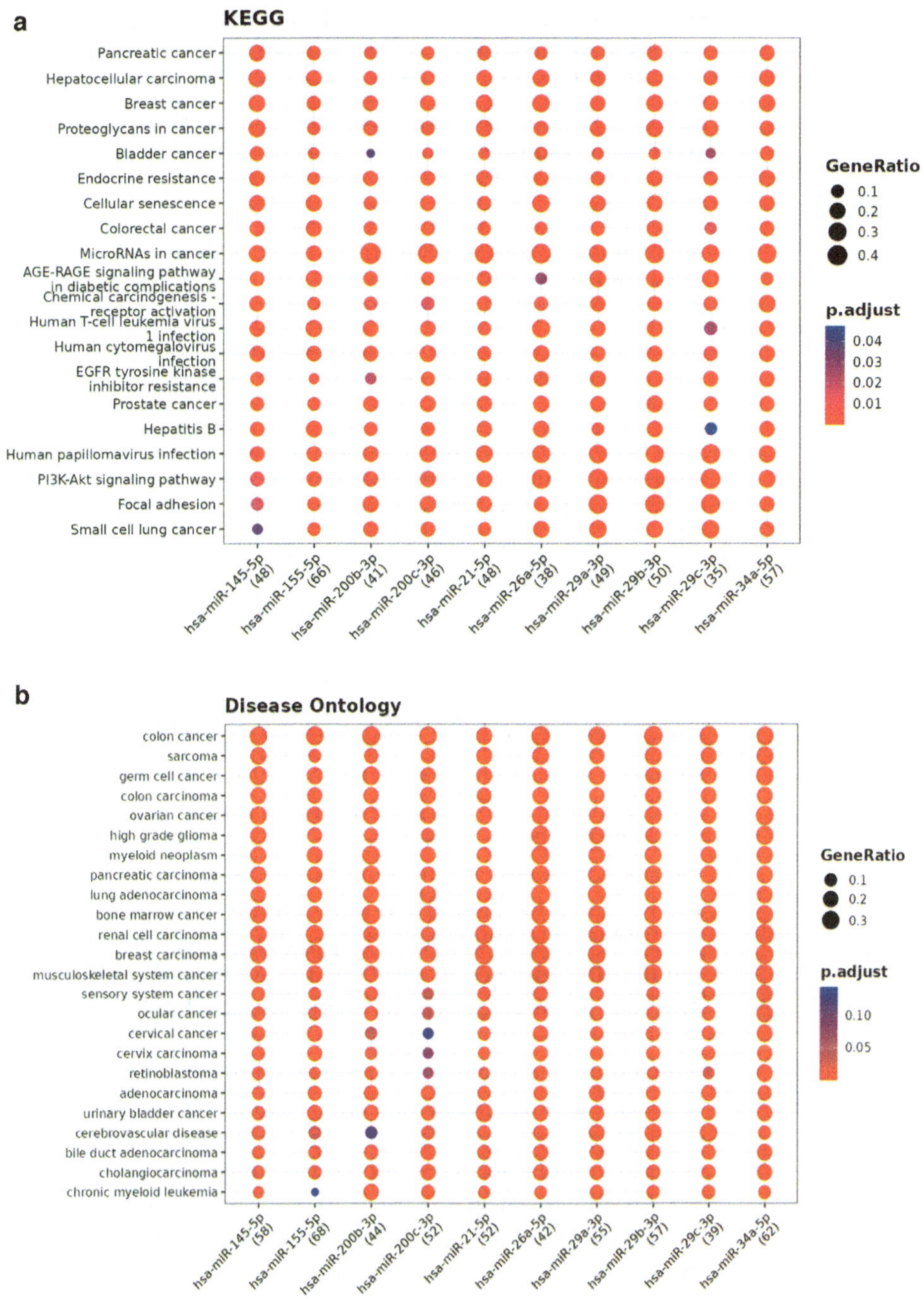

Fig. 5 Functional enrichment analysis. Dot plots of functional enrichment analysis for target genes of selected miRNAs. On the *y*-axis, annotation categories (KEGG pathways and Disease Ontology terms) are reported, while on the *x*-axis, selected miRNAs are depicted. The size of the dots represents gene ratio (i.e., the number of miRNA targets found enriched in each category over the number of total genes associated with that category), while the color of the dots represents the adjusted *p*-values (FDR) in a color scale from red to blue. (**a**) Enriched pathways in the KEGG database. (**b**) Enriched terms in Disease Ontology

switch miRNAs according to miRTarBase. Specifically, the number of targets identified for the switch miRNAs in each disease is reported in Table 2. Regarding miRNA–target interaction predictions derived from TargetScan, MIENTURNET did not show any statistically significant results.

3.4 miRNA–Target Network

To graphically represent MIENTURNET results, a miRNA–target interaction network was constructed (Fig. 4). The nodes represent the switch miRNAs and their target genes, respectively,while the edges represent the interactions identified by MIENTURNET according to miRTarBase. Comprehensively, the number of nodes is 802, comprising 641 target genes and 161 switch miRNAs, while the number of edges is 2239. As expected, diseases that were associated with several common miRNAs shared most of the identified targets, suggesting similar regulation patterns in common biological processes orchestrated by the same players.

3.5 Functional Enrichment Analysis

The targets of the ten switch miRNAs showing the highest degree (i.e., number of interactions) in the miRNA–target network were deeply investigated by performing a functional enrichment analysis for KEGG pathways [41] and Disease Ontology [42]. The selected miRNAs univocally associated with one disease were the following: hsa-mir-155-5p, hsa-mir-200b-3p, and hsa-mir-200c-3p associated with BRCA; hsa-miR-145-5p, hsa-miR-26a-5p, and hsa-miR-29a-3p associated with LUSC; hsa-miR-34a-5p associated with THCA; and hsa-miR-29b-3p associated with diabetes. Among the ones shared by multiple diseases, we found hsa-mir-21-5p associated with BRCA, LUSC, and THCA, and hsa-miR-29c-3p associated with diabetes and LUSC. Notably, the corresponding targets were found to be significantly enriched in several pathways related to cancer, such as miRNAs in cancer, EGFR tyrosine kinases, PI3K/AKT pathway, and cellular senescence (Fig. 5a). These pathways are widely recognized for their activation in several human cancers contributing to carcinogenesis, proliferation, invasion, and metastasis of tumor cells. These findings underscore the fundamental roles of miRNAs as master regulators of common oncogenic pathways. Furthermore, the enrichment of these targets in pathways such as focal adhesion and inhibitor resistance highlights the intricate regulatory interplay between miRNAs and their target genes, which is critical in the dysregulation of key cellular processes shared by several different diseases. Specifically, focal adhesion is a relevant integrator of mechanical and biochemical signals, and its dysregulation is a common hallmark of numerous complex diseases [43, 44]. The enrichment analysis of Disease Ontology (Fig. 5b) revealed several terms related to cancer. Interestingly, these included malignancies such as colon cancer, sarcoma, and pancreatic carcinoma, which were absent in our initial datasets. These observations highlight the involvement of the selected targets in oncogenic pathways shared across diverse tumors, suggesting the broader relevance of miRNA–target regulation in different cancer types.

Acknowledgments

This work was supported by Finalizzata Giovani Ricercatori 2021 to G. F. (grant no.: GR-2021-12372614, CUP: B83C22007560001), by Progetto di Ricerca di Ateneo 2021 (grant no.: RM12117A34663A2C) of Sapienza University of Rome, and by the European Union – NextGenerationEU through the Italian Ministry of University and Research under PNRR – M4C2-I1.3 Project PE_00000019 "HEAL ITALIA" to P. P. – CUP B53C22004000006. The views and opinions expressed are those of the authors only and do not necessarily reflect those of the European Union or the European Commission. Neither the European Union nor the European Commission can be held responsible for them.

Conflict of Interest The authors have no conflicts of interest to declare that are relevant to the content of this chapter.

References

1. Venter JC, Adams MD, Myers EW et al (2001) The sequence of the human genome. Science 291:1304–1351. https://doi.org/10.1126/science.1058040
2. The ENCODE Project Consortium (2007) Identification and analysis of functional elements in 1% of the human genome by the ENCODE pilot project. Nature 447:799–816. https://doi.org/10.1038/nature05874
3. Barrett LW, Fletcher S, Wilton SD (2012) Regulation of eukaryotic gene expression by the untranslated gene regions and other non-coding elements. Cell Mol Life Sci 69:3613–3634. https://doi.org/10.1007/s00018-012-0990-9
4. Pesole G, Mignone F, Gissi C et al (2001) Structural and functional features of eukaryotic mRNA untranslated regions. Gene 276:73–81. https://doi.org/10.1016/S0378-1119(01)00674-6
5. Ding B, Liu Y, Liu Z et al (2021) Noncoding loci without epigenomic signals can be essential for maintaining global chromatin organization and cell viability. Sci Adv 7:eabi6020. https://doi.org/10.1126/sciadv.abi6020
6. Mattick JS, Amaral PP, Carninci P et al (2023) Long non-coding RNAs: definitions, functions, challenges and recommendations. Nat Rev Mol Cell Biol 24:430–447. https://doi.org/10.1038/s41580-022-00566-8
7. Taft RJ, Mattick JS (2003) Increasing biological complexity is positively correlated with the relative genome-wide expansion of non-protein-coding DNA sequences. Genome Biol 5: P1. https://doi.org/10.1186/gb-2003-5-1-p1
8. Taft RJ, Pheasant M, Mattick JS (2007) The relationship between non-protein-coding DNA and eukaryotic complexity. BioEssays 29:288–299. https://doi.org/10.1002/bies.20544
9. Liu G, Mattick J, Taft RJ (2013) A meta-analysis of the genomic and transcriptomic composition of complex life. Cell Cycle 12:2061–2072. https://doi.org/10.4161/cc.25134
10. Mattick JS (2009) The genetic signatures of noncoding RNAs. PLOS Genet 5:e1000459. https://doi.org/10.1371/journal.pgen.1000459
11. Zhang P, Wu W, Chen Q, Chen M (2019) Non-coding RNAs and their integrated networks. J Integr Bioinforma 16:20190027. https://doi.org/10.1515/jib-2019-0027
12. He L, Hannon GJ (2004) MicroRNAs: small RNAs with a big role in gene regulation. Nat Rev Genet 5:522–531. https://doi.org/10.1038/nrg1379
13. Miska EA (2005) How microRNAs control cell division, differentiation and death. Curr Opin Genet Dev 15:563–568. https://doi.org/10.1016/j.gde.2005.08.005

14. Lui P-Y, Jin D-Y, Stevenson NJ (2015) MicroRNA: master controllers of intracellular signaling pathways. Cell Mol Life Sci 72:3531–3542. https://doi.org/10.1007/s00018-015-1940-0
15. Hamidi AA, Taghehchian N, Basirat Z et al (2022) MicroRNAs as the critical regulators of cell migration and invasion in thyroid cancer. Biomark Res 10: 40. https://doi.org/10.1186/s40364-022-00382-4
16. Maegdefessel L (2014) The emerging role of microRNAs in cardiovascular disease. J Intern Med 276:633–644. https://doi.org/10.1111/joim.12298
17. Zhou S, Jin J, Wang J et al (2018) miRNAS in cardiovascular diseases: potential biomarkers, therapeutic targets and challenges. Acta Pharmacol Sin 39:1073–1084. https://doi.org/10.1038/aps.2018.30
18. Quinlan S, Kenny A, Medina M et al (2017) MicroRNAs in neurodegenerative diseases. In: International review of cell and molecular biology. Elsevier, pp 309–343
19. Li S, Lei Z, Sun T (2023) The role of microRNAs in neurodegenerative diseases: a review. Cell Biol Toxicol 39:53–83. https://doi.org/10.1007/s10565-022-09761-x
20. Ha T-Y (2011) MicroRNAs in human diseases: from cancer to cardiovascular disease. Immune Netw 11:135. https://doi.org/10.4110/in.2011.11.3.135
21. Li Y, Kowdley KV (2012) MicroRNAs in common human diseases. Genomics Proteomics Bioinform 10:246–253. https://doi.org/10.1016/j.gpb.2012.07.005
22. Peng Y, Croce CM (2016) The role of MicroRNAs in human cancer. Signal Transduct Target Ther 1:15004. https://doi.org/10.1038/sigtrans.2015.4
23. Ratti M, Lampis A, Ghidini M et al (2020) MicroRNAs (miRNAs) and long non-coding RNAs (lncRNAs) as new tools for cancer therapy: first steps from bench to bedside. Target Oncol 15:261–278. https://doi.org/10.1007/s11523-020-00717-x
24. Li Z, Rana TM (2014) Therapeutic targeting of microRNAs: current status and future challenges. Nat Rev Drug Discov 13:622–638. https://doi.org/10.1038/nrd4359
25. Kim T, Croce CM (2023) MicroRNA: trends in clinical trials of cancer diagnosis and therapy strategies. Exp Mol Med 55:1314–1321. https://doi.org/10.1038/s12276-023-01050-9
26. Chan SY, Loscalzo J (2012) The emerging paradigm of network medicine in the study of human disease. Circ Res 111:359–374. https://doi.org/10.1161/CIRCRESAHA.111.258541
27. Paci P, Colombo T, Fiscon G et al (2017) SWIM: a computational tool to unveiling crucial nodes in complex biological networks. Sci Rep 7: 44797. https://doi.org/10.1038/srep44797
28. Paci P, Fiscon G (2022) SWIMmeR: an R-based software to unveiling crucial nodes in complex biological networks. Bioinformatics 38:586–588. https://doi.org/10.1093/bioinformatics/btab657
29. Licursi V, Conte F, Fiscon G, Paci P (2019) MIENTURNET: an interactive web tool for microRNA-target enrichment and network-based analysis. BMC Bioinformatics 20: 545. https://doi.org/10.1186/s12859-019-3105-x
30. Fiscon G, Conte F, Farina L et al (2019) Identification of disease–miRNA networks across different cancer types using SWIM. In: Laganà A (ed) MicroRNA target identification: methods and protocols. Springer, New York, NY, pp 169–181
31. (2022) The Cancer Genome Atlas Program (TCGA) - NCI. https://www.cancer.gov/ccg/research/genome-sequencing/tcga. Accessed 26 Nov 2024
32. Home - GEO - NCBI. https://www.ncbi.nlm.nih.gov/geo/. Accessed 26 Nov 2024
33. Love MI, Huber W, Anders S (2014) Moderated estimation of fold change and dispersion for RNA-seq data with DESeq2. Genome Biol 15: 550. https://doi.org/10.1186/s13059-014-0550-8
34. Palumbo MC, Zenoni S, Fasoli M et al (2014) Integrated network analysis identifies fight-club nodes as a class of hubs encompassing key putative switch genes that induce major transcriptome reprogramming during grapevine development. Plant Cell 26:4617–35. https://doi.org/10.1105/tpc.114.133710
35. Pecce V, Verrienti A, Fiscon G et al (2021) The role of FOSL1 in stem-like cell reprogramming processes. Sci Rep 11: 14677. https://doi.org/10.1038/s41598-021-94072-0
36. Agarwal V, Bell GW, Nam J-W, Bartel DP (2015) Predicting effective microRNA target sites in mammalian mRNAs. eLife 4:e05005. https://doi.org/10.7554/eLife.05005
37. Huang H-Y, Lin Y-C-D, Cui S et al (2021) miRTarBase update 2022: an informative resource for experimentally validated miRNA–target interactions. Nucleic Acids Res 50:D222. https://doi.org/10.1093/nar/gkab1079
38. Shannon P, Markiel A, Ozier O et al (2003) Cytoscape: a software environment for integrated models of biomolecular interaction networks. Genome Res 13:2498–2504. https://doi.org/10.1101/gr.1239303
39. Diener C, Keller A, Meese E (2022) Emerging concepts of miRNA therapeutics: from cells to clinic. Trends Genet TIG 38:613–626. https://doi.org/10.1016/j.tig.2022.02.006

40. Barabási A-L, Gulbahce N, Loscalzo J (2011) Network medicine: a network-based approach to human disease. Nat Rev Genet 12:56–68. https://doi.org/10.1038/nrg2918
41. KEGG: kyoto encyclopedia of genes and genomes - PubMed. https://pubmed.ncbi.nlm.nih.gov/10592173/. Accessed 28 Nov 2024
42. The DO-KB Knowledgebase: a 20-year journey developing the disease open science ecosystem - PubMed. https://pubmed.ncbi.nlm.nih.gov/37953304/. Accessed 28 Nov 2024
43. Mishra YG, Manavathi B (2021) Focal adhesion dynamics in cellular function and disease. Cell Signal 85:110046. https://doi.org/10.1016/j.cellsig.2021.110046
44. Legerstee K, Houtsmuller AB (2021) A layered view on focal adhesions. Biology 10:1189. https://doi.org/10.3390/biology10111189

Chapter 12

Integration of Omics Data to Identify Cancer-Related MicroRNA

Luciano Cascione

Abstract

MicroRNAs (miRNAs) are small noncoding RNAs that play critical regulatory roles in gene expression by targeting messenger RNAs (mRNAs) for degradation or translational inhibition. Their ability to modulate multiple genes simultaneously places them at the center of many biological processes, including those involved in tumorigenesis. Depending on their targets, miRNAs can act as oncogenes or tumor suppressors, contributing to cancer development and progression. Advances in high-throughput sequencing now allow simultaneous profiling of miRNA and mRNA expression in the same biological samples. When properly integrated, these datasets can uncover functional miRNA–mRNA regulatory networks, reveal cancer-specific signatures, and identify novel biomarkers or therapeutic targets. However, the integrative analysis of miRNA and mRNA expression data remains computationally challenging and requires well-structured pipelines and rigorous statistical approaches. In this chapter, we present updated computational strategies to identify biologically relevant miRNA–mRNA interactions through expression correlation, target prediction, and functional enrichment. The chapter includes step-by-step protocols, example R code, and recommendations for integrating additional data layers such as copy number variation and DNA methylation, with the goal of improving the robustness and interpretability of miRNA-based analyses in cancer.

Keywords MicroRNAs, Cancer, Multi-omics integration, miRNA–mRNA regulatory networks, Bioinformatics pipeline

1 Introduction

MicroRNAs (miRNAs) are a class of small noncoding RNAs, approximately 20–24 nucleotides in length, that control gene expression at the post-transcriptional level [1]. They act predominantly by base-pairing with complementary sequences in the 3′ untranslated regions (3′-UTRs) of messenger RNAs (mRNAs), leading to transcript degradation or translational repression [2]. A single miRNA can regulate multiple genes, and conversely, a single mRNA may be regulated by several miRNAs, forming an intricate network of gene regulation [3]. In the context of cancer, miRNAs are frequently

Alessandro Laganà and Giovanni Nigita (eds.), *MicroRNA Target Identification*, Methods in Molecular Biology, vol. 3056, https://doi.org/10.1007/978-1-0716-5392-0_12,

dysregulated and can act either as oncogenes (oncomiRs) or tumor suppressors, depending on their target genes [4]. OncomiRs are typically overexpressed in tumors, where they downregulate tumor suppressor genes, while tumor-suppressive miRNAs are often deleted, silenced by promoter hypermethylation, or otherwise repressed, resulting in unchecked expression of oncogenes [5]. The expanding availability of high-throughput transcriptomics data, including both mRNA and small RNA sequencing, has made it possible to investigate miRNA–mRNA interactions at a systems level. However, the functional interpretation of these datasets requires careful integration, as miRNA–mRNA interactions are highly context-dependent, non-linear, and regulated by numerous molecular factors beyond expression levels alone. Emerging aspects such as ceRNA networks [6, 7], RNA-binding proteins (RBPs) [8], RNA editing [9], isomiRs [10], and subcellular localization have enriched our understanding of miRNA–mRNA interactions [11]. Integrating miRNA and mRNA data allows researchers to uncover functional regulatory relationships. This chapter offers a full computational workflow for preprocessing, differential analysis, integration, and interpretation of miRNA and mRNA data, with a focus on reproducibility and biological relevance.

2 Material

To successfully carry out an integrative analysis of miRNA and mRNA expression data, it is essential to begin with high-quality genomic and transcriptomic resources, as well as a set of well-maintained tools. This section outlines the datasets, software, and computational tools required to execute the workflow described in the following methods section. Publicly available datasets such as those from The Cancer Genome Atlas (TCGA) or GEO provide a rich resource for such studies. A thoughtful setup of your computational environment and data organization structure will facilitate efficient analysis, reproducibility, and downstream biological interpretation.

2.1 Genomic Data and Annotation

1. Human genome reference (e.g., GRCh38) in FASTA format can be downloaded from http://hgdownload.soe.ucsc.edu/goldenPath/hg38/bigZips/. Other rich genomic resources are Ensembl https://www.ensembl.org/index.html and the NCBI https://www.ncbi.nlm.nih.gov/datasets/genome/GCF_000001405.26/
2. GENCODE gene annotation in GTF format: https://www.gencodegenes.org/human/
3. Mature miRNA sequences from miRBase (v22 or later): https://www.mirbase.org/download/

4. Public datasets (e.g., TCGA, GEO) with matched miRNA/mRNA expression.
5. Clinical data for survival analysis.

2.2 Software and Tools

1. **RNA-seq aligners**
   a. STAR: A fast RNA-seq aligner that performs spliced alignment of transcriptomic reads to the reference genome (https://github.com/alexdobin/STAR)
   b. HISAT2: An efficient alignment program for mapping next-generation sequencing reads to a population of genomes (http://daehwankimlab.github.io/hisat2/download/)
2. **miRNA-seq**:
   a. BWA: A lightweight tool for aligning short sequencing reads to a reference genome, commonly used for small RNAs (https://github.com/lh3/bwa)
   b. miRDeep2: A tool to discover and quantify known and novel microRNAs from small RNA-Seq data (https://github.com/rajewsky-lab/mirdeep2)
3. **Quality control**: FastQC identifies potential issues in raw reads (https://www.bioinformatics.babraham.ac.uk/projects/fastqc/)
4. **Trimming**: Cutadapt removes adapter sequences from high-throughput sequencing reads with flexible options (https://cutadapt.readthedocs.io/en/stable/installation.html)
5. **Quantification**: featureCounts counts reads mapped to genomic features such as genes or exons.
6. **Visualization**: The Integrative Genomics Viewer (IGV) for interactive exploration of large genomic datasets (http://software.broadinstitute.org/software/igv/)
7. **R packages**:
   a. ggplot2: A widely used R package for creating elegant and customizable data visualizations. https://ggplot2.tidyverse.org/
   b. edgeR: An R package for differential expression analysis of RNA-seq data using negative binomial models. https://bioconductor.org/packages/release/bioc/html/edgeR.html
   c. DESeq2: Performs differential expression analysis based on the negative binomial distribution. https://bioconductor.org/packages/release/bioc/html/DESeq2.html

d. limma: An R package for linear modeling of microarray and RNA-seq data. https://bioconductor.org/packages/release/bioc/html/limma.html

e. survival: Provides functions for survival analysis, including Cox models and Kaplan–Meier estimators. https://cran.r-project.org/web/packages/survival/index.html

f. anamiR: An integrated pipeline for miRNA–mRNA correlation and pathway analysis in R. https://bioconductor.org/packages/release/bioc/html/anamiR.html

2.3 R Setup Code

Before starting any computational analysis in R, it is important to begin with a clean environment. Removing all objects currently in memory helps avoid accidental reuse of old variables or data structures that may interfere with the new analysis. This step ensures reproducibility and reduces the chance of introducing hidden bugs due to leftover objects from previous sessions. Also, make sure to set your working directory at the start of your script to avoid ambiguity.

```
# Load libraries library(edgeR) library(DESeq2) library(limma)
library(anamiR) library(survival) library(gplots)
library(clusterProfiler)
library(enrichplot)
library(survival)
library(survminer)
# Set a working directory
setwd("~/project_root/")
```

2.4 Directory Structure

Maintaining a consistent and organized directory structure is crucial for reproducibility, collaboration, and scalability in computational biology projects (*see* **Note 1**). Separating raw data, processed data, scripts, results, and figures into clearly labeled folders makes it easier to navigate complex analyses and track changes over time. Such organization also facilitates automation using workflow managers and ensures that collaborators or reviewers can follow and reproduce the full analysis pipeline without ambiguity. This practice is aligned with the recommendations in "Ten Simple Rules for Structuring Papers" [12], and particularly with "Ten Simple Rules for Reproducible Computational Research" by Sandve et al. [13], which emphasizes the importance of managing files in a clean, reproducible manner with clear documentation and version control.

A typical recommended layout may resemble the following:

```
project_root/
├── data/
│   ├── raw/ # Unmodified original data
│   ├── processed/ # Cleaned, filtered, and normalized data
│   └── annotations/ # Reference files like GTF or FASTA
├── scripts/ # R scripts and notebooks
├── results/ # Output tables, metrics, and results
```

```
├── figures/ # Plots and visual summaries
├── reports/ # Final rendered documents manuscripts
└── qc_reports/ # Quality Control graph and statistics
```

3 Methods

This section provides a step-by-step protocol for analyzing and integrating miRNA and mRNA expression data, including preprocessing, quantification, differential expression analysis, correlation, and survival modeling. All major steps include example R code to facilitate reproducibility. We assume the starting point is high-throughput sequencing data, specifically generated using the Illumina platform, which remains the most widely adopted technology for transcriptomic studies. Raw reads from Illumina sequencers are initially stored in a proprietary binary base call (BCL) format. Therefore, the first preprocessing step involves converting these BCL files into the more universally accepted and human-readable FASTQ format. This conversion is typically performed by the sequencing facility and is outside the scope of this chapter. Once FASTQ files are obtained, the first bioinformatic task involves removing technical sequences such as Illumina adapter contamination and poly(A)/poly(T) tails introduced during cDNA synthesis or library preparation. These artifacts, if not removed, can interfere with read alignment and quantification. We recommend using Cutadapt, a flexible and efficient tool for adapter and quality trimming. Other alternatives such as Trimmomatic or the FASTX-Toolkit may also be used, depending on user preferences or pipeline constraints. Quality control (QC) is a critical next step to assess the integrity and characteristics of the sequencing reads. Tools such as FastQC provide comprehensive summaries including per-base sequence quality, GC content, sequence length distribution, and levels of sequence duplication. Reviewing these metrics ensures that only high-quality reads are retained for downstream analysis. Once reads pass QC, they can be aligned to a reference genome and quantified to generate gene- or transcript-level count matrices for both mRNA and miRNA profiles. All subsequent analyses ranging from differential expression testing to integrative modeling are guided by the study's experimental design and biological goals. An overview of the complete analytical pipeline is presented in Fig. 1, highlighting the essential steps from raw sequence data to functional interpretation.

3.1 Preprocessing and Quality Control

Download and uncompress the human genome sequence file (hg38) and the comprehensive gene annotation (gtf file):

cd /project_dir/annotations/

wget https://ftp.ensembl.org/pub/current_fasta/homo_sapiens/dna/Homo_sapiens.GRCh38.dna.primary_assembly.fa.gz

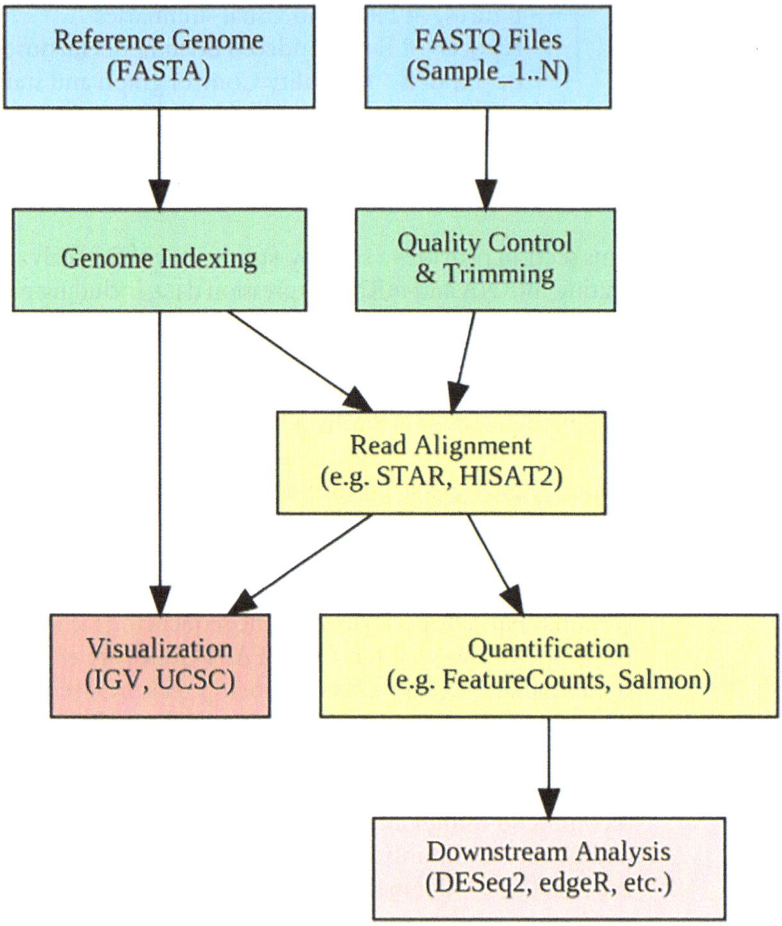

Fig. 1 RNA-seq data analysis pipeline diagrams. These diagrams illustrate alternative layouts of a standard RNA-Seq data analysis pipeline, highlighting key computational steps from raw data input to biological interpretation

wget https://ftp.ebi.ac.uk/pub/databases/gencode/Gencode_human/release_48/gencode.v48.annotation.gtf.gz

gunzip -c gencode.v48.annotation.gtf.gz

Create STAR indexes (*see* **Note 2**): This is an initial and one-time step, which prepares a genome index file to be used during read mapping.

mkdir./star-indexes

cd./star-indexes

STAR --runThreadN 4 --runMode genomeGenerate –-genomeDir./ --genomeFastaFiles /project_dir/annotations/Homo_sapiens.GRCh38.dna.primary_assembly.fa.gz –sjdbGTFfile genomes/hg38/gencode.v48.annotation.gtf

3.2 Quality Control

Before proceeding with read alignment or quantification, it is essential to evaluate the quality of raw sequencing data. Low-quality bases, sequencing artifacts, or residual adapter sequences can significantly affect the accuracy of downstream analyses such as transcript quantification and differential expression. In this step, we use FastQC to perform a comprehensive quality control check on the raw FASTQ files, providing insights into base quality scores, sequence length distribution, GC content, overrepresented sequences, and potential adapter contamination. If adapter sequences or poly(A)/poly(T) tails are detected (*see* **Note 3**), they must be removed using a robust trimming tool. Cutadapt is a widely used, flexible, and efficient tool for removing adapter contamination and other unwanted sequence features. It is compatible with both single-end and paired-end reads and can be easily integrated into pipelines. After trimming, it is good practice to re-run FastQC on the cleaned data to confirm improvement in quality.

Run FastQC on raw FASTQ files (gzip-compressed and example for paired-end data)

cd /project_dir/data/raw/

mkdir./trimmed_reads/

Fastqc sample1_R1.fastq.gz sample1_R2.fastq.gz -o /project_dir/qc_reports/

Trim adapters with Cutadapt (example for paired-end data)

cutadapt -a AGATCGGAAGAGC -A AGATCGGAAGAGC -o./trimmed_reads/sample1_R1_trimmed.fastq.gz -p./trimmed_reads/sample1_R2_trimmed.fastq.gz /sample1_R1.fastq.gz sample1_R2.fastq.gz --minimum-length 20 -q 20

Re-run FastQC on trimmed files

Fastqc./trimmed_reads/sample1_R1_trimmed.fastq.gz trimmed_reads/sample1_R2_trimmed.fastq.gz -o /project_dir/qc_reports/

3.3 Read Alignment and Quantification

To analyze RNA-Seq data, we will align mRNA reads to the reference genome using the STAR aligner, which is optimized for spliced reads and transcriptome mapping. For miRNA reads, we will use miRDeep2, a tool specifically designed to handle short, unspliced sequences characteristic of microRNAs. This step is crucial for downstream differential expression analysis and functional interpretation.

cd /project_dir/processed/

mkdir./star_alignment/

GENOME_DIR=/project_dir/annotations/star-indexes

GTF_FILE=/project_dir/annotations/gencode.v48.annotation.gtf

R1=/project_dir/data/raw/trimmed_reads/sample1_R1_trimmed.fastq.gz

R2=/project_dir/data/raw/trimmed_reads/sample1_R2_trimmed.fastq.gz

STAR alignment (mRNA) STAR --runThreadN 10 --genomeDir $GENOME_DIR –readFilesIn $R1 $R2 --sjdbGTFfile $GTF_FILE --outSAMtype BAM SortedBy-Coordinate --quantMode

TranscriptomeSAM GeneCounts --read-FilesCommand gunzip -c --outFileNamePrefix "SAMPLE_name"

In the next step, we index the BAM file, creating a ".bai" file that enables rapid, random access to specific genomic regions without scanning the entire file. This is essential for efficient visualization in genome browsers (such as IGV) and for tools that need to retrieve alignments from targeted regions, such as during variant calling or coverage analysis.

samtools index Sample_1_Aligned.sortedByCoord.out.bam

The same alignment and quantification procedure should be carried out in parallel for all samples in the cohort to ensure consistency and comparability. Upon completion, each sample will yield a sorted BAM file containing the aligned reads, as well as a gene-level count file named Sample_1.ReadsPerGene.out.tab. This file is generated automatically by STAR when the *--quantMode GeneCounts* option is enabled. This flag instructs STAR to count the number of reads that map to each gene during the alignment process, producing output that is directly comparable to the counts obtained from htseq-count with its default settings. This option simplifies the pipeline by eliminating the need for an additional read-counting step post-alignment. It is important to note that gene annotation is required for STAR to perform this quantification. The annotations must be provided either during genome index generation or at the mapping step using the *--sjdbGTFfile* option. Accepted formats include GTF or GFF files from reliable sources such as Ensembl or GENCODE. Ensuring that the annotation file matches the reference genome build used for alignment is crucial to avoid mismatches and incorrect gene assignments. By collecting the ReadsPerGene.out.tab files from all samples, one can easily construct a count matrix for downstream differential expression analysis using packages such as edgeR, DESeq2, or limma.

3.4 MicroRNA Read Alignment and Quantification

Alignment of microRNA sequencing data requires special considerations due to the short length of the reads (typically 18–30 nucleotides), the presence of closely related miRNA family members, and the possibility of multiple alignments. Unlike the procedure presented in the previous section, where spliced aligner such as STAR is preferred, miRNA reads are best aligned with aligners optimized for short sequences (*see* **Note 4**). Two commonly used tools for miRNA alignment are Bowtie (used within pipelines such as miRDeep2) and BWA for small RNA libraries. Here, we outline a typical alignment workflow using miRDeep2, which wraps both adapter trimming and Bowtie-based alignment steps, followed by miRNA quantification.

Convert genome to index for Bowtie (if not already available)

cd /project_dir/annotations/

mkdir./bowtie_genome_index

bowtie-build Homo_sapiens.GRCh38.dna.primary_assembly.fa.gz bowtie_genome_index

Convert FASTQ to miRDeep2's.arf alignment format

gunzip -c sample_miRNA.fastq.gz | mapper.pl - -e -h -j -l 18 -m -s reads_collapsed.fa -t reads_vs_genome.arf -p bowtie_genome_index

We exclude reads with low complexity (-e), with non-standard letters (-h), shorter than 18 nt (-j 18), and that map to more than 5 loci (-m) (*see* **Note 5**).

Run miRDeep2 core algorithm

GENOME=/project_dir/annotations/ Homo_sapiens.GRCh38.dna.primary_assembly.fa.gz

cd /project_dir/data/

miRDeep2.pl reads_collapsed.fa genome.fa reads_vs_genome.arf mature_miRNA.fa hairpin_miRNA.fa > miRDeep2_output.txt 2> miRDeep2_error.log

This step quantifies known miRNAs and predicts novel ones. The output includes:

- A summary of known and novel miRNA expression.
- PDF plots showing hairpin structures and expression.
- Read count tables ready for downstream normalization.

3.5 Differential Expression Analysis from miRDeep2 Output

To perform differential expression analysis using limma on microRNA (miRNA) data derived from multiple samples processed with miRDeep2, you first need to aggregate the read count results into a matrix format, where rows represent miRNAs and columns represent samples. The main file of interest from miRDeep2 is quantification_*.csv or miRNAs_expressed_all_samples.csv (if using combined analysis). This file contains read counts for all detected miRNAs across all samples in a single table, making it straightforward to perform differential expression directly. Below is a detailed step-by-step guide and R code to perform differential expression analysis between normal and tumor samples using limma.

Load the combined miRDeep2 count matrix

counts <- read.csv("miRNAs_expressed_all_samples.csv", row.names = 1)

Inspect the first few rows and columns

head(counts[, 1:5])

Define sample groups: update according to your experiment

Assume 3 normal samples followed by 3 tumor samples

group <- factor(c(rep("normal",3), rep("tumor",3)) (*see* **Note 6**)

group <- relevel(x=group,ref="tumor")

Create design matrix (*see* **Note 7**)

design <- model.matrix(~group)

colnames(design) <- c("Intercept", "Tumor_vs_Normal")

Prepare metadata and design matrix

Create DGEList object

dge <- DGEList(counts = counts)

Filter out lowly expressed miRNAs (*see* **Note 8**)

keep <- filterByExpr(dge, group = group)

```
dge <- dge[keep,, keep.lib.sizes = FALSE]
```

Filter, normalize, and transform counts

```
# Normalize using TMM (Trimmed Mean of M-values)
dge <- calcNormFactors(dge)
# Transform counts with voom
v <- voom(dge, design, plot = TRUE)
```

Visualizing the miRNome landscape

Before proceeding to formal statistical testing, it is useful to visualize the global expression landscape of the miRNA transcriptome (miRNome) across samples. Such exploratory analysis helps identify sample outliers, batch effects, and the overall structure of variation between biological groups. Two widely used methods are:

- Multi-dimensional scaling (MDS) plots to assess overall sample similarity based on expression profiles.
- Heatmaps to visualize expression patterns of the most variable or significantly differentially expressed miRNAs.

We use the limma::plotMDS() function for MDS and gplots::heatmap.2() for the heatmap, based on log-transformed counts per million (logCPM) values calculated with edgeR.

```
#Assuming 'dge' is your filtered and normalized DGEList object
logCPM <- cpm(dge, log = TRUE)
# MDS plot using the top 500 most variable miRNAs
plotMDS(logCPM, labels = group, col = as.numeric(group),
main = "MDS Plot of miRNA Expression", cex = 1.2)
legend("topright", legend = levels(group),
col = 1:length(levels(group)), pch = 20)
# Select top 50 DE miRNAs from the limma results
top_miRNAs <- rownames(results)[1:50]
# Subset the logCPM matrix
heat_data <- logCPM[top_miRNAs,]
# Optional: scale rows (z-score normalization)
heat_data_scaled <- t(scale(t(heat_data)))
# Plot heatmap
heatmap.2(heat_data_scaled,
trace = "none", scale = "row",
col = colorRampPalette(c("blue", "white", "red"))(100),
margins = c(6, 10),
ColSideColors = as.character(as.numeric(group)),
main = "Top 50 Differentially Expressed miRNAs",
dendrogram = "both", key = TRUE, density.info = "none")
```

Fit linear model and extract differential expression results

```
# Fit model and compute moderated t-statistics
fit <- lmFit(v, design)
fit <- eBayes(fit)
# Extract results
```

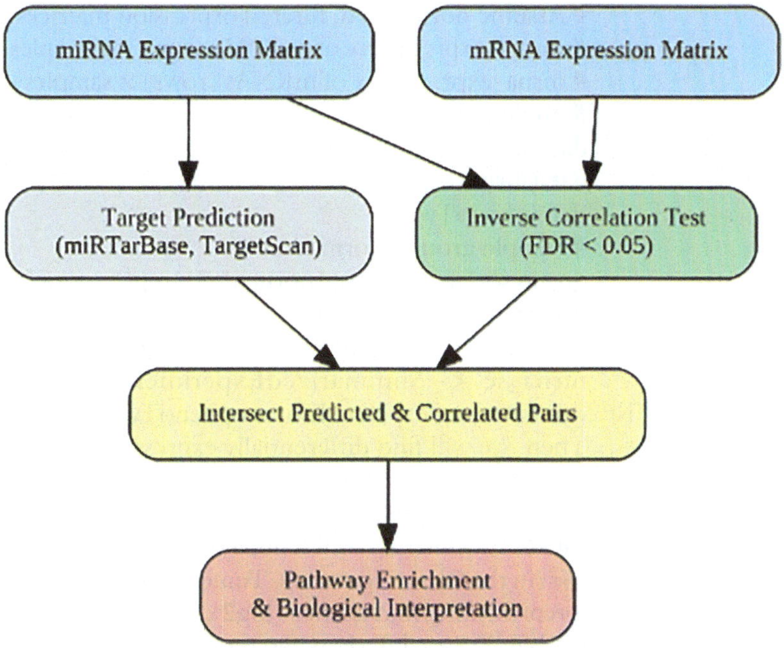

Fig. 2 Workflow for integrative miRNA–mRNA analysis. Overview of the integrative analysis pipeline for identifying functional miRNA–mRNA pairs. The workflow includes independent profiling, correlation analysis, validation using target prediction and literature, and downstream functional annotation to derive biological insights

```
results <- topTable(fit, coef = "Tumor_vs_Normal", number = Inf)
# Display top DE miRNAs
head(results)
```

3.6 miRNAs and mRNAs Integrating miRNA and mRNA Expression to Identify Regulatory Interactions

Once both miRNA and mRNA quantification are complete, a central analytical goal is to identify putative regulatory interactions. The most biologically meaningful miRNA–mRNA interactions typically exhibit inverse expression patterns, reflecting the post-transcriptional repression mechanism by which miRNAs inhibit their targets. A common first step is to perform differential expression analysis separately for miRNAs and mRNAs as described in the previous section. These two result sets are then integrated using target prediction databases (e.g., miRTarBase, TargetScan) to retrieve known or predicted miRNA–gene pairs. The general workflow is summarized in Fig. 2.

The anamiR R package streamlines this process into a coherent workflow that includes:

- Differential expression analysis.
- Correlation testing between miRNAs and mRNAs.
- Intersection with target prediction databases.
- Optional filtering by pathways or gene sets.

# Assume normalized, filtered expression matrices:
# mirna_expr: matrix of miRNAs (rows) × samples (columns)
# mrna_expr: matrix of mRNAs (rows) × samples (columns)
# phenotype
data(mirna_expr)
data(mrna_expr)
data(phenoType)
# Sample group information
group <- factor(c(rep("normal",3), rep("tumor",3)))

mirna_se <- SummarizedExperiment(assays = SimpleList(counts=mirna_expr), colData = phenoType)

mrna_se <- SummarizedExperiment(assays = SimpleList(counts=mrna_expr), colData = phenoType)

Then, we will find differentially expressed genes and miRNAs. The differExp_discrete function has three statistical methods, and we use limma for demonstration purposes. This function outputs genes and miRNAs, which are differentially expressed in the comparison of interest (in the Class: Normal vs. Tumor) and adds another three columns representing fold change (log2), *p*-value, and adjusted *p*-value.

mRNA_DE <- differExp_discrete(se = mrna_se, class = "group", method = "limma", log2 = FALSE, p_value.cutoff = 0.05, logratio = 0.5, p_adjust.method = "BH")

miRNA_DE <- differExp_discrete(se = mirna_se, class = "group", method = "limma", log2 = FALSE, p_value.cutoff = 0.05, logratio = 0.5, p_adjust.method = "BH")

Before comparing results with known interaction databases, it is crucial to convert all miRNA identifiers to the latest miRBase version (v21 or later) (*see* **Note 9**). Inconsistent IDs can lead to loss of annotations and missed interactions. This step ensures that all miRNA names are standardized and compatible with public resources such as miRTarBase, TargetScan, and miRDB.

mirna_21 <- miR_converter(data = mirna_d, remove_old = TRUE, original_version = 17, latest_version = 21)

To identify regulatory relationships, we test for significant negative correlations between miRNA and mRNA expression levels.

cor <- negative_cor(mrna_data = mrna_d, mirna_data = mirna_21,

cut.off = −0.5, method = "pearson")

Each row in the resulting list represents a potential interaction. Only pairs showing a Pearson correlation coefficient below the specified threshold (e.g., < −0.5) are retained (*see* **Note 10**). This is based on the biological premise that miRNAs downregulate their target genes. The next step is to check which predicted interactions are supported by known target prediction or validation databases. This adds confidence to the selected miRNA–mRNA pairs.

sup <- database_support(cor_data = cor, Sum.cutoff = 2, org = "hsa")

Sum.cutoff = 2 filters for interactions supported by at least two databases, and org = "hsa" specifies that the analysis is for human miRNAs.

3.7 Functional Enrichment Analysis

After identifying high-confidence miRNA–target gene interactions, it is often biologically insightful to assess which pathways or functional gene sets are significantly enriched among the target genes. This helps connect molecular interactions to broader cellular processes and disease mechanisms. The anamiR package provides the enrichment() function to perform pathway enrichment analysis on genes involved in the validated miRNA–mRNA interactions (i.e., those in the sup object). The method supports enrichment against several biological databases, including KEGG, GO, and Reactome, and uses a combination of hypergeometric testing and permutation-based empirical *p*-values to ensure robust results.

```
pathEnrich <- enrichment(data_support = sup, org = "hsa",
per_time = 500)
```

per_time = 500 defines the number of permutations used to calculate empirical *p*-values, which estimate how likely the enrichment results are under a random background model. The output includes both the classical *p*-value based on the hypergeometric distribution and an empirical *p*-value representing the proportion of permutations that yielded results as extreme as the observed ones. This dual significance testing increases the robustness of the conclusions, especially when dealing with high-dimensional or noisy data.

3.8 Visualizing and Exporting Pathway Enrichment Results

It is important to visualize the results and optionally export them for use with external tools. Visual representations such as bar plots and dot plots help highlight the most significantly enriched pathways and make it easier to communicate biological insights. To do this, we recommend using the barplot() and dotplot() functions from the clusterProfiler package, which is fully compatible with enrichment objects generated by anamiR.

```
# Convert 'path' object from anamiR to a data.frame if needed
df_path <- as.data.frame(pathEnrich)
# Optional: subset top 10 enriched terms based on adjusted
p-value
df_top <- df_path[order(df_path$p.adjust),][1:10,]
# Bar plot of top enriched pathways
barplot(as.enrichResult(pathEnrich),
showCategory = 10,
title = "Top Enriched Pathways",
font.size = 12)
# Dot plot alternative
dotplot(as.enrichResult(pathEnrich),
showCategory = 10,
title = "Dotplot of Enriched Pathways",
font.size = 12)
```

You may want to export the list of enriched genes or pathways for input into external tools such as Enrichr (https://maayanlab.cloud/Enrichr/) or DAVID (https://david.ncifcrf.gov/).

3.9 Regression

A more powerful and informative method to uncover miRNA–mRNA regulatory relationships is to use multivariable regression models, in which a gene's expression is modeled as a function of its corresponding miRNA levels, while simultaneously adjusting for genomic and epigenomic factors such as copy number alterations (CNAs) and DNA methylation. This approach enables the identification of significant negative associations, consistent with the repressive function of miRNAs, that are robust to confounding variables. Because linear models are part of base R functionality, you do not need specialized software to implement this (*see* **Note 11**). For example, in a linear regression framework:

```
lm_model <- lm(log2(gene_expr + 1) ~ log2(miRNA_expr + 1) + MET + CNA, data = my_data)
summary(lm_model)
```

MET and CNA are additional predictors accounting for methylation and copy number, both known to modulate gene expression independently of miRNAs. Integrating these factors helps isolate the direct effect of miRNAs (*see* **Note 12**). It should be noted that gene regulation is often influenced by additional layers, including transcriptional networks, RNA-binding proteins, and post-transcriptional modifications, making the models simplified representations of biological complexity. While these indirect factors are important (and increasingly modeled in deep learning and graph-based frameworks), they require extensive data and are beyond the intended scope of this chapter. Even so, our regression model provides a transparent and interpretable method to test miRNA-mediated regulation within the context of multi-omics cancer data.

3.10 Survival Analysis of Integrated Pairs

In the final stage of our miRNA–mRNA integrative analysis, we evaluate whether candidate genes or miRNAs identified through differential expression or regulatory modeling are associated with clinical outcomes, such as overall survival (OS) or progression-free survival (PFS). This step provides translational relevance to the bioinformatics findings and helps prioritize biomarkers with potential prognostic or therapeutic utility. The Kaplan–Meier estimator, coupled with log-rank tests, is the most used method to evaluate survival differences across groups defined by molecular expression. For continuous expression values, samples are often stratified into high- and low-expression groups using median or quantile-based thresholds (*see* **Note 13**). The survival curves can then be visualized using the ggsurvplot() function from the survminer package, which offers high-quality plots and flexible customization.

```
# Example: Stratify patients by median expression of a candidate miRNA
miRNA_expr <- miRNA_data["hsa-miR-21",] # Replace with candidate of interest
group <- ifelse(miRNA_expr > median(miRNA_expr), "High", "Low")
# Prepare survival object
surv_obj <- Surv(time = clinical_data$OS_time,
event = clinical_data$OS_event)
# Fit Kaplan-Meier model
fit <- survfit(surv_obj ~ group)
# Plot with ggsurvplot
ggsurvplot(fit,
data = clinical_data,
pval = TRUE,
risk.table = TRUE,
title = "Survival Stratified by miR-21 Expression",
xlab = "Time (days)",
legend.labs = c("Low expression", "High expression"),
palette = c("#1F77B4", "#D62728"))
```

This example evaluates the association between miR-21 expression and survival but can be applied to any miRNA or mRNA of interest. For multiple markers, survival modeling can be extended using multivariate Cox proportional hazards models, which allow the inclusion of clinical covariates (e.g., stage, age, treatment):

```
coxph(Surv(OS_time, OS_event) ~ miR_21 + age + stage,
data = clinical_data)
```

This concludes our comprehensive workflow, from miRNA/mRNA quantification to the identification of functional interactions and their clinical implications. Future advances, including machine learning, single-cell integration, and spatial transcriptomics, will further refine the discovery of robust miRNA–gene biomarkers for precision oncology. Depending on cohort size, the entire analysis may take several days and require moderate computational infrastructure (*see* **Note 14**).

4 Notes

Note 1: Pipeline Reproducibility and Directory Structure
Adhering to a well-structured folder hierarchy (e.g., separating raw, processed, annotation, and results) enables reproducibility and simplifies debugging. Consider version-controlling scripts using git and containerizing environments with Docker for full reproducibility.
Note 2: STAR Genome Indexing
STAR index generation is computationally intensive and must match the annotation version used later in the pipeline. Mismatched genome/annotation versions (e.g., GRCh38 + Gencode v22 vs. v48)

can lead to misaligned reads and incorrect quantification. Save index versions for reproducibility and consistency across samples.

Note 3: Adapter and Quality Trimming

Residual adapter sequences and poly(A)/poly(T) tails can severely compromise read alignment and quantification accuracy. Always inspect FastQC output for adapter contamination and low-quality bases. Cutadapt is preferred for flexibility, but settings must be tuned for short miRNA reads. Re-running FastQC post-trimming is essential to verify improvements.

Note 4: Read Alignment Discrepancies Between mRNA and miRNA

STAR is optimized for spliced long reads; using it for miRNA (short reads) will result in poor alignment rates. Conversely, Bowtie/BWA or miRDeep2 should not be used for spliced mRNA reads. This technical distinction is often overlooked in pipelines combining both data types.

Note 5: Low Complexity and Multi-Mapped miRNA Reads

Small RNAs are prone to multi-mapping due to sequence similarity among miRNA families. miRDeep2's default setting to exclude reads mapping >5 loci is essential but may remove biologically meaningful data in repeat-rich regions. Balance sensitivity with specificity depending on the biological context.

Note 6: Batch Effects in High-Throughput Datasets

When integrating public datasets (e.g., from TCGA or GEO), batch effects can confound differential expression and correlation analysis. Use tools such as sva or limma::removeBatchEffect() to adjust for known batch variables, especially if multiple platforms or institutions are involved.

Note 7: Common Pitfalls in Differential Expression Analysis

Misspecification of the design matrix (e.g., incorrect factor levels or missing intercepts) leads to erroneous results. Always inspect the design matrix and validate contrasts with model.matrix() and contrasts.fit().

Note 8: Sensitivity to Low Abundance miRNAs

Lowly expressed miRNAs can yield unstable statistical results. Filtering out miRNAs with low counts across all samples using filterByExpr() improves robustness but may remove biologically relevant miRNAs. Carefully consider trade-offs, especially for tumor suppressors.

Note 9: Inconsistent miRNA Naming Conventions

Use of outdated miRBase IDs can cause failure in target mapping and database lookups. Always convert miRNA IDs to the latest miRBase version using tools such as miR_converter. Manual inspection is sometimes needed for novel miRNAs predicted by miRDeep2.

Note 10: Inverse Correlation Thresholds for miRNA–mRNA Pairing

The commonly used Pearson threshold (e.g., <-0.5) for identifying miRNA–mRNA interactions may be too strict for weakly repressive but real interactions. Use context-specific or pathway-informed thresholds, and validate with multiple databases (e.g., miRTarBase, TargetScan).

Note 11: Confounding in Regression Models

Multivariable models (miRNA + MET + CNA) can be skewed if predictors are collinear. Standardize predictors and assess variance inflation factors (VIFs). Interpretation should be cautious if data for epigenetic covariates (e.g., methylation) are sparse or from a different platform.

Note 12: Extending with Additional Omics

While the chapter focuses on expression data, the pipeline is extensible to include CNV, methylation, and proteomics. When doing so, make sure all omics layers are normalized and matched at the sample level, and ensure correct scaling for joint modeling.

Note 13: Survival Analysis Stratification Bias

Using the median to stratify expression (high vs. low) can oversimplify continuous expression effects. Explore optimal cut points (e.g., surv_cutpoint in survminer) and validate across cohorts. Avoid overfitting by applying the same cutoff in an independent validation set.

Note 14: Sensitivity and Time Requirements

The full pipeline, from raw data to survival modeling, requires moderate computational resources (8–16 CPU threads, 64 + GB RAM) and can take 1–3 days per cohort depending on size and complexity. Running alignment in parallel for all samples significantly reduces total time.

Conflict of Interest

The authors have no conflicts of interest to declare that are relevant to the content of this chapter.

References

1. Bartel DP (2004) MicroRNAs: genomics, biogenesis, mechanism, and function. Cell. 116(2):281–297.
2. Michlewski G, Cáceres JF (2019) Post-transcriptional control of miRNA biogenesis. RNA. 25(1):1–16
3. Treiber T, Treiber N, Meister G (2019) Regulation of microRNA biogenesis and its crosstalk with other cellular pathways. Nat Rev Mol Cell Biol 20(1):5–20
4. Zhou K, Liu M, Cao Y New Insight into microRNA functions in cancer: oncogene–microRNA–tumor suppressor gene network. Front Mol Biosci 4:46. https://www.frontiersin.org/journals/molecular-biosciences/articles/10.3389/fmolb.2017.00046/full
5. Svoronos AA, Engelman DM, Slack FJ (2016) Oncomir or tumor suppressor? The duplicity of microRNAs in cancer. Cancer Res 76(13):3666–3670.
6. To KKW, Zhang H, Cho WC (2024) Competing endogenous RNAs (ceRNAs) and drug resistance to cancer therapy. Cancer Drug Resist 7:37
7. Aparo A, Avesani S, Parmigiani L, Napoli S, Bertoni F, Bonnici V et al (2025) EasyCircR: detection and reconstruction of circular RNAs post-transcriptional regulatory interaction networks. Comput Biol Med 188:109846
8. Nalbant E, Akkaya-Ulum YZ (2024) Exploring regulatory mechanisms on miRNAs and their implications in inflammation-related diseases. Clin Exp Med 24(1):142

9. Truong VA, Chang YH, Dang TQ, Tu Y, Tu J, Chang CW et al (2024) Programmable editing of primary MicroRNA switches stem cell differentiation and improves tissue regeneration. Nat Commun 15(1):8358
10. Wyman SK, Knouf EC, Parkin RK, Fritz BR, Lin DW, Dennis LM et al (2011) Post-transcriptional generation of miRNA variants by multiple nucleotidyl transferases contributes to miRNA transcriptome complexity. Genome Res 21(9):1450–1461.
11. Jie M, Feng T, Huang W, Zhang M, Feng Y, Jiang H et al (2021) Subcellular localization of miRNAs and implications in cellular homeostasis. Genes. 12(6):856
12. Mensh B, Kording K (2017) Ten simple rules for structuring papers. PLOS Comput Biol 13(9):e1005619
13. Sandve GK, Nekrutenko A, Taylor J, Hovig E (2013) Ten simple rules for reproducible computational research. PLOS Comput Biol 9(10):e1003285

Check for updates

Chapter 13

MicroRNAs and Their Profiling via Single-Cell Sequencing Technologies

Chinmayee Goda, Shaopeng Gu, Anjun Ma, Mario Acunzo, Giulia Romano, Adrienne M. Dorrance, Ramiro Garzon, Qin Ma and Giovanni Nigita

Abstract

MicroRNAs (miRNAs) are small non-coding RNAs that play essential roles in gene regulation, cellular function, and disease pathogenesis. Advances in single-cell RNA sequencing technologies have enabled the profiling of miRNAs at single-cell resolution, providing unprecedented insight into cell-specific regulatory networks and heterogeneity. This chapter presents an overview of miRNA biology, technical approaches for single-cell miRNA sequencing, and recent bioinformatics tools developed for data analysis. We discuss challenges in library preparation, such as adapter biases and low RNA input, and highlight integrative strategies for co-profiling miRNAs with other omics layers. Finally, we outline in the conclusion the potential of single-cell miRNA profiling to contribute to precision medicine and therapeutic development, including its possible use in biomarker discovery, monitoring tumor heterogeneity, and informing personalized treatment strategies. As the field progresses, continued innovation will be critical to overcoming existing barriers and fully harnessing the power of single-cell miRNA analyses.

Keywords miRNAs, ncRNAs, scRNA-seq

1 Introduction

1.1 Overview of microRNAs

MicroRNAs (miRNAs) are a class of small non-coding RNA molecules, typically around 21–25 nucleotides in length, that play a pivotal role in post-transcriptional gene regulation [1]. Their discovery in 1993 marked a significant advancement in understanding gene regulation at the molecular level [2–5]. miRNAs fine-tune gene expression by binding to messenger RNAs (mRNAs) in a sequence-specific manner, influencing various cellular processes [6–9]. They bind to mRNA through sequence complementarity, leading to the inhibition of gene expression via post-transcriptional regulation [10]. Their biological significance lies in their ability to negatively regulate protein-coding genes, thereby impacting diverse cellular processes,

Alessandro Laganà and Giovanni Nigita (eds.), *MicroRNA Target Identification*, Methods in Molecular Biology, vol. 3056,
https://doi.org/10.1007/978-1-0716-5392-0_13,

including development, differentiation, proliferation, and apoptosis [11]. miRNAs have been implicated in various diseases, including cancer, where dysregulation can contribute to tumorigenesis, metastasis, and chemoresistance [12, 13], making them potential therapeutic targets [14–16].

One intriguing aspect of miRNAs is their role in intercellular communication, primarily through vesicle-delivered miRNAs exchanged between different cell types in the tumor microenvironment [17–20]. These vesicle-enclosed miRNAs have been shown to play a role in communication between macrophages and cancer cells, influencing processes such as apoptosis and tumor growth [21–23]. Moreover, miRNAs are involved in host–pathogen interactions, modulating the host's immune response to pathogens, which highlights their diverse regulatory roles in mediating interactions between different organisms [24, 25].

The profiling of miRNAs using next-generation sequencing (NGS) technologies [26–28], including single-cell RNA sequencing, and the development of bioinformatics tools [29] and pipelines [30] has transformed miRNA research by enabling robust cross-cohort comparisons and the discovery of clinically relevant miRNA signatures, with downstream applications in molecular subtyping, non-invasive biomarker development (e.g., liquid biopsy), and prediction of therapeutic response. NGS enables high-throughput sequencing of small RNA transcriptomes, providing insights into miRNA expression profiles across various cell types and conditions, as well as characterizing variations in miRNAs produced during biogenesis, known as miRNA isoforms (e.g., shifted miRNAs, edited miRNAs) [31–36]. Single-cell RNA sequencing technologies have further advanced our understanding of miRNA expression at the single-cell level, enabling the identification of cell-specific miRNA signatures and regulatory networks [37]. These technologies have paved the way for comprehensive studies of miRNA expression patterns, target identification, and functional analyses, thereby enhancing our understanding of miRNA roles in health and disease.

This chapter provides a comprehensive overview of miRNAs, including their biogenesis and mechanisms of action, and details the principles and applications of single-cell sequencing technologies. It discusses the technical approaches for profiling miRNAs at single-cell resolution, highlighting technological advances and future directions in the field.

1.2 Introduction to Single-Cell Sequencing Technologies

Evolution and Importance of Single-Cell Sequencing

Single-cell RNA sequencing (scRNA-seq) is a powerful tool that allows a comprehensive understanding of individual cells that comprise a tissue. The first scRNA-seq study was reported by Tang et al. [38] in

2009, in which they analyzed gene expression profiles of individual cells from a single mouse blastomere. This groundbreaking *Nature* study laid the foundation for the subsequent development of scRNA-seq technologies, and since then, numerous modified and improved scRNA-seq methodologies have been developed. With the introduction of the Fluidigm C1 microfluidics system in 2012, it became possible to profile gene expression patterns of up to 96 cells in a single run within a day [39]. This paved the way for the high-throughput Fluidigm integrated fluidic circuits, which enabled the examination of up to 800 cells simultaneously. In recent years, microdroplet technologies (Drop-seq) [40] have replaced Fluidigm microchambers, enabling the capture and analysis of hundreds of thousands of single cells within microdrops. Currently, 10× Genomics, a type of droplet-based scRNA-seq technology, is the most widely used scRNA-seq method.

Comparison with Bulk Sequencing Techniques

scRNA-seq platforms provide multiple advantages over traditional bulk RNA sequencing. Bulk RNA sequencing provides an average gene expression profile of cells within a tissue, whereas scRNA-seq provides a gene expression profile of individual cells within a tissue. Through scRNA-seq, it is possible to identify heterogeneity among cell types within a tissue, thereby gaining a better understanding of complex, heterogeneous tissues. Since bulk RNA sequencing generates an average gene expression profile, the expression profiles of rare cells cannot be captured without sufficient enrichment of these cells. scRNA-seq enables the profiling of all cells within a tissue, allowing for the identification of rare cell types and their transcriptomes without requiring additional enrichment steps. This is particularly important in the analysis of rare cell populations, such as cancer stem cells [41] and cells of the tumor microenvironment [42], which would otherwise be diluted by the transcriptome of tumor cells in bulk sequencing techniques. Furthermore, by analyzing the transcriptomes of individual cells, lineage tracing and the study of developmental trajectories can be performed, which ultimately provide insights into the mechanisms of cell fate decisions [43].

2 Biogenesis and Mechanism of Action of miRNAs

2.1 Biogenesis of miRNAs

miRNAs are small non-coding RNA molecules essential for post-transcriptional gene regulation. The biogenesis of miRNAs involves multiple steps that lead to the maturation of functional miRNAs (Fig. 1). This process starts with the transcription of primary miRNA (pri-miRNA) by RNA polymerase II in the nucleus [44]. The pri-miRNA is then processed by the Microprocessor complex, which includes Drosha and its partner DiGeorge Critical Region 8 (DGCR8), resulting in the production of precursor miRNA (pre-miRNA) [44]. Drosha, a ribonuclease III enzyme, cleaves the stem-loop structure of the

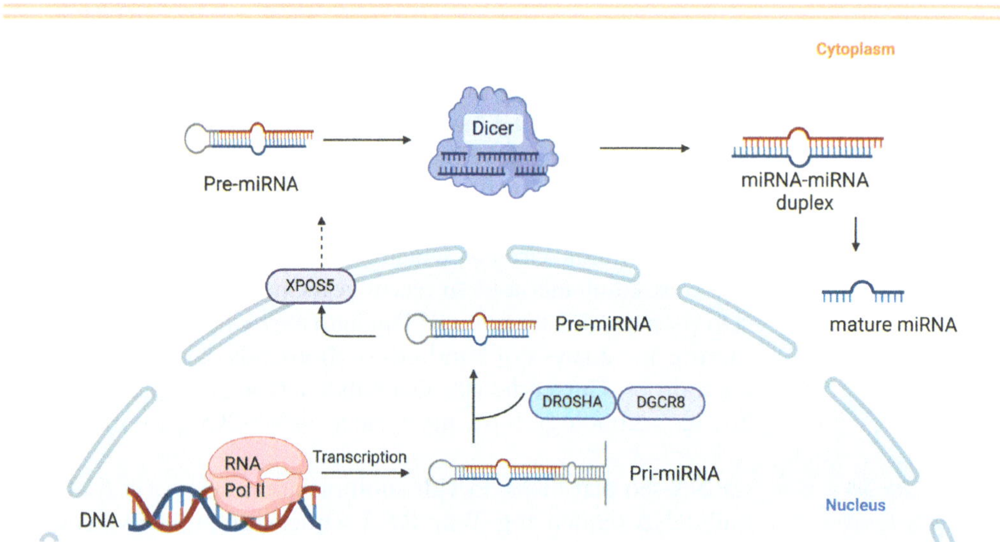

Fig. 1 Schematic representation of the canonical biogenesis of miRNAs

pri-miRNA to generate pre-miRNA in the nucleus [45]. This crucial step prepares the precursor that will eventually mature into miRNAs.

After processing by Drosha, the pre-miRNA is exported to the cytoplasm by Exportin 5 (XPO5) [46]. In the cytoplasm, the pre-miRNA undergoes further processing by Dicer, another ribonuclease III enzyme, to produce mature miRNA molecules [47]. Dicer processes the pre-miRNA into a ~22-nucleotide mature miRNA duplex [48]. This duplex is unwound, and one strand, known as the guide strand, is incorporated into the RNA-induced silencing complex (RISC) [49]. The RISC complex uses the guide strand to target mRNAs through base pairing with complementary sequences, mainly in the mRNA 3′ untranslated region (UTR) [10].

The mechanism of miRNA action involves the binding of the miRNA-loaded RISC complex to the target mRNA, resulting in post-transcriptional gene regulation [10]. The miRNA recognizes its target mRNA through partial complementarity, leading to translational repression or mRNA degradation [10]. Translational repression occurs when the miRNA-RISC complex binds to the mRNA, blocking translation initiation or promoting mRNA decay [10]. Conversely, mRNA degradation involves cleavage of the target mRNA by the RISC complex, leading to its degradation [10]. This post-transcriptional regulation mediated by miRNAs is crucial for fine-tuning gene expression and controlling various cellular processes [49].

2.2 Canonical Function of microRNAs

miRNAs play a crucial role in post-transcriptional gene regulation, influencing target recognition and subsequent regulatory actions. These molecules regulate the translation and stability of their target mRNAs [49]. Canonically, miRNAs modulate protein translation or

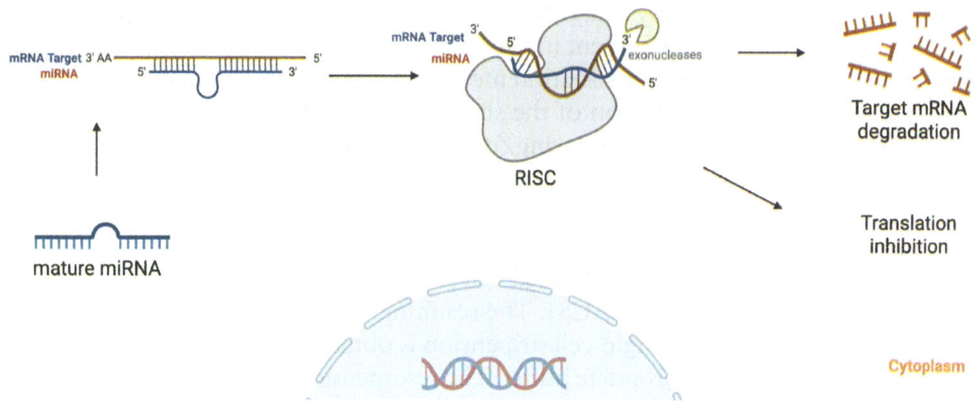

Fig. 2 microRNAs' canonical function

induce mRNA degradation by binding to specific regions, principally the 3′UTRs of target mRNAs (Fig. 2) [10]. Evidence suggests that miRNAs may bind to 5′ UTRs and CDS regions to negatively regulate gene expression [50, 51]. This interaction occurs through base pairing between the miRNA and the mRNA, resulting in translational repression or mRNA degradation [10]. The miRNA's actions at the initiation of translation and in mRNA decay are well-established mechanisms of post-transcriptional regulation [1, 49].

When miRNAs bind to their target mRNAs, they cause translational repression, where the miRNA-RISC complex inhibits translation initiation or promotes mRNA decay [49]. This involves the miRNA-RISC complex binding to the mRNA, suppressing protein translation [49]. Additionally, miRNAs can initiate mRNA degradation by cleaving the target mRNA, thereby reducing gene expression [1]. The negative regulation of gene expression by miRNAs at the post-transcriptional level is crucial for controlling cellular processes, including differentiation, proliferation, and metastasis [11].

3 Single-Cell Sequencing Technologies

This section provides a brief overview of the experimental steps involved in performing single-cell sequencing (Fig. 3). We also highlight the major sequencing platforms currently in use, along with their strengths and limitations.

3.1 Fundamental Principles

Isolation and Preparation of Single Cells

Effective isolation and preparation of single cells from samples is an important first step in single-cell sequencing. An optimal sample for scRNA-seq is a uniform cell suspension with viability greater than 90%. Isolation of single cells from blood and liquid biopsies is relatively straightforward and requires minimal effort, as the cells are already present in a suspension. In the case of solid tissues, however, various dissociation methods must be employed, depending on the composition of the starting tissue. Dissociation typically involves mechanical mincing/dissociation and/or degradation of extracellular matrix using enzymes such as collagenase [52]. Depending on the cellular composition of the sample, enrichment of cells is performed to enhance the chances of capturing cells of interest, either by fluorescence-activated cell sorting (FACS) or magnetic-activated cell sorting (MACS). The resulting single cells are washed thoroughly, and a single-cell suspension is obtained by resuspending the cells in an appropriate buffer. The resuspension buffer should be Ca^{2+}, Mg^{2+}, and EDTA-free since they interfere with the downstream reverse transcription reaction. The cell suspension is then analyzed through a microscope for cell concentration, viability, and morphology using a viability dye such as trypan blue. During the cell preparation process, it is crucial to maintain low temperatures and employ gentle handling techniques to ensure optimal cell viability and minimize cellular damage. The single-cell suspension is then partitioned into thousands of individual cells using various techniques.

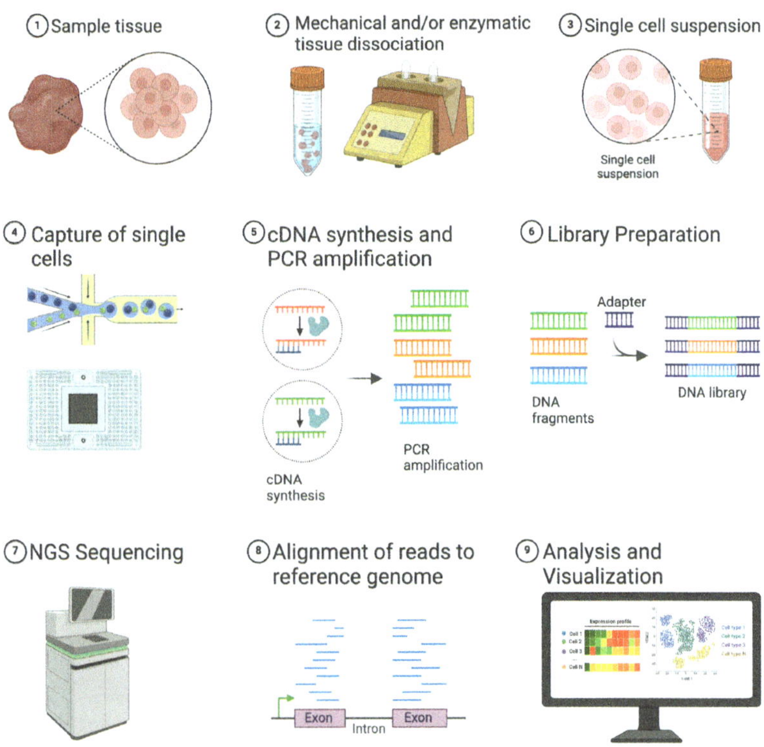

Fig. 3 Overview of scRNA sequencing

Following that, each single cell is lysed individually, and a reverse transcription reaction is carried out to generate. The master mix consists of enzymes and reagents required for the reverse transcription reaction, while the gel beads contain poly(dT) primers. To generate single cells within each droplet, cells are delivered at a limiting dilution, such that most of the generated GEMs contain no cells, while the remaining GEMs contain a single cell each.

Amplification and Sequencing of RNA from Single Cells

Following the partition of single cells, each cell is lysed, and a pool of poly(dT) primers with a unique barcode per cell is used to perform a reverse transcription reaction. This results in the generation of full-length cDNA from polyadenylated mRNA, where cDNA from each cell is barcoded with unique barcodes per cell. Sequencing primers and sample indexes are then added via end repair and adaptor ligation, followed by a second round of PCR amplification. The library is then sequenced using a next-generation sequencer.

3.2 Major Platforms and Techniques

Overview of Technologies (e.g., 10× Genomics, Fluidigm, SMART-seq)

There are various platforms currently available to perform scRNA-sequencing. Depending on the question being addressed and the strengths and limitations of each platform, an appropriate platform can be chosen for a scRNA-seq experiment.

Fluidigm. Fluidigm was one of the first commercially available platforms to utilize a microfluidic chip for partitioning single cells into individual, tiny chambers. The Fluidigm C1 chip uses an integrated microfluidic chip (IFC) that automatically captures individual cells from the prepared single-cell suspension [53]. There are various sizes of IFCs available for use, depending on the size and shape of the cells to be captured. Furthermore, since the IFCs are optically clear, they can be examined under a light microscope to evaluate the number and quality of captured cells. The cells are stained with a viability dye prior to examination under a light microscope to determine the number of live cells. Once the number and quality of cells are checked, lysis and reverse transcription reagents are delivered to each cell. Each cell is lysed, and first-strand cDNA is synthesized using a reverse transcriptase (RT) within the micro-chambers. This is followed by a polymerase chain reaction (PCR), resulting in the amplification of the cDNA. The cDNA is then tagged with primers and cut with transposons before being tagged with dual indices that provide a unique barcode to libraries prepared from each cell. These individually barcoded cDNA libraries from individual cells are then combined, cleaned, and sequenced using a next-generation sequencer.

10× Genomics. 10× Genomics is the most widely used single-cell technology today. The basic principle involves loading a single-cell suspension onto a microfluidic chip, where it is partitioned into thousands of cells within gel beads-in-emulsion (GEMs) containing single cells. For the generation of GEMs, barcoded gel beads and partitioning oil are loaded onto a microfluidic chip along with a master mix containing the cell suspension. The master mix consists of enzymes and reagents required for the reverse transcription reaction, while the gel beads contain the poly(dT) primers. To generate single cells within each droplet, cells are delivered at a limiting dilution, such that most of the generated GEMs contain no cells, while the remaining GEMs contain a single cell each. Following the generation of GEM, the gel bead is dissolved, primers are released, and the enclosed cell is lysed. The primers and unique barcodes are mixed with the cell lysate along with the master mix containing the reverse transcription reagents. GEMs are incubated to generate barcoded, full-length cDNA from polyadenylated mRNA, where cDNA from each cell is barcoded with unique barcodes per cell. After incubation, GEMs are dissociated, and pooled fractions of first-strand cDNA from all cells are recovered using silane magnetic beads. Barcoded, full-length cDNA is then amplified via PCR to generate sufficient DNA concentration for library construction. Enzymatic fragmentation and size selection are used to optimize the cDNA amplicon size. Sequencing primers and sample indexes are then added via end repair and adaptor ligation, followed by a second round of PCR amplification. The library is then sequenced using a next-generation sequencer.

SMART-seq. Most scRNA-sequencing platforms preferentially amplify either the 5′ or 3′ end of mRNAs, resulting in coverage of reads at either end of the transcript. Most mammalian genes are made of multiple exons and undergo alternative RNA processing [54], which would not be detected by read coverage of only a part of the mRNA. SMART-seq (Switching Mechanism at the 5′ End of RNA Template) is a plate-based single-cell sequencing technology that generates read coverage across the full length of transcripts, which thus allows the detection and quantification of transcript isoforms, gene fusions, and point mutations [55]. Single cells in individual wells of a plate are lysed to release the RNA, followed by binding of oligo(dT) primers to the 3′ ends of polyA mRNAs. The primed mRNAs are then reverse transcribed using the Moloney Murine Leukemia Virus (M-MLV) RT. This RT has terminal transferase activity that adds non-templated nucleotides to the 3′ end of the cDNA ends. This is followed by the annealing of a template-switching oligo (TSO), which enables the newly synthesized cDNA to switch templates and continue extending, thereby facilitating the generation of full-length cDNA. The resulting full-length cDNA contains the complete 5′ end of the mRNA as well as an anchor sequence that serves as a universal priming site for second-strand synthesis. The cDNA is then amplified using PCR to generate sufficient material for sequencing. The amplified cDNA is then ligated with sequencing primers, and libraries are sequenced using a next-generation sequencer.

Strengths and Limitations of Different Platforms

Each platform has strengths and limitations. The greatest strength of the 10×Genomics platform is its ease in carrying out high-throughput

Table 1
Strengths and limitations of various scRNA-sequencing platforms

Platform	Strengths	Limitations
10× Genomics	• High throughput • Ease of use • Lower cost per cell	• Lower read depth • Less quality control • Potential for doublets
Fluidigm	• High read depth • Visual quality control • Precise cell capture	• Low throughput • Higher cost per cell • Complex workflow
SMART-seq	• High sensitivity • Isoform analysis • Rare cell population analysis	• Technical complexity • Low throughput • Longer processing time

studies. A single 10× Genomics experiment can capture up to 10,000 cells, thereby drastically reducing the cost per cell. Fluidigm and SMART-seq are low-throughput platforms and struggle with high costs per cell. The 10× Genomics workflow is relatively straightforward compared to the complex workflows of the IFCs in Fluidigm and the technical complexity of SMART-seq. However, due to the high-throughput nature of 10× Genomics, each cell captured has fewer reads per transcript, which may result in missing genes that have low expression. High read depth and high sensitivity can be achieved with the Fluidigm and SMART-seq platforms. Fluidigm enables the visual inspection of cells within the IFCs prior to sequencing, thereby allowing for proper quality control and the precise capture of cells. 10× Genomics does not allow for the inspection of individual cells for quality before sequencing, and as a result, has a higher chance of capturing multiple cells in a single droplet, thereby lowering the data quality compared to Fluidigm. SMART-seq captures full-length transcripts, enabling detailed analysis of isoforms and alternative splicing events within the transcriptome. This ultimately enables the study of rare cell types within a heterogeneous population with greater sensitivity compared to other platforms. However, the SMART-seq workflow is very complex, and the protocol requires greater technical expertise to be performed accurately. The full-length sequencing protocol can also take longer than the other platforms that focus on a smaller portion of the transcript. The strengths and limitations of the platforms have been summarized in Table 1.

4 Profiling MicroRNAs at Single-Cell Resolution

4.1 Technical Approaches

Efficient single-cell isolation is essential for studying small RNAs, including miRNAs, at the single-cell level. Live, single cells of interest are typically sorted into each well of a multi-well plate using methods such as FACS. The lysis of individual cells is then followed by the addition of lysis buffer to each well, allowing for the extraction of RNA from the cells. Ligation-based protocols or polyadenylation-based protocols are then used to add adaptors to the 3′ and 5′ ends of RNAs using T4 RNA ligases, followed by RT using RT primers and an RT enzyme [56]. Free, unbound adaptors are removed using exonucleases, followed by PCR amplification of libraries, adapter ligation, and ultimately NGS .

Methods for MicroRNA Isolation and Sequencing from Single Cells

The conventional method for generating small RNA libraries involves ligation-based protocols that sequentially ligate short oligonucleotides (adapters) to miRNA using T4 RNA ligase. This is followed by reverse transcription and PCR amplification to generate the DNA libraries, which are ultimately sequenced using Illumina sequencing. The protocol typically involves enzymatic digestion of 3′ adapters after ligation, prior to 5′ adapter ligation. This ensures that the excess 3′ adapters are blocked by RT primers after ligation and then degraded by exonucleases [57]

Ligation-based protocols have several challenges, as discussed in the next section. Hence, another strategy involves a polyadenylation-based miRNA-seq protocol. Most miRNAs lack a poly(A) tail or other consensus 3′ end sequences, making it impossible to use oligo-dT primers typically used in sequencing mRNAs. Poly(A) polymerase (PAP) is an enzyme that can perform polyadenylation to the 3′ end of RNAs [58] Utilizing this enzyme, several protocols have been developed, including polyadenylation ligation-mediated sequencing (PALM-seq) and vast transcriptome analysis of single cells by dA-tailing (VASA-seq) [59, 60] They involve addition of poly(A/C) tails to purified small RNAs, followed by capture using oligo-d(T/G) primers.

Challenges in MicroRNA Quantification and Strategies to Overcome Them

One of the significant challenges faced in miRNA quantification is the formation of adapter dimers. To ensure efficient adapter ligation, a high concentration of adapters is required relative to the small amounts of small RNAs. This results in high levels of adapter dimers, which impede the amplification of small RNA-cDNA libraries. One strategy to avoid this involves the use of chemically modified adapters, such as 3′ CleanTag, which inhibits the formation and reverse transcription of adapter dimers [61]. Another protocol uses random splint ligation, in which a double-stranded adapter anneals to RNAs via a short, single-stranded, degenerate extension. Following 3′ adapter ligation, excess non-ligated splint adapters are cleaved by lambda exonucleases, thereby preventing adapter dimer formation [62] Cas9-assisted small RNA sequencing (CAS-seq) is another strategy that was developed to detect less abundant small RNAs and prevent adapter self-ligation [63] Following the ligation of 3′ and 5′ adapters, this method uses biotin-labeled RT primers, followed by Cas9/single-guide RNA cleavage to degrade adapter heterodimers. Streptavidin-bead-mediated selection and further library size selection by polyacrylamide gel electrophoresis enable the enrichment of small RNAs prior to NGS.

Another challenge faced in miRNA quantification is adapter ligation biases. This occurs due to the base preferences of RNA ligases and the ligation preferences resulting from the formation of stable secondary structures post-ligation between adapters and small RNAs of similar length [64, 65]. To prevent ligation biases, one strategy is to include 4–8 nucleotide randomized bases within adapters at the ligation ends [64]. Another strategy to increase adapter ligation efficiency is the parallel single-cell small RNA sequencing protocol (PSCSR-seq) [66]. This method involves microinjection of a 3′ ligation reaction, resulting in a reduction of reaction volume to nanoliter-scale, which ultimately inhibits the formation of side products. Furthermore, library yield is enhanced by a heating step, which possibly allows the release of small RNAs from RISC.

4.2 miRNA and mRNA Co-Abundance

miRNAs regulate the expression of target mRNA, which ultimately impacts cell function. Hence, the simultaneous quantification of mRNAs and miRNAs from the same cell adds value because miRNA activity is context- and substrate-dependent: the functional impact of a given miRNA depends on which target (and competing) transcripts are expressed at that moment, and may therefore differ across cellular states even within the same cell type. To this end, several strategies have been developed to quantify mRNAs and miRNAs from the same cell. Holo-seq involves the isolation of single cells through microscope-supervised pipetting and cell lysis, followed by conventional bulk total RNA-seq and small RNA holo-sequencing [67]. The co-seq strategy involves lysing single cells and then splitting the RNA from each cell into two half-cell fractions. Each half-cell fraction is then subjected to small RNA and mRNA sequencing, yielding a paired half-cell miRNA–mRNA co-profile [68]. VASA-seq is another technology that can profile both miRNA and mRNA from single cells by capturing both non-polyadenylated and polyadenylated transcripts.

4.3 Bioinformatics Tools for Single-Cell MicroRNA Sequencing Data

Analyzing and interpreting sc-miRNA-seq data involves balancing robust computational methodologies with the unique biological details of small RNA measurements. Raw reads from single-cell small RNA sequencing (sc-sRNA-seq) experiments are typically preprocessed by trimming barcodes, unique molecular identifiers (UMIs), and adapter sequences, followed by quality filtering, and then aligning the filtered reads to reference sequences. This preprocessing ensures the accurate quantification of miRNAs and other classes of small non-coding RNAs (e.g., fragments of tRNAs or ribosomal RNAs), [35, 69–71] setting the stage for downstream analyses, such as clustering and differential expression (DE) analyses. Moreover, the inherent sparsity and technical noise in single-cell assays require specialized normalization and batch-correction strategies to harmonize data across different experiments and platforms.

Various specialized bioinformatics tools have been developed to address the unique challenges presented by single-cell microRNA sequencing (sc-miRNA-seq) data. These tools incorporate pipelines that begin with correcting sequencing errors (such as removing UMIs and adapter contaminants) to ensure that the counts reflect true biological variation rather than technical artifacts. For instance, tools such as SingmiR offer web-based platforms that automate many of these preprocessing steps, including alignment to small RNA databases and normalization against established annotations [72]. Once the preprocessing is complete, both canonical miRNAs and their modified miRNAs (isomiRs) will be quantified. Dedicated software modules perform differential expression analyses similar to those used in scRNA-seq gene expression studies, applying dimensionality reduction, clustering, and statistical inference methods to distinguish between cell types and states.

To fully capture the complexity of cellular regulation, integrative analyses that combine sc-miRNA-seq with other omics modalities are increasingly essential [73, 74]. By merging miRNA profiles with transcriptomic, proteomic, or epigenomic data, researchers can dissect the multilayered regulatory networks that govern cell function and identity. A typical integrative workflow begins by aligning miRNA expression data with corresponding scRNA-seq measurements, thereby enabling correlation between miRNA abundance and gene expression patterns. This step is critical for inferring miRNA–mRNA regulatory relationships, particularly when using approaches such as cell-specific network inference or statistical resampling methods to overcome the limitations associated with smaller cell populations. Moreover, combining miRNA data with epigenetic or proteomic profiles allows for the construction of dynamic regulatory models that capture both transcriptional and post-transcriptional layers. While integrating multimodal data can reveal critical insights (such as tissue-specific regulatory circuits involving miRNAs), it also introduces new challenges related to data normalization, batch effects, and aligning disparate data types. Nonetheless, advances in computational methods and statistical modeling are progressively overcoming these obstacles, facilitating a more comprehensive interpretation of the intricate biological interplay that defines cellular phenotypes.

5 Conclusion

The integration of single-cell sequencing technologies into miRNA research profoundly transforms our ability to resolve gene regulation at an unprecedented cellular resolution. This chapter outlined the critical role of miRNAs in post-transcriptional regulation and detailed the experimental and computational innovations enabling their study at the single-cell level. Despite challenges related to technical noise, adapter ligation biases, and low input material, continued refinements in molecular protocols and analytical tools are steadily overcoming these limitations. Looking ahead, the future of single-cell miRNA profiling is promising. Emerging methods that simultaneously capture multiple RNA species from the same cell, along with integrative multi-omics data, are expected to deepen our understanding of cell-specific regulatory networks. Such approaches hold significant potential for precision medicine, enabling the discovery of novel diagnostic biomarkers, tracking tumor heterogeneity, and identifying cell type-specific therapeutic targets. Moreover, miRNA-based interventions may eventually be tailored to individual patients by leveraging insights gained from single-cell data. Nonetheless, the field must also confront key challenges, including data sparsity, cross-platform variability, and the need for scalable, user-friendly bioinformatics tools. Addressing these issues will unlock new opportunities to map dynamic regulatory landscapes and translate single-cell insights into tangible clinical applications. As technologies continue to mature, single-cell miRNA profiling is poised to become an indispensable resource in molecular and translational medicine.

Acknowledgments

G.N. was supported in part by the National Cancer Institute (National Institute of Health) grant (R21CA287180).

Conflict of Interest The authors have no conflicts of interest to declare that are relevant to the content of this chapter.

References

1. Gebert LFR, Macrae IJ (2018) Regulation of microRNA function in animals. Nature Rev Mol Cell Biol 21:743. https://doi.org/10.1038/s41580-018-0045-7
2. Lee RC, Feinbaum RL, Ambros V (1993) The C. elegans heterochronic gene lin-4 encodes small RNAs with antisense complementarity to lin-14. Cell 75:843–854
3. Reinhart BJ, Slack FJ, Basson M et al (2000) The 21-nucleotide let-7 RNA regulates developmental timing in Caenorhabditis elegans. Nature 403:901–906. https://doi.org/10.1038/35002607
4. Lagos-Quintana M, Rauhut R, Lendeckel W, Tuschl T (2001) Identification of novel genes coding for small expressed RNAs. Science (New York, NY) 294:853–858. https://doi.org/10.1126/science.1064921
5. Lau NC, Lim LP, Weinstein EG, Bartel DP (2001) An abundant class of tiny RNAs with probable regulatory roles in Caenorhabditis elegans. Science (New York, NY) 294:858–862. https://doi.org/10.1126/science.1065062
6. Lewis BP, Burge CB, Bartel DP (2005) Conserved seed pairing, often flanked by adenosines, indicates that thousands of human genes are microRNA targets. Cell 120:15–20. https://doi.org/10.1016/j.cell.2004.12.035
7. Grimson A, Farh KK-H, Johnston WK et al (2007) MicroRNA targeting specificity in mammals: determinants beyond seed pairing. Mol Cell 27:91–105. https://doi.org/10.1016/j.molcel.2007.06.017
8. Shin C, Nam J-W, Farh KK-H et al (2010) Expanding the MicroRNA targeting code: functional sites with centered pairing. Mol Cell

38:789–802. https://doi.org/10.1016/j.molcel.2010.06.005
9. Auyeung VC, Ulitsky I, McGeary SE, Bartel DP (2013) Beyond secondary structure: primary-sequence determinants license Pri-miRNA hairpins for processing. Cell 152:844–858. https://doi.org/10.1016/j.cell.2013.01.031
10. Bartel DP (2009) MicroRNAs: target recognition and regulatory functions. Cell 136:215–233. https://doi.org/10.1016/j.cell.2009.01.002
11. O'Brien J, Hayder H, Zayed Y, Peng C (2018) Overview of microRNA biogenesis, mechanisms of actions, and circulation. Front Endocrinol 9:843–12. https://doi.org/10.3389/fendo.2018.00402
12. Calin GA, Croce CM (2006) MicroRNA signatures in human cancers. Nature Rev Cancer 6:857–866. https://doi.org/10.1038/nrc1997
13. Croce CM (2009) Causes and consequences of microRNA dysregulation in cancer. Nat Rev Genet 10:704–714. https://doi.org/10.1038/nrg2634
14. Christopher AF, Kaur RP, Kaur G et al (2016) MicroRNA therapeutics: discovering novel targets and developing specific therapy. Perspect Clin Res 7:68–74. https://doi.org/10.4103/2229-3485.179431
15. Menon A, Abd-Aziz N, Khalid K et al (2022) miRNA: a promising therapeutic target in cancer. Int J Mol Sci 23:11502. https://doi.org/10.3390/ijms231911502
16. Diener C, Keller A, Meese E (2022) Emerging concepts of miRNA therapeutics: from cells to clinic. Trends Genet 38:613–626. https://doi.org/10.1016/j.tig.2022.02.006
17. O'Brien K, Breyne K, Ughetto S et al (2020) RNA delivery by extracellular vesicles in mammalian cells and its applications. Nat Rev Mol Cell Biol 21:585–606. https://doi.org/10.1038/s41580-020-0251-y
18. Conti I, Varano G, Simioni C et al (2020) miRNAs as influencers of cell–cell communication in tumor microenvironment. Cells 9:220. https://doi.org/10.3390/cells9010220
19. Mansoori B, Baradaran B, Nazari A et al (2022) MicroRNAs in the cancer cell-to-cell communication: an insight into biological vehicles. Biomed Pharmacother 153:113449. https://doi.org/10.1016/j.biopha.2022.113449
20. Buzas EI (2023) The roles of extracellular vesicles in the immune system. Nat Rev Immunol 23:236–250. https://doi.org/10.1038/s41577-022-00763-8
21. Jiang J, Mei J, Ma Y et al (2022) Tumor hijacks macrophages and microbiota through extracellular vesicles. Exploration 2:20210144. https://doi.org/10.1002/exp.20210144
22. Zhou M, He X, Mei C, Ou C (2023) Exosome derived from tumor-associated macrophages: biogenesis, functions, and therapeutic implications in human cancers. Biomark Res 11:100. https://doi.org/10.1186/s40364-023-00538-w
23. Kumar MA, Baba SK, Sadida HQ et al (2024) Extracellular vesicles as tools and targets in therapy for diseases. Signal Transduct Target Ther 9:27. https://doi.org/10.1038/s41392-024-01735-1
24. Zhou X, Li X, Wu M (2018) miRNAs reshape immunity and inflammatory responses in bacterial infection. Signal Transduct Target Ther 3:14. https://doi.org/10.1038/s41392-018-0006-9
25. Acuña SM, Floeter-Winter LM, Muxel SM (2020) MicroRNAs: biological regulators in pathogen–host interactions. Cells 9:113. https://doi.org/10.3390/cells9010113
26. Motameny S, Wolters S, Nürnberg P, Schumacher B (2010) Next generation sequencing of miRNAs – strategies, resources and methods. Genes 1:70–84. https://doi.org/10.3390/genes1010070
27. Kolanowska M, Kubiak A, Jażdżewski K, Wójcicka A (2018) MicroRNA analysis using next-generation sequencing. Methods Mol Biol (Clifton, NJ) 1823:87–101. https://doi.org/10.1007/978-1-4939-8624-8_8
28. Benesova S, Kubista M, Valihrach L (2021) Small RNA-sequencing: approaches and considerations for miRNA analysis. Diagnostics 11:964. https://doi.org/10.3390/diagnostics11060964
29. Chen L, Heikkinen L, Wang C et al (2019) Trends in the development of miRNA bioinformatics tools. Brief Bioinform 20:1836–1852. https://doi.org/10.1093/bib/bby054
30. Veneziano D, Nigita G, Ferro A (2015) Computational approaches for the analysis of ncRNA through deep sequencing techniques. Front Bioeng Biotechnol 3:77. https://doi.org/10.3389/fbioe.2015.00077
31. Telonis AG, Loher P, Jing Y et al (2015) Beyond the one-locus-one-miRNA paradigm: microRNA isoforms enable deeper insights into breast cancer heterogeneity. Nucleic Acids Res 43:9158–9175. https://doi.org/10.1093/nar/gkv922
32. McCall MN, Kim M-S, Adil M et al (2017) Toward the human cellular microRNAome. Genome Res 27:1769–1781. https://doi.org/10.1101/gr.222067.117
33. Desvignes T, Loher P, Eilbeck K et al (2019) Unification of miRNA and isomiR research: the mirGFF3 format and the mirtop API. Bioinformatics 36:698–703. https://doi.org/10.1093/bioinformatics/btz675
34. Tomasello L, Distefano R, Nigita G, Croce CM (2021) The microRNA family gets wider: the isomiRs classification and role. Front Cell Dev

Biol 9:668648. https://doi.org/10.3389/fcell.2021.668648
35. Distefano R, Tomasello L, Vinciguerra GLR et al (2022) Pan-cancer analysis of canonical and modified miRNAs enhances the resolution of the functional miRNAome in cancer. Cancer Res 82:3687–3700. https://doi.org/10.1158/0008-5472.can-22-0240
36. Patil AH, Baran A, Brehm ZP et al (2022) A curated human cellular microRNAome based on 196 primary cell types. Gigascience 11:giac083. https://doi.org/10.1093/gigascience/giac083
37. Chakraborty C, Bhattacharya M, Agoramoorthy G (2020) Single-cell sequencing of miRNAs: a modified technology. Cell Biol Int 44:1773–1780. https://doi.org/10.1002/cbin.11376
38. Tang F, Barbacioru C, Wang Y et al (2009) mRNA-seq whole-transcriptome analysis of a single cell. Nat Methods 6:377–382. https://doi.org/10.1038/nmeth.1315
39. DeLaughter DM, Bick AG, Wakimoto H et al (2016) Single-cell resolution of temporal gene expression during heart development. Dev Cell 39:480–490. https://doi.org/10.1016/j.devcel.2016.10.001
40. Macosko EZ, Basu A, Satija R et al (2015) Highly parallel genome-wide expression profiling of individual cells using nanoliter droplets. Cell 161:1202–1214. https://doi.org/10.1016/j.cell.2015.05.002
41. Bhaduri A, Lullo ED, Jung D et al (2020) Outer radial glia-like cancer stem cells contribute to heterogeneity of glioblastoma. Cell Stem Cell 26:48–63. https://doi.org/10.1016/j.stem.2019.11.015 e6
42. Lambrechts D, Wauters E, Boeckx B et al (2018) Phenotype molding of stromal cells in the lung tumor microenvironment. Nat Med 24:1277–1289. https://doi.org/10.1038/s41591-018-0096-5
43. Galen Pvan, Hovestadt V, MHW II et al (2019) Single-cell RNA-seq reveals AML hierarchies relevant to disease progression and immunity. Cell 176:1265–1281. https://doi.org/10.1016/j.cell.2019.01.031 e24
44. Treiber T, Treiber N, Meister G (2019) Regulation of microRNA biogenesis and its crosstalk with other cellular pathways. Nat Rev Mol Cell Biol 20:5–20. https://doi.org/10.1038/s41580-018-0059-1
45. Han J, Lee Y, Yeom K-H et al (2004) The Drosha-DGCR8 complex in primary microRNA processing. Genes Dev 18:3016–3027. https://doi.org/10.1101/gad.1262504
46. Wu K, He J, Pu W, Peng Y (2018) The role of exportin-5 in MicroRNA biogenesis and cancer. Genom, Proteom Bioinform 16:120–126. https://doi.org/10.1016/j.gpb.2017.09.004
47. Hutvágner G, McLachlan J, Pasquinelli AE et al (2001) A cellular function for the RNA-interference enzyme Dicer in the maturation of the let-7 small temporal RNA. Science (New York, NY) 293:834–838. https://doi.org/10.1126/science.1062961
48. Cullen BR (2004) Transcription and processing of human microRNA precursors. Mol Cell 16:861–865. https://doi.org/10.1016/j.molcel.2004.12.002
49. Jonas S, Izaurralde E (2015) Towards a molecular understanding of microRNA-mediated gene silencing. Nat Rev Genet 16:421–433. https://doi.org/10.1038/nrg3965
50. Lytle JR, Yario TA, Steitz JA (2007) Target mRNAs are repressed as efficiently by microRNA-binding sites in the 5′ UTR as in the 3′ UTR. Proc Natl Acad Sci USA 104:9667–9672. https://doi.org/10.1073/pnas.0703820104
51. Tay Y, Zhang J, Thomson AM et al (2008) MicroRNAs to Nanog, Oct4 and Sox2 coding regions modulate embryonic stem cell differentiation. Nature 455:1124–1128. https://doi.org/10.1038/nature07299
52. Bustos Y, Dorrance AM, Goda C (2025) Stem cell niche, methods and protocols. Methods Mol Biol (Clifton, NJ) 2939:55–63. https://doi.org/10.1007/7651_2025_630
53. Durruthy-Durruthy R, Ray M (2018) Disease gene identification, methods and protocols. Methods Mol Biol 1706:199–221. https://doi.org/10.1007/978-1-4939-7471-9_11
54. Pan Q, Shai O, Lee LJ et al (2008) Deep surveying of alternative splicing complexity in the human transcriptome by high-throughput sequencing. Nat Genet 40:1413–1415. https://doi.org/10.1038/ng.259
55. Ramsköld D, Luo S, Wang Y-C et al (2012) Full-length mRNA-seq from single-cell levels of RNA and individual circulating tumor cells. Nat Biotechnol 30:777–782. https://doi.org/10.1038/nbt.2282
56. Pfeffer S, Lagos-Quintana M, Tuschl T (2005) Cloning of small RNA molecules. Curr Protoc Mol Biol 72:26.4.1–26.4.18. https://doi.org/10.1002/0471142727.mb2604s72
57. Faridani OR, Abdullayev I, Hagemann-Jensen M et al (2016) Single-cell sequencing of the small-RNA transcriptome. Nat Biotechnol 34:1264–1266. https://doi.org/10.1038/nbt.3701
58. Yehudai-Resheff S, Schuster G (2000) Characterization of the E.coli poly(A) polymerase: nucleotide specificity, RNA-binding affinities and RNA structure dependence. Nucleic Acids Res 28:1139–1144. https://doi.org/10.1093/nar/28.5.1139
59. Liu Z, Wang T, Yang X et al (2022) Polyadenylation ligation-mediated sequencing

(PALM-Seq) characterizes cell-free coding and non-coding RNAs in human biofluids. Clin Transl Med 12:e987. https://doi.org/10.1002/ctm2.987

60. Salmen F, Jonghe JD, Kaminski TS et al (2022) High-throughput total RNA sequencing in single cells using VASA-seq. Nat Biotechnol 40:1780–1793. https://doi.org/10.1038/s41587-022-01361-8
61. Khnouf R, Shore S, Han CM et al (2018) Efficient production of on-target reads for small RNA sequencing of single cells using modified adapters. Anal Chem 90:12609–12615. https://doi.org/10.1021/acs.analchem.8b02773
62. Maguire S, Lohman GJS, Guan S (2020) A low-bias and sensitive small RNA library preparation method using randomized splint ligation. Nucleic Acids Res 48:e80. https://doi.org/10.1093/nar/gkaa480
63. Yang Q, Li R, Lyu Q et al (2019) Single-cell CAS-seq reveals a class of short PIWI-interacting RNAs in human oocytes. Nat Commun 10:3389. https://doi.org/10.1038/s41467-019-11312-8
64. Zhuang F, Fuchs RT, Sun Z et al (2012) Structural bias in T4 RNA ligase-mediated 3′-adapter ligation. Nucleic Acids Res 40:e54. https://doi.org/10.1093/nar/gkr1263
65. Raabe CA, Tang T-H, Brosius J, Rozhdestvensky TS (2014) Biases in small RNA deep sequencing data. Nucleic Acids Res 42:1414–1426. https://doi.org/10.1093/nar/gkt1021
66. Li J, Zhang Z, Zhuang Y et al (2023) Small RNA transcriptome analysis using parallel single-cell small RNA sequencing. Sci Rep 13:7501. https://doi.org/10.1038/s41598-023-34390-7
67. Xiao Z, Cheng G, Jiao Y et al (2018) Holo-Seq: single-cell sequencing of holo-transcriptome. Genome Biol 19:163. https://doi.org/10.1186/s13059-018-1553-7
68. Wang N, Zheng J, Chen Z et al (2019) Single-cell microRNA-mRNA co-sequencing reveals non-genetic heterogeneity and mechanisms of microRNA regulation. Nat Commun 10:95. https://doi.org/10.1038/s41467-018-07981-6
69. Ferlita AL, Alaimo S, Nigita G et al (2024) tRFUniverse: a comprehensive resource for the interactive analyses of tRNA-derived ncRNAs in human cancer. iScience 27:108810. https://doi.org/10.1016/j.isci.2024.108810
70. Cherlin T, Jing Y, Shah S et al (2024) The subcellular distribution of miRNA isoforms, tRNA-derived fragments, and rRNA-derived fragments depends on nucleotide sequence and cell type. BMC Biol 22:205. https://doi.org/10.1186/s12915-024-01970-6
71. Patil AH, Halushka MK (2021) miRge3.0: a comprehensive microRNA and tRF sequencing analysis pipeline. NAR Genom Bioinform 3:lqab068. https://doi.org/10.1093/nargab/lqab068
72. Engel A, Rishik S, Hirsch P et al (2024) SingmiR: a single-cell miRNA alignment and analysis tool. Nucleic Acids Res 52:W374–W380. https://doi.org/10.1093/nar/gkae225
73. Baysoy A, Bai Z, Satija R, Fan R (2023) The technological landscape and applications of single-cell multi-omics. Nat Rev Mol Cell Biol 24:1–19. https://doi.org/10.1038/s41580-023-00615-w
74. Maji RK, Leisegang MS, Boon RA, Schulz MH (2025) Revealing microRNA regulation in single cells. Trends Genet 41:522–536. https://doi.org/10.1016/j.tig.2024.12.009

Chapter 14

The Yenos/MinION Multi-Cancer Early Detection Liquid Biopsy Test: An Innovative Approach to MicroRNA Quantification

Anastassia Kanavarioti

Abstract

The MinION from Oxford Nanopore Technologies was repurposed to detect and quantify microRNAs (miRNAs) in total RNA isolated from serum or urine samples. Detection and quantification were enabled with a proprietary probe, complementary to the target miRNA, enhanced with osmium tags. The assay yields absolute miRNA copies, at the femtomolar (fM) level, per 1 μL of total RNA isolate with up to ± 20% accuracy. This unprecedented accuracy is the result of a bracketing experimental protocol where one experiment is designed to include about 20% more probe and the other experiment about 20% less probe than the target miRNA. Several miRNAs, quantified in a commercial combined serum from healthy men and in the urine of healthy volunteers, exhibited comparable copy numbers after normalization to the same total RNA content. These observations refute the belief that the collection method and the subject's demographics cause sample variability. Our ready-to-use platform discriminated healthy from cancer samples based on miRNA overexpression. Preliminary results suggested that known cancer biomarkers, such as miR-21, miR-375 and miR-141, are approximately 1.8-fold elevated in prostate, pancreatic, breast, ovarian, colorectal, and lung cancer indications than in healthy samples. The latter led us to exploit 1.5-fold the healthy level (1.5 HL) as the threshold to discriminate healthy from cancer samples with zero data overlap (the Yenos test). Materials and methods to conduct this test are described below.

Keywords The Yenos test, MicroRNA quantification, Liquid biopsies, MinION, Nanopore-array, Single molecule ion-conductance, Osmylated ss nucleic acid probes

1 Introduction

Biological fluids, called liquid biopsies, have attracted worldwide attention and are viewed as potential non-invasive sources for early cancer/disease detection [1, 2]. Liquid biopsies are known to carry short single-stranded (ss) nucleic acids, such as miRNAs [3–5], circulating tumor DNA (ctDNA), fragmented tRNA, long noncoding RNA, which have been evaluated in numerous medical studies as cancer/disease biomarkers. However, their low abundance in biological fluids presented an analytical challenge, which has prevented

Alessandro Laganà and Giovanni Nigita (eds.), *MicroRNA Target Identification*, Methods in Molecular Biology, vol. 3056,
https://doi.org/10.1007/978-1-0716-5392-0_14,

their validation and utilization in diagnostics. Current analytical technologies to detect and quantify fM-level miRNAs rely mostly on PCR amplification, which works well for detection but not for quantification [6]. The problem is exacerbated due to the absence of a reliable normalization method. Medical studies comparing miRNA levels in the serum of cancer patients and of healthy individuals yielded mean values to suggest miRNA overexpression in the cancer samples [7]. Nevertheless, the actual data exhibited strong overlap between healthy and cancer samples [7], which brought uncertainty to the miRNA field and prohibited FDA approvals.

To address these issues, we developed, optimized, and internally validated a novel analytical platform (see Fig. 1) [8, 9]. We used this platform to detect and quantify selected miRNAs in serum and urine samples from healthy individuals and treatment-naïve cancer patients [10–12]. This novel technology is amplification-free and exploits the MinION device from Oxford Nanopore Technologies (ONT),

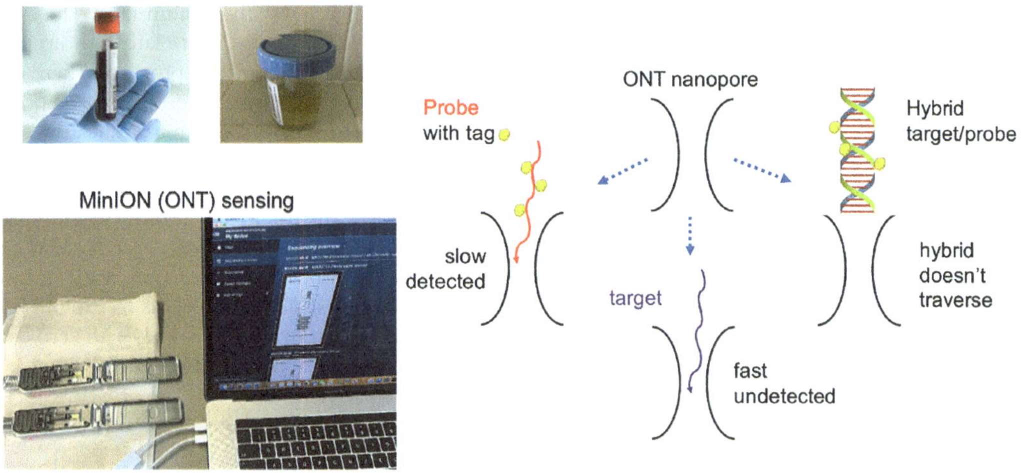

Fig. 1 Graphical abstract of the Yenos test. Left, top: collection of the biospecimen, blood or urine. Left, bottom: Running two nanopore experiments simultaneously using one laptop and two MinION devices/flow cells. Right, overview of the detection/quantification using nanopore detection and a Yenos probe, which is an oligonucleotide complementary to the miRNA target and enhanced with osmium tags. The experiment measures the ion current (I) in picoAmperes (pA) as a function of time (t) in milliseconds (ms). In practice, I is constant at I_o, which is the open nanopore ion current (I_o) due to the presence of the electrolyte supplied by the ONT buffer. When a single molecule traverses the nanopore, I_o is reduced to a new value, I_r, because the molecule occupies the space that would have been occupied by the electrolyte. Ion current reduction (dip in this platform) lasts for a time, τ. Both I_r and τ depend on the characteristics of the traversing molecule. During the experiment, data were acquired by the MinKNOW software and stored in a *fast-5* file. Upon experiment completion, the *fast-5* file is read/analyzed by *OsBp_detect* software [9]. The analysis determines whether the free probe is in excess and detected (left on the scheme above) or if the probe is not detected because it is fully hybridized with the target (right on the scheme above). The Yenos probes traverse much slower compared to intact ss nucleic acids and are selectively detected by the relatively slow acquisition rate of this platform. The hybrid is too big, cannot fit through these nanopores and does not get detected. This platform reverses voltage automatically to prevent the nanopores from being blocked by the hybrids

which is currently the only commercially available nanopore-array device [13]. The MinION is promoted and supported by ONT for enzyme-assisted sequencing of long DNA/RNA. We repurposed it to conduct unassisted, voltage-driven, ion-conductance experiments and used it to measure absolute copies of miRNAs in serum and urine. This was accomplished by first isolating total RNA from the biospecimen using commercial kits. Notably, the concentration of many miRNAs is at the fM level in serum, while these miRNAs are about 50-fold lower in urine. However, the corresponding total RNA isolation kits for serum and urine yield similar RNA content. Detection and quantification (sensing) of a target miRNA is enabled via an osmium-tagged proprietary probe complementary to the miRNA target. The Yenos proprietary probes were developed and optimized for efficient hybridization with the target and preferential nanopore detection [8, 10]. The osmium label is responsible for the preferential detection of the probe by the ONT nanopores over the RNA background. It was observed that liquid biopsies vary substantially in total RNA content, but miRNA copies were proportional to the total RNA content. Hence, we instituted normalization of the observed miRNA copies in a biospecimen to the same RNA content (16 ng/μL), which happens to be close to the RNA content of several lots of a commercial combined serum (see Materials) used by us extensively during the development stage [10–12].

The MinKNOW software, downloaded from the ONT website, is used in conjunction with a state-of-the-art laptop to run the experiments. The data, ion current (I) as a function of time, are acquired as a *fast-5* file. The latter is analyzed by our publicly available software tool, *OsBp-detect* [9] to select the events corresponding to single-molecule full translocations. Quantitative comparison between control and sample experiments answers the question of whether the probe copies are more or less than the target copies in the tested sample. Quantification is based on the known concentration of the probe and the 1:1 hybridization between target miRNA and its probe. Notably the MinION proprietary nanopores permit the translocation of ss nucleic acids, as well as the Yenos probes, but prohibit the translocation of ds nucleic acids, including the hybrid between miRNA target and its complementary probe (see scheme in Fig. 1). Probe detection implies that probe copies are more than the target miRNA copies, while no probe detection (silencing) implies that probe copies are fewer than the target miRNA copies in the tested sample. Experiments for miRNA copy number determination are designed to differ by about 40% in one of the components, RNA isolate or probe. This bracketing experimental design returns miRNA copy number from the average of two such experiments, one yielding detection and the other yielding silencing. This experimental approach is low throughput, and more than two experiments are required to quantify a single miRNA. It may be used to measure overexpression or underexpression and validate a miRNA cancer/disease biomarker. The technology may be extended to other ss nucleic acids, such as ctDNA, long ncRNA, when a part of the target sequence is known so that the complementary probe can be designed. To the best of our knowledge, this is the first analytical platform using a bracketing approach with a protocol-defined accuracy; it yields the same accuracy across all miRNA concentrations at the levels found in liquid biopsies.

2 Materials and Equipment

2.1 Materials

Register with ONT and purchase MinION device(s), DNA flow cells and the Flow cell Priming Kit XL (EXP-FLP004-XL). RNA flow cells are not suitable for this method.

Download the MinKNOW software and test the flow cells upon arrival.

Order custom-made probes, complementary to target miRNAs, from Yenos Analytical LLC.

Purchase the Monarch T2010S kit from New England BioLabs and use it for the isolation of total RNA from blood/serum, and the slurry kit No. 29600 from Norgen Biotek Corp. and use it for the isolation of total RNA from urine. Follow manufacturer's kit instructions.

Use Ambion nuclease-free water, not treated with DEPC.

Use sterile, RNase-free test tubes for sample preparation.

Practice conducting nanopore experiments using human serum, combined from healthy men, purchased from Sigma–Aldrich (cat no H6914) (see Table 1 and *see* **Notes 1–4**).

2.2 Equipment

Use a Laptop MacBook Pro, preferably M4 or M5, loaded with the MinKNOW™ software from the ONT website and a MinION Mk1D with flow cells FLO-MIN114, both purchased from ONT. The Oxford Nanopore Technologies® MinION™ is an electronic device that provides the interface between the user's computer and the nanopore sensor array (see Fig. 1). The MinION powers the application-specific integrated circuit (ASIC), performs temperature control and transfers data to the laptop through a single USB 3.0 port. A state-of-the-art laptop can conduct simultaneously two nanopore experiments. The MinKNOW™ software controls the device, experimental scripts and acquires the data, as a *fast*-5 file.

Table 1
miRNA copies (HL_{miRNA}) per 1 μL of total RNA [8, 10–12] or 2 μL serum in H6914

miR-15b	Let7-b	miR-21-5p	miR-375-3p	miR-141-3p
17,710	12,150	10,494	9,240	6,096

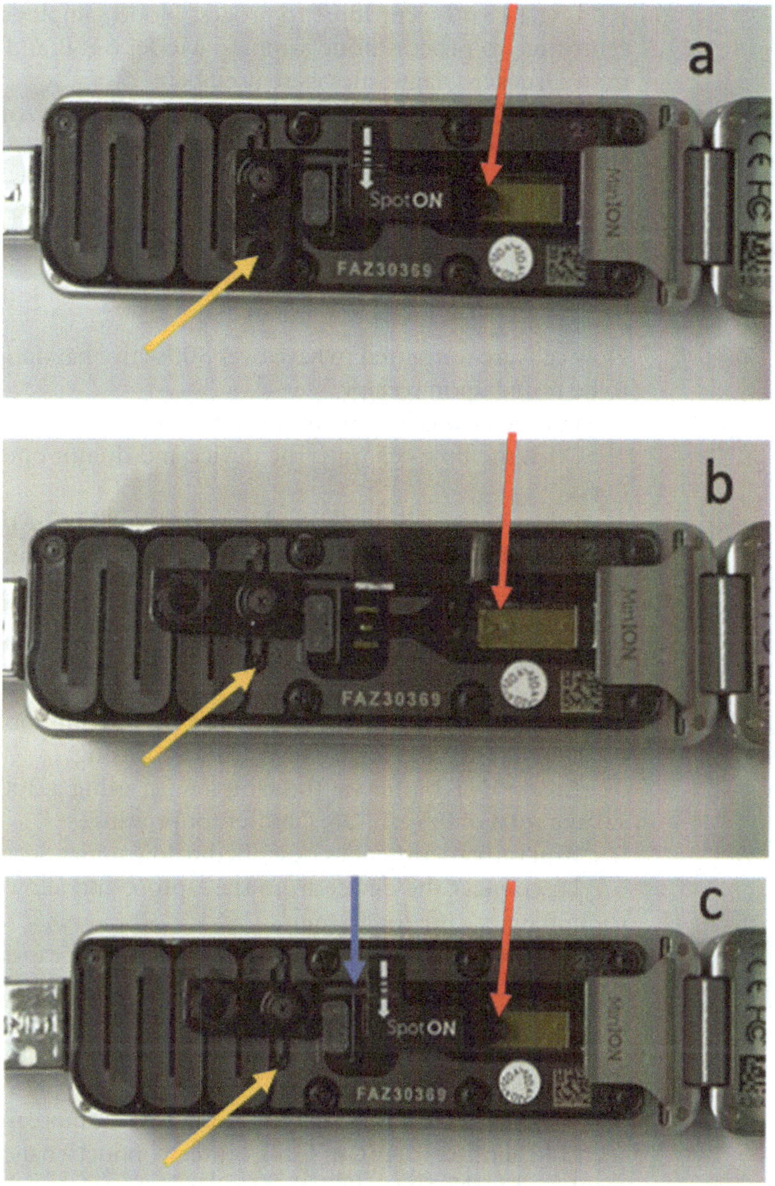

Fig. 2 Flow cell usage to conduct a sensing experiment. Sample port (red arrow), priming port (yellow arrow) and point up to which excess buffer/sample is removed (blue arrow). (**a**) Both ports are closed when flow cell is not in use. (**b**) Both ports are open to load buffer or sample. (**c**) Sample port is closed, while priming port is open and used to remove excess solution up to the point indicated by the blue arrow. See text under flow cell priming and conducting experiment

Load the *OsBp_detect* software from Github [9] to analyze the *fast-5* file.

Use a microcentrifuge in conjunction with the total RNA isolation kits.

Use a 15 mL centrifuge to process blood samples and a 50 mL centrifuge to process urine samples. Model 642B and Horizon24 from Drucker Diagnostics were used here.

Use a DS-11 Nanodrop spectrophotometer from DeNovix Corp., or comparable, to measure total RNA in the isolate.

3 Methods

3.1 Testing Flow Cell Warranty

Flow cells are warranted to have over 800 active nanopores and need to be tested upon receipt.

Let the flow cell come to room temperature before testing.

Load the flow cell onto the device and do not open any of the ports.

Select "flow cell check" and flow cell type FLO-MIN114.

If the flow cell check reports over 800 active nanopores, it is acceptable.

If the flow cell check reports below 800 active nanopores, contact ONT and request a replacement.

3.2 Priming a New Flow Cell (see Fig. 2 and see Notes 1–4)

Thaw ONT buffer and centrifuge it at 14,000 rpm for 1 min ahead of using it. Take the supernatant without mixing.

Load ONT buffer via the sample port using a 200 μL pipette. Load a total of 0.5 mL ONT buffer for priming.

Run the experiment as described below under Sect. 3.3.

Disconnect the device from the laptop, and allow it to stay at room temperature for 15 min.

Connect the device with the laptop and run the experiment again. Do not open the ports.

Analyze the *fast-5* files as described below under Sect. 3.3.

If the second experiment reports total events less than 700,000, the flow cell is ready for use.

If the second experiment reports total events more than 700,000, then add another 0.1 mL of ONT buffer, conduct a nanopore experiment, and analyze it to show that total events are fewer than 700,000.

3.3 Conducting a Yenos/MinION Sensing Experiment (see Fig. 2 and see Notes 1–4)

Keep both sample port and priming port closed when not in use.

When ready to use, open both ports (Fig. 2).

Load the buffer or sample via the sample port using a 200 μL pipette. Close the sample port.

Remove excess of solution using the priming port and a 1 mL pipette up to the blue arrow level.

Close the priming port.

Use "start sequencing" mode.

Select flow cell type FLO-MIN114, select "ligation sequencing kit (SQK-LSK114), PCR Barcoding Expansion 1–12(EXP-PBC001).

These kits are only selected to satisfy computer requirements, and other selections may work too. None of the kits, besides the flow cell priming kit, is required for the Yenos/MinION platform to work.

Disable "base calling," select run length 45 min (0.75 h) and select "output bulk file Raw (1–512).

Data Analysis of a Sensing Experiment Using the OsBp_detect Algorithm

Upon completion of the experiment, a *fast-5* file is automatically generated and found in Library/MinKNOW/data.

Load the *OsBp_detect* commands sequentially into the laptop's terminal:

cd osbp_detect/

source.venv/bin/activate (starts virtual environment)

just run-gui (Starts the graphical user interface)

The window displays the threshold parameters: event duration (in tps): 4–1200 (1.3–400 ms); lowest *Ir/Io* <0.55; all *Ir/Io* <0.6; select channels 1–512 and maximum events 100,000.

Load the desired *fast-5* file and run the script. The output is three separate tsv files, which can all be opened

in Microsoft Excel (xlxs) and save it as such. One tsv file contains all the channels, the second tsv file contains only the channels that reported more than 100,000, and the third tsv file contains only the channels that reported less than 100,000 events. Channels that report more than 100,000 events are considered malfunctioning and may be excluded. Use the third file for the analysis of the sample. Group the data in the fourth column (*Ir/Io*) in a histogram with 0.05 bins from 0.05 to 0.55 and graph the data for better visualization.

Calculate the R value, where R = ratio of late $(Ir/Io)_{max}$ at 0.3 ± 0.05 over early $(Ir/Io)_{max}$ at 0.15 ± 0.05.

3.4 Conducting a Yenos test, i.e., a Total of Four Nanopore Experiments on a Single Flow Cell, Preferably the Same Day

Sample Preparation (see Notes 5–10)

Isolate the total RNA of the desired biospecimen and measure total RNA content (Z) in ng/μL.

Select the miRNA target and the complementary probe, and note the miRNA copy number (HL_{miRNA}) from Table 1, defined for a total RNA content $C = 16$ ng/μL.

Select a probe aliquot (Υ in μL) so that probe copies (P) are around 100,000.

$$\Upsilon = P/(\text{probe concentration (in fM)} \times 600) \quad (1)$$

To test a sample at the 1.5 HL miRNA threshold

Use X μL of total RNA isolate so that expected miRNA target copies are 1.5-fold of the reference miRNA copy number (see Table 1) normalized to 16 ng/μL RNA content

$$X = P \times C/(Z \times 1.5\ \mathrm{HL}_{\mathrm{miRNA}}) \quad (2)$$

See Fig. 3d as an example: Biospecimen with total RNA Z = 12.1 ng/μL, testing for miR-375 with $\mathrm{HL}_{\mathrm{miR\text{-}375}}$ = 9,240 copies per 1 μL total RNA at C = 16 ng/μL and using 4 μL probe375 at 41.6 fM concentration requires $X = (4 \times 41.6 \times 600) \times 16/(12.1 \times 1.5 \times 9240)$ = 9.53 μL of the total RNA of this biospecimen.

Run ONT Buffer Twice

Load 75 μL of ONT buffer as described under Sect. 3.3

Between runs disconnect the MinION and let it stand at room temperature for 15 min. Do not open the ports between duplicate experiments.

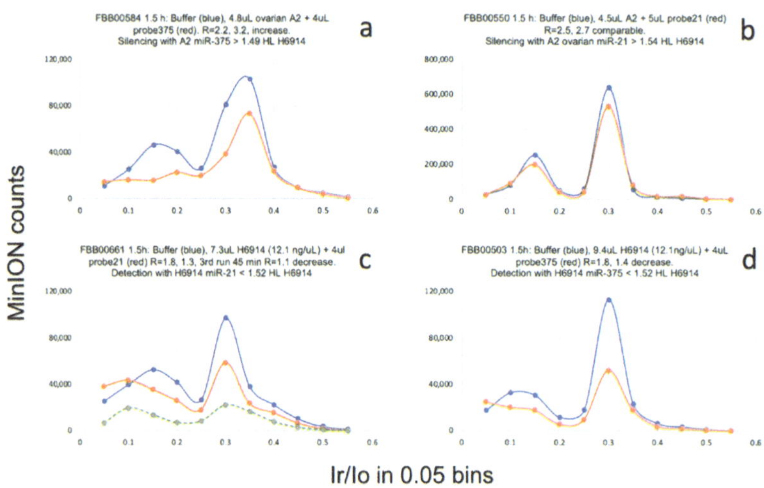

Fig. 3 Yenos tests analyzed by *OsBp_detect. Graph of Minion counts (selected events) as a function of Ir/Io bin.* Each experiment is run for 45 min at −180 mV on the same flow cell. Two consecutive experiments are run with buffer only, analyzed using *OsBp_detect* and the data summed up to make for a 1.5 h experiment (see blue trace in figures). The test sample is added next and run for 45 min at −180mV twice. Similarly, the data of the two experiments with the test sample are summed up to mimic a 1.5 h experiment (red trace). values of late/early $(Ir/Io)_{max}$ are calculated and compared (see under Methods). For confirmation, a third experiment can be conducted while the sample is still in the flow cell (green dashed trace in panel c). (**a**) Sample: 4.8 μL of total RNA isolated from a serum sample of a patient with ovarian cancer (A2, 24.2 ng/μL) mixed with 4 μL of probe375 and 75 μL ONT buffer. (**b**) Sample: 4.5 μL of total RNA isolated from a serum sample of a patient with ovarian cancer (A2, 24.2 ng/μL) mixed with 5 μL of probe21 and 7 5μL ONT buffer. (**c**) Sample: 7.3 μL of total RNA isolated from serum H6914 (12.1 ng/μL) mixed with 4 μL probe21 and 75 μL ONT buffer. (**d**) Sample: 9.4 μ L total RNA isolated from serum H6914 (12.1 ng/μL) mixed with 4 μL probe375 and 75 μL ONT buffer. The *R* values and the result of the test (detection or silencing) are included in the title of each experiment (see under Methods)

Run the Sample Twice (see *Notes 11 and 12*)

Prepare the sample as described under Sect. 3.4 and add 75 µL ONT buffer.

Wait 30 min after running the second buffer experiment (see Sect. 3.4) while keeping the flow cell at room temperature.

Open both ports and load all the sample from the sample port with a 200 µL pipette.

Close the sample port.

Remove excess solution from the priming port with a 1 mL pipette up to the blue arrow level (Fig. 2).

Close both ports, connect the device to the laptop and run the first experiment with the sample.

Upon completion of the first sample run, disconnect the device, keep it at room temperature for 15 min, reconnect and conduct the second sample run. Do not open the ports.

Analyze the *fast-5* files as described under Sect. 3.3.

The Yenos Test Result (see Fig. 3 and see *Note 13)*

If the four above experiments (the Yenos test) yield a probe detection conclusion, then target miRNA measures less than 1.5 HL, while if they yield probe silencing, the target miRNA measures more than 1.5 HL.

Compare the four *R* values obtained from the two ONT buffer experiments and the two sample experiments. Lower or higher means here that values, $R(\text{sample})/R(\text{buffer})$, differ by more than ±17%.

- If all four *R* values are comparable, the test suggests silencing.
- If both buffer *R* values are comparable and at least one of the sample experiments exhibits an *R* value lower than the *R* values of the buffer, the test suggests detection.
- If the buffer *R* values are comparable and one sample *R* value is significantly higher than the buffer *R* values, but the second sample *R* value is significantly lower than the buffer *R* values, then the test is inconclusive.
- If both buffer *R* values are comparable and one of the sample's *R* values is comparable to the buffer *R*, and the other significantly higher than the buffer *R* value, then the test suggests silencing.
- If the buffer *R* values are significantly different from each other, combine the two buffer experiments as one 1.5 h experiment and the two-sample experiment also as one 1.5 h experiment and compare the resulting two *R* values. Comparable *R* values, or sample *R* value higher than buffer *R* value, suggest silencing. If sample *R* value is significantly lower than buffer *R* value, then the test is a detection one.
- If the four experiments do not yield a firm conclusion, conduct a third sample run, the next day is ok, and compare the new *R* value to the other four.

Quality Control of the Yenos Test

- Flow cells with less than 200 active nanopores do not yield reliable experiments.
- A Yenos test where a sample run exhibits significantly more events than the buffer experiment run ahead of it should be disregarded.
- Similarly, a Yenos test where the sample run exhibits dramatically, let us say more than 50%, fewer events compared to the buffer run ahead of it, should be disregarded.
- Experiments where the $(Ir/Io)_{max}$ values differ by more than ± 0.05 from 0.3 and/or 0.15 should be disregarded.

3.5 Confirm miRNA Copy Number in a Total RNA Sample Isolated from a Biospecimen

Conduct, at least, two Yenos tests, as described under Sect. 3.4.

If the sample is healthy or if the sample is from a cancer patient but the miRNA target is not a cancer biomarker, use Eq. 2 at a threshold of 1.3 HL_{miRNA} for the first Yenos test and Eq. 2 at a threshold of 0.7 HL_{miRNA} for the second test.

If the sample is obtained from a cancer patient and the target miRNA is a cancer biomarker, use Eq. 2 at a threshold of 1.5 HL_{miRNA} for the first Yenos test and Eq. 2 at a threshold of 2.2 HL_{miRNA} for the second test.

3.6 Measure miRNA Copy Number in a Total RNA Sample Isolated from any Biospecimen

Conduct multiple Yenos tests, as described under Sect. 3.4, using a broad range of HL_{miRNA}, before narrowing the range in order to obtain miRNA copy number with ±20% accuracy.

Note that the total RNA sample should be diluted with water, if the specific miRNA target measures higher than 30,000 copies per 1 μL of total RNA. Similarly, biospecimens that exhibit high total RNA content may be diluted with water to about 25 ng/μL.

4 Notes

1. It is imperative to practice loading sample from the sample port without entering air into the nanopore array. Loading of the sample for the ion-conductance nanopore experiment follows a different protocol than the one proposed for the ONT-promoted sequencing experiments. Air/bubbles cannot be removed and deactivate the nanopores. The best way to add a solution is to hold the pipette on top of the sampling port and let three to four drops come out before setting the pipette into the groove of the sampling port. Loading should be relatively slow but take less than 15 s for a 75 μL load. Loading buffer or sample into the flow cell should be done while the flow cell is in the device but disconnected from

power. Care should be taken that spilling does not happen while loading. Flow cells may be used until they have about 200 active pores. Only sequential experiments are to be compared. Flow cells come with the nanopore array filled with a liquid that contains buffer as well as proprietary molecules that traverse the nanopores and give detectable events, which we call noise for simplification. Loading the ONT proprietary buffer reduces this noise.

2. Ports tend to dry out and get plugged. Usually, the problem disappears when one persists by keeping the pressure without forcing the liquid through. This wets the dry port and permits loading.
3. The ONT buffer is not optimized for sensing. The ONT buffer has a diluting effect on the solution that is present inside the flow cell. Passing too much ONT buffer depletes the "noise," but also reduces the pore activity. Therefore, cleaning the flow cell after a sample is strongly discouraged. Currently, the flow cell can support up to 12 (45 min) sensing experiments. This is sufficient for priming a flow cell and running 2–3 Yenos tests.
4. It is recommended to keep a flow cell in the device, disconnected from the laptop, at room temperature, and use it for up to 10 days. If experiments are not planned accordingly, the flow cell can stay in the refrigerator at 4°C–8°C. Flow cells should be primed a day or two before running experiments. If the priming is satisfactory (see under Sect. 3.2.), the first Yenos test can be done by using a single buffer experiment instead of two.
5. It is recommended to use the same 10 μL pipette to prepare the sample by aliquoting total RNA and probe and to keep the aliquots between 3 and 10 μL to reduce sampling errors.
6. The Yenos probes are stable and should be stored at −20°C for up to a year. Thawing and freezing is not an issue, so aliquoting should be avoided.
7. Calculations are based on the actual copies of the probe. Therefore, the total volume of the sample mixture tested in the nanopore experiment is not taken into consideration. It is strongly recommended that the aliquots of the RNA isolate and probe are kept below 15 μL total volume, while the volume of ONT buffer is kept at 75 μL.
8. Contamination is not a major issue considering that samples are all made comparable by keeping the probe molecules to a certain level of about 100,000. This level is a compromise between having enough probe copies for detection above the

noise and not a lot of probe copies hybridized with the target. The hybrid is doubly negatively charged, moves toward the pores, and prevents ss molecules from translocating. The hybrid does not permanently block the nanopores because the device reverses voltage regularly.

9. Since the probe molecules appear at the early $(Ir/Io)_{max} = 0.15$, it is necessary to design the experiment by assuring that the anticipated free probe copies will be detectable over the counts observed at this early $(Ir/Io)_{max}$ in the buffer/control sample.

10. A biospecimen that yields less than 8 ng/μL total RNA is a lower limit for this technology, primarily because the RNA measurement is inaccurate. Isolates with more than 30 ng/μL total RNA should be diluted.

11. The experiment duration is 45 min. This yields a *fast*-5 file of about 10 GB and a corresponding Excel file with less than a million entries, which is the upper limit of Excel. Nevertheless, 45 min is not sufficient to let all the sample translocate. Sometimes the first 45 min experiment does not show probe translocation, but the second 45 min experiment does. It is not uncommon for a third experiment to show evidence for probe translocation. A silencing experiment looks more like a buffer experiment, only that total number of events is fewer because the hybrid blocks the nanopores partially during the duration of the experiment.

12. From the point of view of cleaning the flow cell after running a sample and using as little ONT buffer as possible, one buffer experiment would have been sufficient. However, there are 2048 nanopores and only 512 detection channels. When a new experiment is initiated, the system selects randomly one nanopore per channel. Nanopores exhibit differences among them, so that selecting one group of nanopores for one experiment and another group of nanopores for the next often yields distinct data. Therefore, it is strongly recommended to run two buffer experiments as a control, followed by, at least, two sample experiments. This strategy, together with deleting any channel that exhibits more than 100,000 events, yields the most reliable results.

13. Typically, $(Ir/Io)_{max}$ is a sharp peak; then the R value is obtained from the events at a single bin value. When $(Ir/Io)_{max}$ is a broad peak, then the R value should be obtained from the sum of the events at the bins with comparable event counts.

Conflict of Interest

Anastassia Kanavarioti is the founder, inventor and director of Yenos Analytical LLC, a company that delivers custom analytical solutions for native, synthetic, or transcribed nucleic acids and engages in the development and manufacturing of labeled/tagged nucleic acids for use in conjunction with nanopore detection and nucleic acid quantification, including miRNAs, with awarded patents for *Nanopore platform for DNA/RNA oligo detection using an osmium tagged complementary probe*, Patent No: US 11,111,527 B1, issued 9-7-2021 and US 11,427,859, issued 9-30-2022 and *Osmium Tagged Probes for Nucleic Acid Detection*, US 11,884,968, issued 01-30-2024. International patents are pending in Europe, Canada, Australia, Japan (awarded No 7857933), China, India, Brazil, and South Korea. There is no competing interest or relationship (commercial, financial, or nonfinancial) between the two companies, namely, Oxford Nanopore Technologies and Yenos Analytical LLC, and their employees.

References

1. Connal S, Cameron JM, Sala A, Brennan PM, Palmer DS, Palmer JD, Perlow H, Baker MJ (2023) Liquid biopsies: the future of cancer early detection. J Transl Med 21(1):118. https://doi.org/10.1186/s12967-023-03960-8
2. Armakolas A, Kotsari M, Koskinas J (2023) Liquid biopsies, novel approaches and future directions. Cancers 15(5):1579. https://doi.org/10.3390/cancers15051579
3. Ambros V (2004) The functions of animal microRNAs. Nature 431(7006):350–355. https://doi.org/10.1038/nature02871
4. Mitchell P et al. (2008) Circulating microRNAs as stable blood-based markers for cancer detection. Proc Natl Acad Sci U S A 105:10513–10518. https://doi.org/10.1073/pnas.0804549105.
5. Iacomino G (2023) miRNAs: the road from bench to bedside. Genes (Basel) 14(2):314. https://doi.org/10.3390/genes14020314
6. Godoy PM, Barczak AJ, DeHoff P et al. (2019) Comparison of reproducibility, accuracy, sensitivity, and specificity of miRNA quantification platforms. Cell Rep 29(12):4212–4222. https://doi.org/10.1016/j.celrep.2019.11.078 e5
7. Hindson CM, Chevillet JR, Briggs HA et al. (2013) Absolute quantification by droplet digital PCR versus analog real-time PCR. Nat Methods 10(10):1003–1005. https://doi.org/10.1038/nmeth.2633 (see Fig. 2c for miR-141 data spread and overlap)
8. Kang ASW, Bernasconi JG, Jack W, Kanavarioti A (2020) Ready-to-use nanopore platform for the detection of any DNA/RNA oligo at attomole range using an Osmium tagged complementary probe. Sci Rep 10(1):19790. https://doi.org/10.1038/s41598-020-76667-1
9. Kanavarioti A, Kang ASW See RNA(OsBp) event detection python package in a public repository. Third revision by Garret D. Grant, see https://github.com/garretdgrant/osbp_detect?tab=readme-ov-file#quick-start
10. Kanavarioti A Femtomolar-level PCR-free quantification of microRNA cancer biomarkers in serum, posted in BioRxiv, https://doi.org/10.1101/2022.12.30.522268
11. Kanavarioti A, Rehman MH, Qureshi S, Rafiq A, Sultan M (2024) High sensitivity and specificity platform to validate microRNA biomarkers in cancer and human diseases. Non-Coding RNA 10(4):42. https://doi.org/10.3390/ncrna10040042
12. Rafiq A, Kanavarioti A (2025) The potential and limitations of the MinION/Yenos platform for miRNA-enabled early cancer detection. Int J Mol Sci 26(8):3822. https://doi.org/10.3390/ijms26083822
13. Oxford Nanopore Technologies, Inc. https://nanoporetech.com

Check for updates

Chapter 15

How to Validate miRNA Targeting: Western Blot Analysis

Roma Karna and Federica Calore

Abstract

MicroRNAs (miRNAs) are non-coding RNA molecules, usually 19–22 nucleotides long, that modulate gene expression by binding to specific mRNA targets. This interaction involves both the translation process and the stability of the mRNAs, thereby affecting gene expression after transcription. Their regulatory activity is carried out by binding to complementary sequences on target mRNAs, usually at their 3′ UTR level, and resulting in their degradation or inhibition of translation. Given their ability to modulate the expression of targeted molecules, miRNAs can act either as oncogenes or tumor suppressor genes in cancer cells. Aberrant miRNA expression can lead to irregular gene expression, contributing to cancer development. Changes in miRNA expression profiles have been documented across multiple cancer types and are linked to clinical factors, such as tumor progression, metastasis, and patient survival rates. Certain miRNAs have potential as diagnostic biomarkers for early cancer detection and can serve as prognostic markers to predict patient outcomes and response to treatments. This chapter outlines the experimental procedures for validating microRNA targets through Western blotting. Western blotting, a "gold-standard" technique for protein analysis, offers a reliable approach for validating the impact of miRNAs on their target genes by assessing protein expression levels [1]. This validation is essential for understanding the roles miRNAs play in health and disease.

Keywords microRNA (miRNA), Western blotting, Protein expression analysis, Post-transcriptional regulation

1 Introduction

Western blotting, also known as protein blotting, immunoblotting, or simply Western blot, is a foundational technique in cell and molecular biology for assessing protein expression levels. Originally introduced by George Stark at Stanford University in 1979, this technique comprises several key steps: (1) *Protein separation*: Proteins are separated on a polyacrylamide gel, typically SDS-PAGE. (2) *Transfer to Membrane*: The proteins that have been separated are subsequently transferred from the gel onto a membrane, which is usually composed of either nitrocellulose or polyvinylidene fluoride (PVDF). (3) *Blocking*: To prevent non-specific interactions, the membrane undergoes a blocking step. (4) *Primary Antibody Incubation*: Next,

Alessandro Laganà and Giovanni Nigita (eds.), *MicroRNA Target Identification*, Methods in Molecular Biology, vol. 3056, https://doi.org/10.1007/978-1-0716-5392-0_15,

the membrane is incubated with a primary antibody that specifically binds to the target protein. (5) *Secondary Antibody Incubation*: A secondary antibody, designed to bind to the primary antibody, is then added; this secondary antibody is linked to either horseradish peroxidase (HRP) or a fluorescent marker. (6) *Detection*: Finally, an HRP substrate like Luminol is added to detect chemiluminescence, or fluorescence is observed if fluorescently labeled antibodies are used.

This chapter presents a standard protocol for confirming miRNA targets by analyzing target protein levels. Since miRNA regulation occurs at the post-transcriptional level, examining the expression of the proteins encoded by the target mRNAs provides biochemical evidence of specific miRNA interactions.

Several techniques are available to verify miRNA targeting, each effective in its approach. This experiment, especially when combined with methods, such as in vitro luciferase assays or quantitative real-time PCR (qRT-PCR), helps clarify mRNA–miRNA interactions and advances research in this area.

2 Materials

2.1 Transfection

1. Cell Lines and Samples: Includes cell lines, primary cells, or tissues.
2. Complete culture media preparation: To a bottle (500 mL) of culture media, add 50 mL of heat-inactivated fetal bovine serum (FBS), along with 5 mL of L-glutamine (if not already present in the basal medium), and 5 mL of a solution containing 10,000 units/mL of penicillin and 10 mg/mL of streptomycin (*see* **Note 1**).
3. Transfection Reagent: Use a cationic liposome-based transfection agent.
4. Pre-miRNA Precursor and Negative Control/miRNA Inhibitor and corresponding Negative Control: Dissolve 5 nmoles of lyophilized oligonucleotides in 100 μL of DNase- and RNase-free ddH_2O to create a 50 μM stock solution.
5. Class II Laminar Flow Hood: For maintaining sterility.
6. CO_2 Incubator: For optimal cell culture conditions.
7. Sterile 24-Well Plates.
8. Pipette Aid.
9. Pipettes and Sterile Filtered Tips.
10. 15-mL Conical Tubes.
11. 1.7-mL Eppendorf Tubes.

2.2 Cell Lysis and Protein Extraction

Materials required for cell lysis may vary depending on the chosen protocol to prepare protein lysates (*see* **Note 2**), though detergent-based lysis is commonly used.

1. Lysis Buffer Stock Solution: Combine 25 mL of 1 M Tris–HCl (pH 7.4), 75 mL of 5 M NaCl, and 50 mL of glycerol, then add ddH_2O to reach a final volume of 500 mL (*see* **Note 3**). Sterilize the solution using a filter unit. On the experiment day, prepare a working solution by combining 965 μL of lysis buffer, 25 μL of 20% NP-40 (nonionic detergent) (*see* **Note 4**), and 10 μL of 100× EDTA-free protease inhibitor cocktail (*see* **Note 5**) in a pre-chilled 1.7 mL tube.
2. 5× Bradford Solution: Dissolve 100 mg of Coomassie Blue G250 in 50 mL of methanol (*see* **Note 6**), add 100 mL of 85% phosphoric acid (H_3PO_4), then dilute with ddH_2O to a final volume of 200 mL. Filter the solution through a sterile filter to remove any precipitates and store at 4°C.
3. Bovine Serum Albumin (BSA) Solution: 1 mg/mL Dissolve 0.1 g of lyophilized BSA in 100 mL of ddH_2O. Aliquot the solution and store at −20°C.
4. Microcentrifuge.
5. Pipette Aid.
6. Pipettes and Tips.
7. 15-mL Conical Tubes.
8. 1.7-mL Eppendorf Tubes.
9. Plastic Cuvettes.
10. Spectrophotometer.

2.3 Protein Separation Using via Polyacrylamide Gel: SDS-PAGE

Caution: Many of the reagents listed below are hazardous and potentially carcinogenic. Conduct all work within a fume hood and wear appropriate personal protective equipment (PPE).

1. **4× Protein Loading Buffer (Sample Buffer):** To prepare, mix 50 mL of ddH_2O, 8 mL of Tris–HCl (pH 6.8), 20 mL of 10% SDS, 10 mL of glycerol, 2.5 mL of 2-mercaptoethanol, and approximately 50 ng of bromophenol blue. After thorough mixing, aliquot and store the buffer at −20°C for long-term use. Keep it at room temperature after opening.
2. **Polyacrylamide Gel Setup** (*see* **Note** 7)**:** For SDS-PAGE, you may either prepare polyacrylamide gels manually or use commercially available precast gels. Follow the manufacturer's instructions for precast gels.
3. **Running Buffer Preparation:** To prepare a 10× stock solution, dissolve 30 g of Tris and 145 g of glycine in approximately 800 mL of ddH_2O. Add 50 mL of 20% SDS, mix

thoroughly until solids are dissolved, and adjust to a final volume of 1 L with a volumetric flask. Store the buffer at room temperature.

4. Pipettes and tips.
5. Pipette aid.
6. 50-mL conical tubes.
7. Suction filtration flasks.
8. Volumetric flasks.
9. Glass plates and spacers.
10. Stands and gaskets.
11. Frames and combs.
12. Power supply.
13. Microcentrifuge.
14. Thermal block.
15. Electrophoresis apparatus.

2.4 Protein Transfer

Caution: Methanol is highly hazardous. Minimize exposure by using appropriate PPE, including gloves, a lab coat, and safety goggles. Always handle methanol in a fume hood with adequate ventilation.

1. **1× Transfer Buffer:** Prepare by combining 100 mL of 10× Running buffer, 200 mL of methanol (*see* **Note 12**), and 700 mL of ddH_2O. Cool the buffer before use.
2. **Ponceau Red Solution:** Dissolve 0.5 g of Ponceau S dye in 5 mL of acetic acid, then dilute with ddH_2O to a final volume of 500 mL.
3. **Nitrocellulose Blotting Membrane**: Precut to match the dimensions of the gel.
4. **Blotting Cellulose Chromatography Papers (3 MM)**: Precut to the gel's dimensions.
5. **Filter Pads or Sponges**: Use for building the transfer stack.
6. **Gel Holder Cassette**: For assembly of the transfer sandwich.
7. **Electrotransfer Apparatus**: Use either a wet or semi-dry transfer system, or the Trans-Blot Turbo Transfer System for automated transfer.
8. **Power Supply**: Required for controlled power output during transfer.
9. **Orbital Shaker**: To facilitate membrane blocking and antibody incubation.

2.5 Blocking, Antibody Binding, and Signal Visualization

1. 10× TBS-T Buffer: Prepare by dissolving 100 g of NaCl and 62 g of Tris in about 700 mL of ddH_2O. Adjust to pH 7.5 with 60–80 mL of 25% HCl, then add 5 mL of Tween® 20. Bring the volume to 1 L with ddH_2O. Store at room temperature and dilute as needed with ddH_2O.
2. Blocking Solution (*see* **Note 13**): Mix 2.5 g of BSA or nonfat dry milk with 50 mL of TBS-T. Store at 4°C. Freshly prepare nonfat dry milk blocking solution for optimal results. Filter Pads or Sponges: Use for building the transfer stack.
3. Primary Antibodies (*see* **Note 14**): Store at 4°C or −20°C. If glycerol is absent in the storage buffer, aliquot the antibodies to prevent freeze–thaw cycles. Dilute in 5% BSA or 5% nonfat dry milk in TBS-T, TBS (without Tween® 20), or PBS on the day of the experiment as directed on the datasheet. Store diluted solutions at −20°C.
4. Secondary Antibodies: Store at 4°C. Dilute according to datasheet instructions using 1% BSA or 1% nonfat dry milk in TBS-T, TBS (without Tween® 20), or PBS solution.
5. ECL Solution: Prepare the solution fresh by mixing luminol and hydrogen peroxide in a 1:1 ratio immediately before use to ensure optimal chemiluminescent activity. Although individual components can be stored at 4°C or room temperature, the mixed solution should not be stored as it may lose effectiveness over time.
6. Chemiluminescent Signal Detection:
   - *Traditional Method*: Use X-ray film in a darkroom setup for signal detection. Film processing is necessary to capture chemiluminescent signals.
   - *Digital Method*: Utilize CCD or CMOS camera-based imaging systems for direct membrane signal detection. and quantification. These systems provide rapid results, higher sensitivity, and do not require a darkroom [2].
7. Orbital Shaker.
8. Pipette Aid.
9. Pipettes and Tips.
10. 15-mL Conical Tubes.
11. Dark Room or Biomolecular Imager: Required for film development in the traditional method or imaging in the digital method.

3 Methods

3.1 Transfection (see *Note 15*)

Maintaining sterile conditions in a class II laminar flow hood is essential when performing transfection experiments. Conduct each condition in at least three replicates for statistical robustness.

1. Cell Seeding (Day Before Transfection): Seed HEK293 cells (*see* **Note 16**) or another suitable cell line (*see* **Note 17**) in a 24-well plate, targeting 60%–70% confluency by the following day. Plate approximately 90,000–100,000 cells per well in 500 μL of culture medium.
2. Medium Replacement (Day of Experiment): Replace the cell medium with 500 μL of fresh medium containing only L-glutamine (*see* **Notes 18–20**) to prepare for transfection.
3. Transfection Mix Preparation: For each condition, prepare transfection mixtures to distribute across three wells. This process requires two separate solutions: one containing medium with the lipo-reagent [3] (*see* **Note 21**) and the other containing medium with a miRNA inhibitor or pre-miR precursor (either experimental or control).
4. Preparation of Lipo-reagent Mix (First Mix): Mix 3 μL of lipo-reagent with 747 μL of serum-free medium, allowing it to sit for 5 min at room temperature. Prepare individual mixtures for each condition using 1.7 mL sterile microcentrifuge tubes and sterile, filtered pipette tips.
5. Preparation of miRNA/Precursor Mix (Second Mix): Combine the miRNA inhibitor or precursor with the medium, ensuring a final volume of 750 μL per condition. Generally, add 100 nM of miRNA inhibitor or pre-miRNA precursor per well, though different concentrations can be tested. Prepare these mixtures in sterile 1.7 mL microcentrifuge tubes using filtered pipette tips.
6. Combining the Mixes: Mix the first and second solutions to yield a final volume of 1.5 mL (750 μL of each). Allow the mixture to sit for 20 min at room temperature to facilitate complex formation. Mix gently without vortexing to maintain transfection complexes.
7. Transfection:
   - *Standard Transfection Protocol*: Follow the manufacturer's guidelines for transfection mix preparation. Dispense 500 μ L of the transfection mixture into each well. After 4–6 h, replace the medium with fresh medium containing FBS and antibiotics. Optimal results can be achieved by lysing cells at 24, 48, and 72 h post-transfection (*see* **Note 22**).

- *Reverse Transfection Protocol*: For reverse transfection, add the transfection mix directly to the wells before cell seeding. After allowing the mix to interact with the plate surface for a few minutes, add the cell suspension. Place cells in a CO_2 incubator, replace the medium after 4–6 h, and collect samples at the designated time points.

3.2 Cell Lysis and Protein Extraction and Quantification

1. Wash Cells: Rinse transfected cells with PBS to eliminate residual FBS from the culture medium. For suspension cells, centrifuge at 300 x *g* for 5 min, discard the supernatant, and wash the pellet with PBS.
2. Add Lysis Buffer: Add approximately 1 mL of chilled lysis buffer (containing detergent and protease inhibitors) per 20 million cells. For suspension cells, apply the buffer directly to the pellet; for adherent cells, add lysis buffer directly to the culture surface or detach cells with PBS, centrifuge, and then lyse as suspension cells (*see* **Note 23**).
3. Incubate on Ice: Keep cells in lysis buffer on ice for at least 20 min, vortexing two to three times during this period.
4. Centrifugation: Centrifuge at 15,000 × *g* for 10 min at 4°C to remove cellular debris.
5. Protein Collection and Storage: Carefully collect the supernatant, containing soluble proteins, and aliquot it to avoid freeze–thaw cycles. Store at −80°C until analysis.
6. Protein Quantification: Use the Bradford assay to measure protein concentration [4]. Dilute 5× Bradford reagent 1:5 with deionized water, and add 1 mL of the diluted reagent to plastic cuvettes.
7. Prepare a Standard Curve: Make serial dilutions of a 1 mg/mL BSA solution (*see* **Note 24**) to create a calibration curve. Add 1 μL of each protein sample to separate cuvettes, allowing the mixture to incubate at room temperature for 5–30 min. Use one cuvette as a blank.
8. Measure Absorbance: Vortex briefly and measure absorbance at 595 nm using a spectrophotometer.

3.3 Protein Separation Using Polyacrylamide Gel: SDS-PAGE

1. Place the protein samples on ice to thaw.
2. Determinethevolumeneededtoachieveapproximately20–30 μg of protein lysate to be loaded onto the gel and combine it with the appropriate amount of loading buffer. For instance, if 20 μ g of protein corresponds to 7.5 μL, then you would add 2.5 μL of loading buffer (*see* **Note 25**).
3. Heat the samples at 95°C for 5–10 min to denature them (*see* **Note 26**).

4. **Preparation of Polyacrylamide Gels** (*refer to* **Note 8**): In a vacuum filtration flask, mix ddH_2O, acrylamide-bisacrylamide solution, 1.25 mL of 1.5 M Tris–HCl (pH 8.9), and 50 μL of 10% SDS (*see* **Note 9**). Adjust ddH_2O and acrylamide-bisacrylamide volumes to achieve the desired gel concentration: for a 12% gel, add 1.68 mL of ddH_2O and 2 mL of acrylamide-bisacrylamide (*see* **Note 10**); for a 10% gel, use 2 mL of ddH_2O and 1.67 mL of acrylamide-bisacrylamide; for a 7.5% gel, use 2.43 mL of ddH_2O and 1.67 mL of acrylamide-bisacrylamide. Add 25 μL of 10% APS and 15 μL of TEMED, swirl gently to prevent air bubbles, connect the flask to a vacuum, and degas the solution (*see* **Note 11**). Pour the mixture between the prepared plates.
5. **Overlay Solution:** To control polymerization, overlay the gel solution with 30% isopropyl alcohol or ddH_2O.
6. **Polymerization Time:** Wait approximately 20 min for the gel to polymerize fully.
7. **Stacking Gel Preparation** (*see* **Note 8**): Combine 1.68 mL of ddH_2O, 2 mL of 30% acrylamide-bisacrylamide solution, 1.25 mL of 0.5 M Tris–HCl (pH 6.8), 50 μL of 10% SDS, 25 μ L of 10% APS, and 15 μL of TEMED. Swirl gently to avoid air bubbles. Carefully remove the overlay from the running gel, add the stacking gel solution, and insert the comb.
8. **Polymerization of the Stacking Gel:** Allow approximately 30 min for the stacking gel to polymerize completely.
9. **Gel Preparation for Electrophoresis:** Remove the gel plates from the frame, gently take out the comb, and prepare the gel for electrophoresis.
10. Position the gel in the electrophoresis chamber and add 1× running buffer as directed by the manufacturer.
11. Carefully take out the comb and introduce the protein samples into the wells of the gel. Do not forget to include a molecular weight marker to serve as a reference for protein sizes and to monitor the progress of the electrophoresis.
12. Connect the apparatus to the power supply and adjust the settings to 200 V and 14 mA for mini-gels, or 50 mA for midi-gels.
13. Terminate the electrophoresis when the band of the molecular weight marker corresponding to your protein of interest is centered in the gel, or when the lowest marker band approaches the bottom of the gel.

3.4 Protein Transfer

To perform protein transfer, create a "sandwich" structure with sponges, blotting cellulose papers (3 MM), the gel, and the membrane. Ensure that each layer of the transfer "sandwich" is completely soaked in transfer buffer to promote efficient protein movement onto the membrane.

1. Position a damp filter pad along with three sheets of blotting paper on the negative side of the transfer cassette.
2. Gently remove the gel from the electrophoresis apparatus and place it on the blotting paper.
3. Place the membrane directly onto the gel, ensuring no air bubbles form between them.
4. Layer three additional sheets of blotting paper and a sponge on top, then secure the cassette.
5. Insert the "sandwich" into the transfer apparatus, ensuring that the negative side of the cassette is oriented toward the anode. This configuration allows negatively charged proteins to move from the gel to the positively charged membrane on the cathode side.
6. Add transfer buffer to the apparatus and set the transfer conditions to 100 V and 400 mA for 1 h and 30 min (*see* **Note 27**). For longer transfers, consider adding ice packs to the apparatus or placing it on ice to keep the solution from overheating.
7. (Optional) Once the transfer is complete, stain the gel with Coomassie Brilliant Blue, a protein-specific dye, to assess how well the transfer was performed.

3.5 Staining and Membrane Blocking

1. Carefully remove the membrane from the transfer assembly, rinse it with ddH_2O, and stain it with 1× Ponceau Red solution on an orbital shaker at room temperature for 5 min to make the proteins visible [5].
2. Capture an image before rinsing off the stain in TBS-T, as the image may be helpful for quantification purposes.
3. Incubate the membrane in a blocking solution on an orbital shaker for 1 h at room temperature.

3.6 Blocking, Antibody Binding, and Signal Visualization

With the membrane now blocked, it is prepared for incubation with primary antibodies that selectively bind to the target proteins.

1. Prepare a dilution of the primary antibody in a 5% solution of nonfat milk or BSA using TBS-T (or TBS or PBS), as per datasheet instructions (*see* **Note 28**).
2. Incubate the membrane in the primary antibody solution using an orbital shaker for 1 h at room temperature, or leave it overnight at 4°C.
3. Rinse the membrane three times with TBS-T for 10 min each on the orbital shaker to eliminate any unbound antibodies.

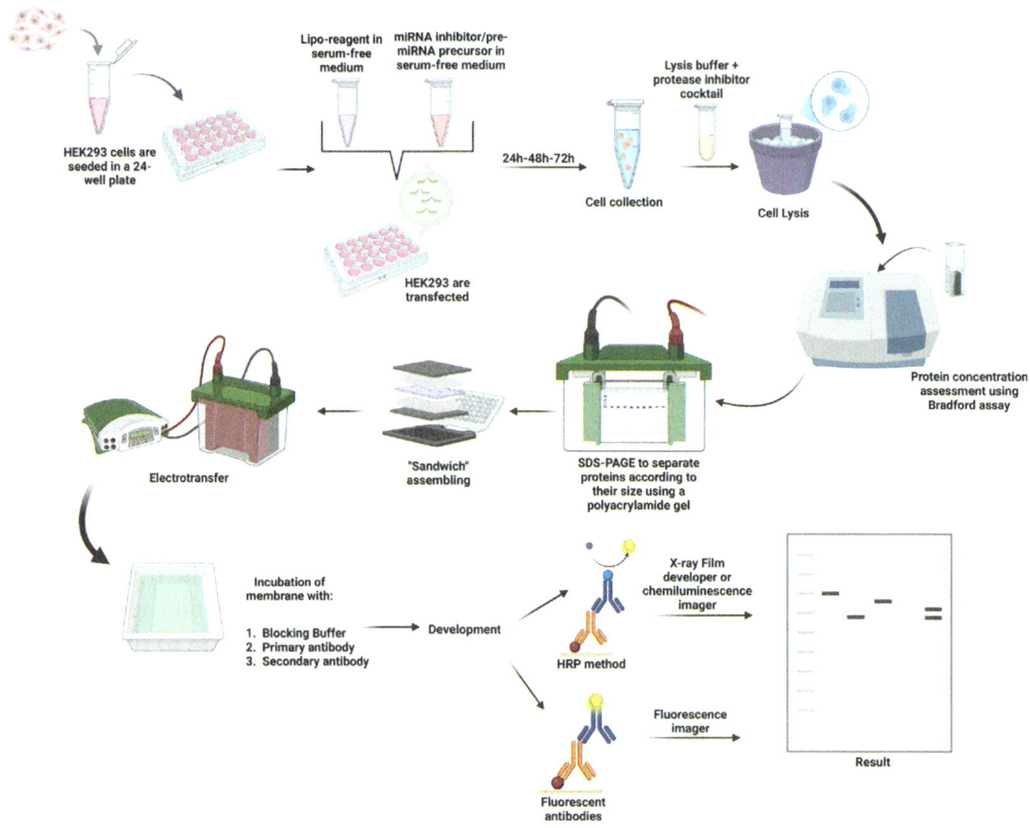

Fig. 1 This figure outlines the various steps described in this chapter for assessing protein expression, beginning with cell transfection. The figure was created using BioRender

4. Incubate the membrane in the diluted secondary antibody solution for 1 h at room temperature on the orbital shaker (*see* **Notes 28** and **29**).
5. Rinse the membrane three times with TBS-T for 10 min each on the orbital shaker to remove any excess secondary antibody.
6. **Protein Detection:**
   - For fluorescent antibodies, use a Fluorescence Imager to read and analyze the fluorescence on the membrane.
   - For the HRP method, apply ECL solution to the membrane and incubate for 1 min at room temperature (*see* **Note 30**).
7. Capture the signal from the ECL and HRP reaction using a chemiluminescence imager or by exposing it onto X-ray film in a darkroom (*see* **Note 31**).
8. Wash membrane thoroughly.

9. Incubate the membrane with an antibody targeting one or more proteins that are universally expressed to normalize the data for the target protein and reduce potential inaccuracies in sample loading or quantification.

This chapter's workflow is summarized in Fig. 1.

4 Notes

1. Cell Culture Medium: Choose a medium suitable for your cell line. Refer to the ATCC website (https://www.atcc.org/) to find the recommended medium for your specific cell culture.
2. Protein Extraction: Various proteins necessitate distinct extraction methods based on their type (such as nuclear or cytoplasmic) and the specific cell type. For instance, neutrophils need nitrogen cavitation for extraction because they are particularly vulnerable to proteases [6].
3. Detergent Lysis: A widely used technique for breaking open mammalian cells, this method utilizes saline to induce cell lysis through osmotic effects. Furthermore, the inclusion of 10% glycerol enhances protein solubility and promotes correct folding by minimizing hydrophilic interactions.
4. Nonionic Detergent: In the lysis buffer, Triton X-100 or NP-40 are used to disrupt cell membranes.
5. Protease Inhibitors: To prevent protein degradation during the extraction process, protease inhibitors are crucial. Additionally, phosphatase inhibitors may be included when studying phosphorylated proteins.
6. Coomassie Staining: Before incorporating any other components, Coomassie should first be dissolved in methanol.
7. Acrylamide Safety: Acrylamide is classified as a Group 2A carcinogen by the International Agency for Research on Cancer (IARC) and presents health hazards. To minimize risk, it is advisable to use precast gels rather than working with liquid acrylamide.
8. Gel Structure: The gel used for electrophoresis has two parts: a stacking gel with lower polyacrylamide concentration and a running gel with higher concentration. The stacking gel organizes proteins into bands, which are separated according to their size in the running gel.
9. SDS Role: SDS, a surfactant, ensures proteins acquire a uniform negative charge, enabling them to migrate based on their size.

10. Gel Concentration: The concentration of the gel is determined by the molecular weight of the protein. Gradient gels (4%–20% polyacrylamide) enable effective separation of proteins across a range of molecular weights.
11. Oxygen Inhibition: Oxygen inhibits radical polymerization during gel formation.
12. Protein Transfer: Methanol in the transfer buffer facilitates protein transfer by removing SDS, enhancing protein adherence to the membrane.
13. Blocking Buffers: Blocking buffers minimize nonspecific binding by occupying available sites on the membrane. Common solutions include 5% nonfat milk or BSA, with BSA preferred for phosphorylated proteins to prevent interference from casein present in milk.
14. Antibody Dilutions: Ideally, antibody dilutions should be made using the same solution as the blocking buffer, although there can be exceptions to this guideline.
15. MiRNA Targeting: Investigating miRNA targeting involves modifying miRNA expression levels to assess differences between samples with various expression levels. There are two primary strategies: the first involves increasing the expression of the miRNA of interest through the use of pre-miRNA precursors or synthetic miRNA oligonucleotides; the second strategy focuses on decreasing miRNA levels using a specific inhibitor. The choice of method depends on the endogenous expression levels of the miRNA in the selected cell model. For cell lines with high levels of miRNA, it is advisable to use a miRNA inhibitor. In contrast, for cells exhibiting low or absent miRNA expression, boosting its levels can help identify downregulation of target proteins in comparison to the control group.
16. HEK293 Cell Line: HEK293 cells, easily transfectable, are often used in miRNA studies.
17. Transfection in Studied Cells: Perform transfections in both HEK293 cells and the cell model under study. The interaction of miRNAs with mRNA can be tissue-specific, so performing the experiments also in the appropriate cell model is crucial.
18. Serum Effects: Serum contains albumin, which negatively impacts transfection efficiency by interfering with liposome reagents.
19. Serum-Free Medium: Serum-free media can be used for sensitive cells, providing necessary nutrients even without FBS.

20. Antibiotic Removal: Removing antibiotics is recommended during transfection, as liposome reagents increase cell permeability, potentially introducing more antibiotics and causing cytotoxicity.
21. Cationic Lipid Structure: Cationic lipids consist of hydrocarbon chains along with a positively charged head group that interacts with nucleic acids. This interaction promotes endocytosis, allowing genetic material to enter cells effectively.
22. Time-Course: Ensure the time-course is long enough to observe changes in target protein expression. This depends on cellular uptake, transcription, and protein turnover.
23. Avoid Trypsin: When working with adherent cells, avoid trypsin as it may degrade extracellular transmembrane protein epitopes.
24. Spectrophotometry: Use a spectrophotometer to measure absorbance and compare known BSA dilutions to determine protein concentrations (mg/mL) in samples.
25. Sample Volume: Adjust sample volumes with ddH_2O or lysis buffer to ensure equal loading in each well.
26. Protein Denaturation: Denaturation is promoted by reducing agents like 2-mercaptoethanol or dithiothreitol (DTT) in the loading buffer, alongside SDS present in the loading buffer, gel, and running buffer.
27. Electric Field for Transfer: In the transfer process, the electric field drives negatively charged proteins from the gel to the positively charged membrane, where they are retained through electrostatic and hydrophobic interactions.
28. Antibody Dilution Optimization: Empirical testing is needed to determine the optimal antibody dilution conditions.
29. Secondary Antibodies: Secondary antibodies are designed to recognize and bind to immunoglobulins from a specific target species, allowing detection of the primary antibody. These antibodies are linked to detection molecules, including fluorophores or horseradish peroxidase (HRP).
30. ECL Solution: Enhanced Chemiluminescence (ECL) involves the preparation of a solution containing the HRP substrate Luminol along with hydrogen peroxide. The reaction is catalyzed by HRP, resulting in the emission of blue light at a wavelength of 450 nm.
31. Signal Intensity: The intensity of the chemiluminescent signal corresponds to the protein's expression level, which can be quantified using image analysis tools like ImageJ.

Conflict of Interest

The authors have no conflicts of interest to declare that are relevant to the content of this chapter.

References

1. Janga SC, Vallabhaneni S (2011) MicroRNAs as post-transcriptional machines and their interplay with cellular networks. Adv Exp Med Biol 722:59–74
2. Khoury MK, Parker I, Aswad DW (2010) Acquisition of chemiluminescent signals from immunoblots with a digital single-lens reflex camera. Anal Biochem 397:129–131
3. Liu D, Ren T, Gao X (2003) Cationic transfection lipids. Curr Med Chem 10(14):1307–1315
4. Bradford MM (1976) A rapid and sensitive method for the quantitation of microgram quantities of protein utilizing the principle of protein-dye binding. Anal Biochem 72:248–254
5. Stochaj WR, Berkelman T, Laird N (2006) Staining membrane-bound proteins with ponceau s. CSH Protoc 2006(5):pdb.prot4543
6. Simpson RJ (2010) Disruption of cultured cells by nitrogen cavitation. Cold Spring Harb Protoc 10(11):pdb.prot5513

Chapter 16

Luciferase Reporter Assay for MicroRNA Target Validation

Soumya Sharma, Roma Karna and Federica Calore

Abstract

MicroRNAs (miRNAs) are small, non-coding RNA molecules, typically 22–25 nucleotides in length, that modulate gene expression after transcription. They exert their function by binding to complementary sequences within the 3′ untranslated regions (3′ UTRs) of target messenger RNAs (mRNAs), leading to mRNA degradation or translational repression. Over the past decade, miRNAs have emerged as critical key mediators of essential cellular functions, such as cell growth, lineage commitment, programmed cell death, and adaptation to stress. Dysregulated miRNA expression has been linked to the development of multiple diseases, including malignancies, heart conditions, and neurological disorders. Precise identification of miRNA target genes is essential for elucidating their functional roles. Although many computational models forecast miRNA–mRNA interactions by analyzing sequence complementarity and additional contextual features, these predictions require rigorous experimental validation to establish their biological relevance. Among the available validation methods, the luciferase reporter assay is one of the most widely employed due to its high sensitivity and specificity. This assay leverages the bioluminescent properties of the luciferase enzyme to quantitatively assess miRNA-mediated gene regulation. A commonly used strategy involves cloning the 3′ UTR of the predicted target gene, including the miRNA recognition site, downstream of the luciferase gene in a plasmid vector. Upon co-transfection with a miRNA mimic into cultured cells, a functional miRNA–mRNA interaction results in reduced luciferase expression, which is quantitatively measured as a decrease in luminescence. Control experiments, such as site-directed mutagenesis of the miRNA binding site, are essential to confirm the specificity of the observed regulatory effects. This chapter provides a detailed methodological framework for performing luciferase assays, encompassing assay design, cell line selection, and data interpretation. By enabling the validation of computational predictions, the luciferase assay serves as a powerful tool for investigating miRNA regulatory networks and their mechanistic contributions to disease pathophysiology.

Keywords miRNA regulation, Luciferase reporter assay, Post-transcriptional gene regulation, 3′ Untranslated region (3′ UTR), miRNA–mRNA interactions

1 Introduction

Experimental validation of microRNA (miRNA) interactions requires a combination of molecular biology techniques to confirm predicted miRNA–mRNA regulatory relationships. These include reporter assays, site-directed mutagenesis, and protein expression analysis, all of which contribute to a more comprehensive understanding of

Alessandro Laganà and Giovanni Nigita (eds.), *MicroRNA Target Identification*, Methods in Molecular Biology, vol. 3056,
https://doi.org/10.1007/978-1-0716-5392-0_16,

miRNA-mediated post-transcriptional control. Among the available approaches, luciferase reporter assays and Western blotting remain widely utilized due to their sensitivity, efficiency, and ability to provide quantitative data on miRNA function.

The luciferase assay capitalizes on the enzymatic activity of luciferase, a bioluminescent enzyme originally isolated from the firefly *Photinus pyralis*. This enzyme catalyzes the oxidation of D-luciferin in an ATP- and oxygen-dependent reaction, yielding oxyluciferin, carbon dioxide, pyrophosphate, adenosine monophosphate (AMP), and light emission:

LUCIFERASE

$$\begin{aligned} &\mathrm{D-Luciferase+ATP+O_2} \\ &\quad \rightarrow \mathrm{Oxyluciferin+CO_2+PPi+AMP+Light} \end{aligned}$$

This bioluminescence serves as a quantifiable output to assess gene regulation. A typical luciferase reporter construct consists of the luciferase coding sequence fused to the 3′ untranslated region (3′ UTR) of the putative miRNA target. Because miRNAs predominantly exert their regulatory effects by binding to sequences in the 3′ UTR [1], inserting this region downstream of the luciferase gene allows for direct evaluation of miRNA-induced repression. A reduction in luminescence indicates target regulation, while unchanged or increased luminescence suggests weak or absent interaction [2].

The luciferase assay is favored in miRNA research for its rapid workflow, high sensitivity, and cost-effectiveness. To enhance data reliability, normalization controls involving alternative luciferases are commonly used. One widely implemented strategy is the dual-luciferase system, in which firefly luciferase serves as the experimental reporter, while Renilla luciferase, which utilizes coelenterazine as a substrate rather than D-luciferin, functions as an internal control. This dual-reporter approach corrects for variations in transfection efficiency and cell viability, thereby improving result consistency.

To facilitate experimental design, several commercial luciferase reporter vectors are available. For instance, Promega provides psiCHECK™-1, which contains Renilla luciferase as a single reporter, and psiCHECK™-2, which integrates both Renilla and firefly luciferases for simultaneous transcriptional analysis and normalization. The latter is widely preferred because it allows for internal normalization within a single plasmid, minimizing variability across experimental conditions.

This chapter details a step-by-step protocol for employing luciferase reporter assays to experimentally validate miRNA target interactions, utilizing optimized reporter constructs and reagents from Promega, a leader in this field.

2 Materials

2.1 Material and Reagents for PCR and Restriction Enzyme Analysis (See Note 1)

1. Template DNA or cDNA: Isolated from cells expressing the target gene.
2. Polymerase Chain Reaction (PCR) Buffer (10× Solution): Contains 15 mM $MgCl_2$, 500 mM KCl, and 100 mM Tris–HCl (pH 8.3 at 25°C).
   Preparation: Dissolve 0.142 g $MgCl_2$ and 3.72 g KCl in 10 mL of 1 M Tris–HCl (pH 8.3) and bring the final volume to 100 mL with DNase- and RNase-free water (sterile ultrapure water). Aliquot and store at −20°C (*see* **Note 2**).
3. Deoxynucleotide Triphosphates (dNTPs) mix: A 10 mM solution is made by diluting the 100 mM standard stock (provided in PCR kits) tenfold using DNase- and RNase-free water. Store at −20°C.
4. High-Fidelity DNA Taq Polymerase: Maintain enzyme stock at −20°C for long-term stability (*see* **Note 3**).
5. Primers (Oligonucleotides): Stock solutions are prepared at 100 μM concentration by dissolving lyophilized primers in DNase- and RNase-free water. For working solutions, dilute the stock tenfold to achieve 10 μM (*see* **Note 4**).
6. Restriction Enzymes: PmeI NotI, XhoI, and SgfI should be stored at −20°C. Each enzyme has specific requirements for optimal activity (*see* **Notes 5** and **6**).
7. Restriction Buffer 1: Dissolve 95.21 mg of $MgCl_2$ and 100 mg of BSA in 70 mL of DNase- and RNase-free water. Add 1 mL of 1 M Bis–Tris propane-HCl (pH 7.0), then bring the total volume to 100 mL with DNase- and RNase-free water. Aliquot the buffer and store at −20°C (*see* **Note 2**).
8. Restriction Buffer 2: Mix 292 mg NaCl, 95.21 mg $MgCl_2$, and 100 mg BSA in 70 mL of DNase- and RNase-free water. Add 1 mL of 1 M Tris–HCl (pH 7.9), then adjust to 100 mL with water. Aliquot and freeze at −20°C (*see* **Note 2**).
9. Restriction Buffer 3: Dissolve 584 mg NaCl, 95.21 mg $MgCl_2$, and 100 mg BSA in 70 mL of DNase- and RNase-free water. Add 5 mL of 1 M Tris–HCl (pH 7.9), and bring the total volume up to 100 mL. Store in aliquots at −20°C (*see* **Note 2**).
10. Restriction Buffer 4: Combine 490 mg potassium acetate, 142.4 mg magnesium acetate, and 100 mg BSA with 70 mL of DNase- and RNase-free water. Add 10 mL of 0.2 M Tris–acetate buffer (pH 7.9), total volume should be100 mL. Aliquot for storage at −20°C (*see* **Note 2**).
11. psiCHECK™-2 Vector: A dual-luciferase reporter vector suitable for cloning and analysis.

12. Agarose Gel Preparation: Dissolve 0.6 g or 1.2 g of agarose in 60 mL of 1× TBE (prepared from 10.8 g Tris, 5.5 g boric acid, and 4 mL of 0.5 M Na_2EDTA in 1 L water). Microwave until the agarose dissolves completely. After cooling slightly, add ethidium bromide (0.2–0.5 μg/mL) under a fume hood. Pour into a mold with a comb and allow to set for 20 min (*see* **Note 7**).
13. Molecular weight markers, stored at −20°C before opening and at 4°C after use.
14. Loading Buffer (10×): includes 10 mM Tris–HCl (pH 7.6), 60 mM EDTA, 0.03% bromophenol blue, and 60% glycerol.
15. Gel Extraction Kit and PCR Cleanup Kit: for nucleic acid purification.
16. T4 DNA Ligase Buffer (10×): Contains 300 mM Tris–HCl (pH 7.8 at 25°C), 100 mM $MgCl_2$, 100 mM DTT, and 10 mM ATP (*see* **Notes 2** and **8**).
17. T4 DNA Ligase: Store at −20°C.
18. Size-Exclusion Spin Columns: for rapid nucleic acid purification.
19. Competent *E. coli* Cells: Aliquot and store at −80°C to prevent freeze-thaw damage.
20. Luria Bertani (LB) Broth: Dissolve 10 g tryptone, 10 g NaCl, and 5 g yeast extract in 800 mL water. Adjust pH to 7.5 with NaOH, then sterilize. Add 100 μg/mL ampicillin before use (*see* **Note 9**).
21. SOC Medium: Enriched growth medium for bacterial recovery post-transformation. Prepare using 20 g tryptone, 5 g yeast extract, 10 mM NaCl, 2.5 mM KCl, 10 mM $MgCl_2$, 10 mM $MgSO_4$, and 20 mM glucose in 1 L of water. Sterilize by autoclaving.
22. Agar Plates with Selective Antibiotics: Dissolve 20 g agar, 10 g tryptone, 10 g NaCl, and 5 g yeast extract in 800 mL water. Adjust pH to 7.5, sterilize, and add 100 μg/mL antibiotic before pouring plates.
23. Plasmid Miniprep Kit [3].
24. Centrifuge: necessary for bacterial pelleting.
25. Microcentrifuge: for nucleic acid isolation.
26. Thermal Cycler: used for PCR amplification.
27. UV Transilluminator: required for gel visualization.
28. Incubators (Bacterial and Heated): necessary for cell growth and enzymatic reactions.
29. Bacterial Shaker: required for liquid culture agitation.
30. Pipettes and Sterile Tips: for precise reagent handling.
31. 15-mL Conical Tubes: used for bacterial centrifugation.

32. 1.5-mL Microcentrifuge Tubes: for sample processing.
33. Erlenmeyer Flasks: required for bacterial culture preparation.
34. Microwave: for agarose gel preparation.
35. Digital Scale: used for the preparation of reagents.
36. Spectrophotometer (NanoDrop or equivalent): necessary for nucleic acid quantification.
37. Sterile Petri Dishes: for bacterial culture plating.

2.2 Reagents and Equipment for Transfection Procedure

1. Cell Lines: Select appropriate cell lines based on the experimental requirements.
2. Growth Medium: Supplement base medium with 50 mL of prewarmed, heat-inactivated fetal bovine serum, 5 mL of L-glutamine, and 5 mL of an antibiotic solution containing 10,000 units/mL penicillin and 10 mg/mL streptomycin (*see* **Note 10**).
3. Transfection Reagent: Utilize a cationic lipid-based formulation to enhance the delivery of nucleic acids into cells. Pre-miRNA Oligonucleotides and Controls: Rehydrate 5 nmol of lyophilized pre-miRNA molecules and non-targeting control oligonucleotides in 100 μL of nuclease-free water to achieve a final concentration of 50 μM.
4. Plasmid DNA: Utilize the plasmid DNA prepared earlier in the protocol.
5. Biosafety Cabinet: Perform all transfection steps under a Class II biosafety cabinet to ensure a sterile working environment.
6. Incubator: Maintain transfected cells in a CO_2-controlled incubator set to optimal conditions for cell growth.
7. Pipetting Equipment: Use a pipette controller for accurate liquid handling.
8. Pipettes and Sterile, Disposable Tips: Use adjustable pipettes with sterile, disposable tips to prevent contamination.
9. Centrifuge Tubes: Prepare 15-mL sterile conical centrifuge tubes for reagent preparation and mixing.
10. Microcentrifuge Tubes: Use 1.5-mL microcentrifuge tubes for small-volume sample preparations.
11. 24-Well Tissue Culture Plates: for transfection experiments.

2.3 Reagents and Equipment for Dual-Luciferase® Reporter Assay

1. Dual-Luciferase® Reporter Assay System: Essential for quantifying Renilla and firefly luciferase activities (*see* **Note 11**).
2. Phosphate-Buffered Saline (PBS): Prepare by dissolving 8 g of NaCl, 0.2 g of KCl, 1.44 g of Na_2HPO_4, and 0.24 g of KH_2PO_4 in 800 mL of distilled water. Adjust the pH to 7.4 using HCl and bring the volume to 1 L with distilled water. Sterilize by autoclaving or using a 0.22 μm filter.

3. Luminometer: Use a manual luminometer or one equipped with a reagent injector for optimal assay performance.
4. Orbital Shaker: used for gentle mixing of assay components.
5. Pipette Controller: necessary for accurate reagent handling.
6. Adjustable Pipettes and Sterile Tips: for precise liquid handling.
7. Sterile Conical Tubes: Use 15-mL conical tubes for reagent preparation and storage.
8. Microcentrifuge Tubes: Use 1.5-mL tubes for small-volume reagent handling and sample storage.
9. White-wall 96-Well Plates: required for luminescence measurements.

3 Methods

3.1 Insert PCR (See Note 12)

Proper preparation is essential when setting up a PCR procedure. Always wear gloves to prevent contamination of the reaction mixture and reagents. Keep all reagents and reaction tubes on ice throughout the procedure. Before starting, ensure that all reagents are fully thawed and properly prepared. Set up the workspace under molecular biology-grade conditions and arrange all necessary lab equipment on the workbench in advance.

1. Utilize 100 ng of DNA or ctDNA as a template to perform amplification of the 3′ UTR region.
2. Prepare a 20 μL total PCR reaction volume for each sample. Assemble a master mix accounting for the number of samples plus one additional reaction. The master mix should contain:

   - 2 μL of 10× homemade or commercial buffer per reaction (including the extra sample).
   - 0.4 μL of each primer at a 10 μM working concentration per reaction (including the extra sample).
   - 0.4 μL of 10 mM dNTPs per reaction (including the extra sample).
   - 0.4 μL of Taq Polymerase per reaction (including the extra sample) (*see* **Note 13**).
   - Adjust the volume with PCR-grade water to reach a total of 20 μL.

Distribute equal aliquots of the master mix into each PCR tube, then add the DNA template individually. Mix gently by vortexing, then briefly centrifuge. Keep the tubes on ice until starting the PCR reaction.

3. Configure the thermal cycler based on the specifications provided in the enzyme datasheet. A typical protocol involves the following steps:
   - Initial denaturation: 95°C for 3 min (1 cycle).
   - PCR amplification:

i. Denaturation at 95°C for 30 s (40 cycles).

ii. Annealing at ~50°C–68°C (depending on the primer sequences) for 30 s (40 cycles).

iii. Extension at 68°C or 72°C (extension temperature depending on the polymerase enzyme) for 1 min per kb (40 cycles).

   - Final extension: 68°C or 72°C for 7 min (1 cycle).
   - Hold: 4°C.

4. Preheat the thermal cycler lid to approximately 105°C.
5. Insert the PCR tubes into the thermal cycler and allow the amplification reaction to proceed until completion.
6. Upon completion, briefly spin down the samples and mix with 2 μL of 10× loading buffer. Prepare the electrophoresis chamber by filling it with 1× TBE buffer and positioning the agarose gel. Carefully load the samples and molecular weight marker into the gel wells, ensuring no spillage. Run the gel at 120 V and monitor the progress until the dye front reaches about 75%–80% of the gel's length.
7. Shut down the electrophoresis system and examine the bands under a transilluminator.
8. Gently excise the gel fragment containing the desired band (*see* **Note 14**) and weigh the excise agarose portion. Follow the manufacturer's guidelines to purify the band using a gel extraction kit.

3.2 Restriction Digestion

It is important to be well-prepared when setting up restriction digestion. Wear gloves to prevent contamination of the reaction mixture or reagents. The reaction tubes should be kept on ice throughout the experiment, and ensure they are fully thawed before starting. Work in molecular biology-grade conditions and organize all necessary lab equipment on the workbench in advance.

1. Prepare two separate tubes: one with the gel-purified insert and the other containing at least 2 μg of psiCHECK™-2 vector, to compensate for losses during purification.

2. Add 5 μL of the suitable 10x buffer (*see* **Note 15**), 1 μL of each enzyme, and DNAse- and RNAse-free ddH_2O to each tube, adjusting the total volume to 50 μL.
3. Allow the digestions to incubate at 37°C for a period of 2–16 h, according to the requirements of the restriction enzyme.
4. Purify the digested insert and vector using a PCR purification kit to remove salts and enzymes, following the manufacturer's instructions (*see* **Note 16**).

3.3 Ligation

Wear gloves to prevent contamination. Keep all reagents and tubes on ice throughout the experiment and ensure they are fully thawed before starting. Work in molecular biology-grade conditions and organize all necessary lab equipment on the workbench in advance.

1. Determine the plasmid concentration using a NanoDrop micro-volume spectrophotometer. Configure the device for dsDNA measurements and analyze 2 μL of the digested psiCHECK™-2 vector (*see* **Note 17**). Calibrate using 2 μL of ddH_2O.
2. Combine 100 ng of purified plasmid with the insert using the specified formula [4] (*see* **Note 18**):

$$\frac{\text{kb of insert}}{\text{kb of vector}} \times \text{ng of vector} = \text{ng of insert required for a } 1:1 \text{ molar ratio}$$

3. Include 2 μL of T4 DNA 10× Ligase Reaction Buffer (*see* **Note 8**) and 2 μL of T4 DNA Ligase in the reaction mix. Adjust the volume to 20 μL with DNAse- and RNAse-free ddH_2O. Include a negative control by replacing the insert DNA with purified ddH_2O.
4. Allow the reaction to incubate for 3 h at room temperature or for 4–18 h at 15°C (*see* **Note 19**).

3.4 Transformation (See Note 20)

1. Keep the necessary aliquots of competent bacterial cells on ice, allowing them to thaw gradually (*see* **Note 21**).
2. Set the thermal block to 42°C in preparation for heat shock of bacterial cells.
3. Plasmid uptake: Add approximately 20–30 ng of the ligation product or negative control to 50 μL of thawed competent bacterial cells. Keep the mixture on ice for 30 min, gently flicking the tube one to two times during incubation. Transfer the tube to a 42°C thermal block for 30 s, then immediately return it to ice for 2 min to enhance plasmid uptake by altering membrane permeability.

4. Resuspend the transformed bacteria in 250 μL of SOC medium. Place the tube in a shaker incubator at 37°C using a speed of 220–250 rpm and allow bacteria to recover for 1 h.
5. Spread appropriate volumes of competent cells (ligation product or negative control) onto 100-mm Petri dishes at room temperature, each containing nutrient agar supplemented with the selective antibiotic (*see* **Note 9**). Work under sterile conditions to avoid contamination.
6. Turn the plates upside down and incubate at 37°C for 12–16 h to promote the formation of visible bacterial colonies on the agar surface (*see* **Note 22**).
7. Using a sterile pipette tip, pick five to ten colonies (or more if background growth is observed in the negative control) and transfer each into a 15 mL tube containing 5 mL of LB broth with the selective antibiotic. Incubate overnight at 37°C in a shaking incubator (220–250 rpm) to amplify the selected colonies.
8. Centrifuge bacterial cultures at ~350–400 × *g* for 20 min to pellet the cells.
9. Extract plasmid DNA from the bacterial cell pellets using the method described by Zhang et al. [3] (*see* **Note 23**).

3.5 Transfection (See Note 24)

The transfection procedure should be performed in a sterile environment within a class II laminar flow hood. It is essential to carry out each experimental condition in at least three replicates to ensure reproducibility.

1. On the day prior to transfection, plate HEK293 cells (*see* **Note 25**) in a 24-well plate at a density sufficient to reach 60%–70% confluence (approximately 90,000–100,000 cells per well) by the time of transfection. Add 500 μL of complete culture medium to each well.
2. On the day of transfection, gently remove the complete culture medium and replace it with 500 μL of medium supplemented exclusively with L-glutamine (*see* **Notes 26–28**).
3. Prepare a mixture with the transfection complexes to be distributed evenly across three wells. For each condition, the mixture should consist of a combination of two distinct preparations:
   a. Preparation 1: Mixture of serum-free medium and the lipofection reagent [5] (*see* **Note 29**),
   b. Preparation 2: Mixture of serum-free medium containing either the plasmid vector, the pre-miR precursor (specific or negative control), or both (*see* **Note 30**).

4. For Preparation 1, combine 3 μL of lipofection reagent with 747 μL of serum-free medium. Allow the mixture to incubate at room temperature for 5 min. Prepare a unique mixture for each condition, using sterile 1.5 mL microcentrifuge tubes and pipette tips.
5. For Preparation 2, combine plasmid DNA and/or pre-miR precursor with serum-free medium to a final volume of 750 μL. Prepare a separate mixture for each condition. Standard transfection conditions include ~1 μg of plasmid DNA per well and a final pre-miR precursor concentration of 100 nM per well. Optimization may require testing different concentrations. Maintain sterility by using dedicated microcentrifuge tubes and pipette tips.
6. Mix equal volumes (750 μL each) of the lipofection reagent solution and the nucleic acid solution, yielding a final transfection mixture volume of 1.5 mL.
7. Let the mixture sit at room temperature for 20 min to facilitate the formation of transfection complexes. Gently invert the tube several times, avoiding vortexing to prevent complex disruption.
8. Dispense 500 μL of the transfection mixture into each well. Incubate cells for 4–6 h before replacing the medium with fresh culture medium containing fetal bovine serum (FBS) and antibiotics. Allow cells to incubate for 48 h before proceeding with luminometer-based analysis.

3.6 Luciferase Assay

Before performing the assay, ensure all frozen reagents are completely thawed at room temperature and mixed thoroughly to prevent the formation of concentration gradients. The dual-luciferase assay measures both firefly and Renilla luciferase activities within the same well, allowing for normalization of transfection efficiency. The assay can be conducted using a manual luminometer or a automated luminometer equipped with a reagent injector.

Traditional luciferase assays require cell lysis to quantify luminescence, which limits the real-time monitoring of gene expression in the same population of cells. However, recent advancements in live-cell luciferase assays now allow non-invasive, real-time luminescence assessment in intact cells. These new methods offer a dynamic assessment of gene expression, overcoming the limitations of end-point microplate-based measurements, by preserving spatial and temporal resolution.

Firefly luciferase-based imaging is widely used due to its high sensitivity, enabling detection of gene expression, signal transduction, and drug responses in a physiologically relevant context. The enzymatic reaction between firefly luciferase with luciferin in the presence of ATP and oxygen generates bioluminescence, which can be quantified using an imaging system, such as CCD cameras or bioluminescent microscopes [6].

1. To prepare the lysis buffer, dilute the 5× passive lysis buffer from the DLR™ Assay Kit with double-distilled water (ddH_2O) (*see* **Note 31**). Mix thoroughly to achieve uniformity. The prepared solution can be stored at 4°C for up to 1 month.
2. Discard the medium from transfected cells and rinse each well with phosphate-buffered saline (PBS) to eliminate residual medium. Aspirate PBS completely.
3. Add 100 μL of passive lysis buffer working solution to each well, ensuring complete coverage of the cell monolayer.
4. Incubate the plate at room temperature on an orbital shaker for 15 min or until complete cell lysis is achieved. This step ensures uniform lysis across all wells.
5. For short-term use, store the lysates at −20°C. Keep the lysates at −80°C for long-term storage in appropriately labeled tubes or vials. Alternatively, the luciferase assay may be performed directly in a 96-well plate.
6. Reconstitute the lyophilized Luciferase Assay Reagent II (LAR II) with 10 mL of Luciferase Assay Buffer II, as provided in the DLR™ Assay Kit (*see* **Note 32**). The reagent is stable for 1 month at −20°C or up to 1 year at −70°C. Avoid repeated freeze–thaw cycles by aliquoting as necessary.
7. If using a luminometer with a reagent injector, dispense 100 μ L of LAR II solution into each well of a 96-well plate. If using a manual luminometer, transfer the solution into the appropriate luminometer tubes (*see* **Note 33**).
8. Add 20 μL of cell lysate to 100 μL LAR II solution and mix gently by pipetting. Avoid vortexing, as this may disrupt the luciferase reaction.
9. Configure the luminometer for a 2-s delay followed by a 10-s luminescence measurement for each sample.
10. Insert the plate or tube into the luminometer and initiate measurement. This step records firefly luciferase activity only.
11. Allow the first reading to complete. If the luminometer is not connected to a computer, manually document the firefly luciferase activity values.

12. Manually dispense or use a reagent injector to add 100 µL of Stop & Glo® Reagent (Substrate + Buffer) (this is the second buffer) into each well or tube (*see* **Note 34**). This reagent allows the assessment of the Renilla luciferase activity only.
13. Allow the second reading to complete, then manually record the values in the case of no computer connection. The experimental procedure is now complete.
14. Data analysis: establish a background luminescence value using measurements from untransfected cells. Subtract this baseline from all sample readings.
15. For tests in triplicate—calculate the average firefly and Renilla luciferase activity from triplicate samples. Since the psiCHECK™-2 Vector system uses firefly luciferase as an internal control, its activity should remain consistent across samples (*see* **Note 35**). Normalize Renilla luciferase activity by dividing each sample's Renilla luminescence value by the corresponding firefly luminescence value. Figure 1 illustrates the expected results if the candidate miRNA targets the 3′ UTR of interest.

This chapter's workflow is summarized in Fig. 2.

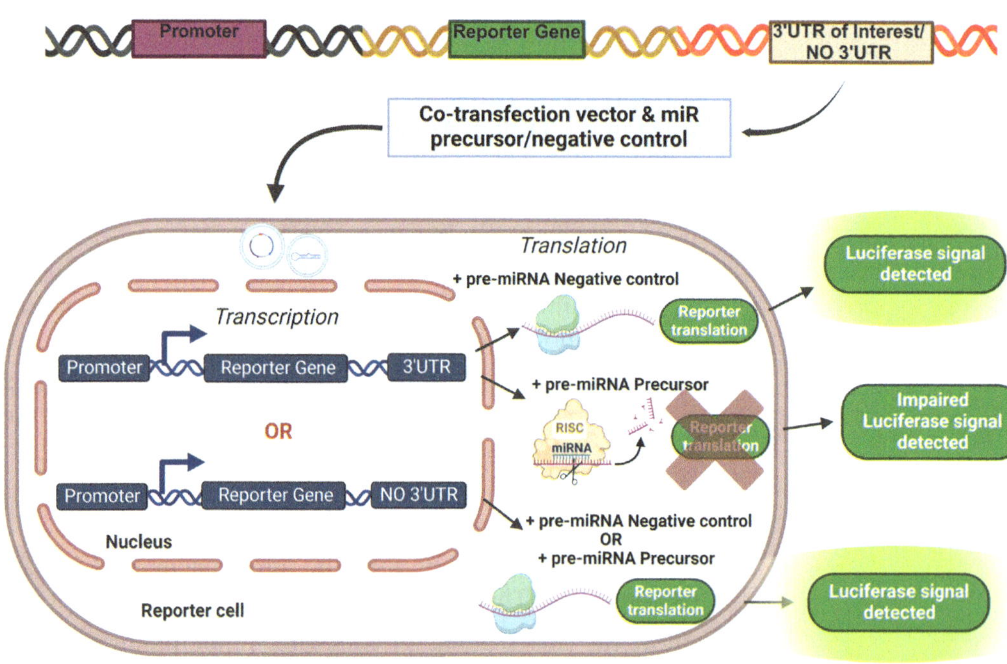

Fig. 1 This figure provides an overview of the key steps outlined in this chapter for measuring firefly and Renilla luciferase luminescence. This figure was generated using BioRender

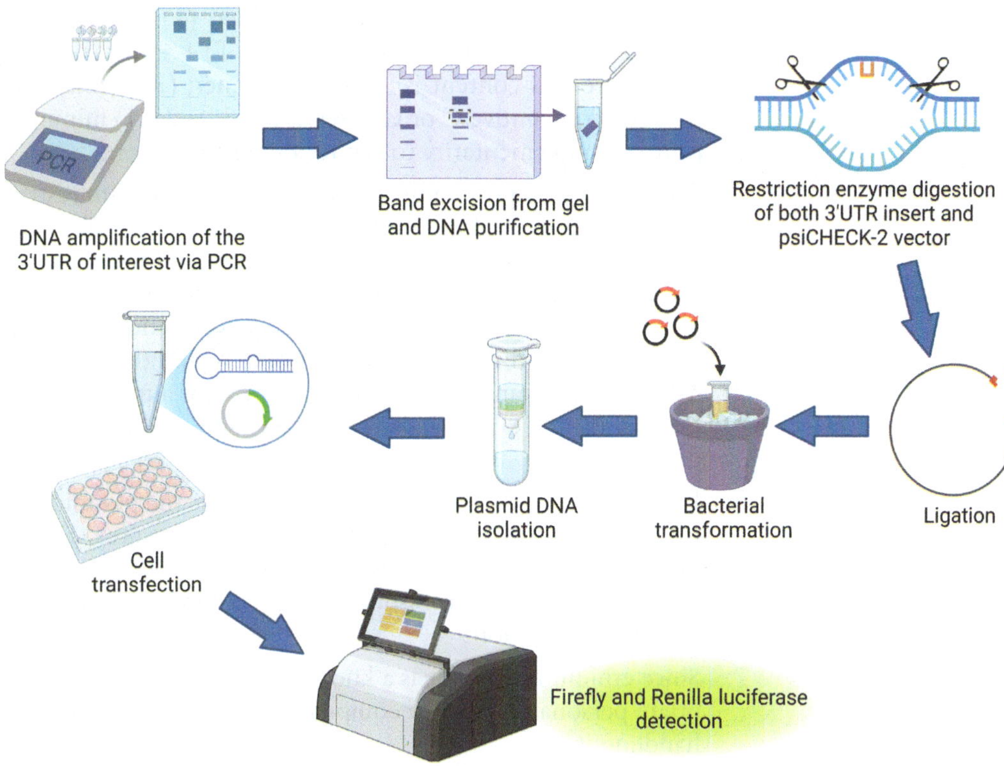

Fig. 2 This figure depicts the varying firefly and Renilla luciferase detection outcomes based on different combinations of pre-miR or negative control miRNA and the vector containing the 3′ UTR of interest or an empty vector. This figure was generated using BioRender

4 Notes

1. MicroRNAs bind to mRNA targets, reducing gene expression by hindering protein synthesis. The initial step in this process involves cloning the 3′ untranslated region (UTR) of the target gene downstream of the open reading frame of a luciferase reporter plasmid. This allows researchers to investigate how microRNAs regulate gene expression.
2. Enzymes are typically sold together with their corresponding buffers. Preparing a "homemade" buffer is uncommon, as companies optimize the buffer's saline composition to enhance enzyme performance and ensure long-term stability.
3. Use a high-fidelity Taq polymerase to ensure accurate introduction of desired mutations. After cloning, sequence the recipient vector to confirm the insert sequence and verify cloning accuracy.

4. For optimal PCR primer design, consider: (1) melting temperature calculation formula: 2°C × (A + T) + 4°C × (G + C); (1) ensure CG content is ≤50% per primer; (3) avoid three or more consecutive C or G bases at the 3′ end; (4) minimize primer complementarity to reduce dimer formation.
5. The vector contains multiple cloning sites (MCS) with restriction sites for NotI, PmeI XhoI, and SgfI. These sites facilitate insertion of the 3′ UTR of the gene of interest downstream of the Renilla stop codon. Ensuring the correct orientation and avoiding antisense insertion is crucial. Restriction enzymes should surround the insert and cut only within the plasmid's MCS, avoiding any other regions. To prevent antisense insertion of the 3′ UTR, verify restriction site positions within the MCS.
6. Use two different restriction enzymes to cut the plasmid and the insert. This prevents self-ligation and ensures correct insert orientation. If two enzymes cannot be used, treat the plasmid to remove a phosphate group before purification to prevent self-ligation.
7. Since ethidium bromide is a known mutagen, wearing appropriate personal protective equipment (PPE) is essential when handling it.
8. T4 ligase is an enzyme that joins DNA fragments, and its activity depends on a specific buffer containing ATP. Do not thaw the buffer at 37°C, as this could degrade ATP and compromise enzyme function.
9. The psiCHECK™-2 plasmid includes a gene that confers resistance to ampicillin.
10. Choose the appropriate culture medium for your cell line using the ATCC website: https://www.atcc.org/.
11. Firefly and Renilla luciferases are enzymes that require distinct substrates for their respective bioluminescent reactions. To accurately measure and differentiate between these reactions, specialized assay kits have been developed. Promega, a leading company in this field, designed the Dual-Luciferase® Reporter (DLR™) Assay System, which sequentially measures luminescence produced by firefly luciferase before suppressing its signal and activating luminescence from Renilla luciferase. This allows for the precise detection of both signals within the same sample. Firefly luciferase generates light by oxidizing beetle luciferin, a process that requires ATP, magnesium ions (Mg^{2+}), and O_2. To enhance the stability and intensity of the luminescence signals, Coenzyme A (CoA) has been incorporated into the latest luciferase reagents. In contrast, the luminescent reaction catalyzed by Renilla luciferase uses O_2 and coelenterazine as substrates.

12. PCR-based cloning generates the insert using standard PCR [7]. This process involves primers designed to amplify the target region, incorporating restriction sites at the ends of the sequence. The PCR reaction, thus, produces an amplicon of the gene's 3′ UTR, surrounded by restriction sites compatible with the multiple cloning site (MCS) of the host vector.
13. Enzyme volume is based on its activity, expressed in U/μL.
14. The transilluminator uses a UV lamp to visualize DNA. Wear UV-blocking safety glasses for protection.
15. Buffers contain varying salt and BSA concentrations, with some optimized for endonuclease activity. These buffers must maintain at least 75% activity for each enzyme. When performing double digestion, discrepancies between the two buffers may occur. To overcome this issue, "Smart" buffers are available for compatibility with multiple enzymes.
16. It is possible to evaluate plasmid digestion by using a "control gel" and comparing it to an uncut vector. A double-digested vector fragment appears smaller than the uncut linear plasmid because it represents the linear vector with the interstitial region removed by the two restriction enzyme cuts.
17. It is recommended to use 100 ng of plasmid DNA for the reaction.
18. Typically, the optimal plasmid-to-insert molar ratio for ligation is 1:3. However, in some instances, variations, such as 1:1 or 3:1, may be beneficial. Testing different concentrations can help establish ideal conditions for the cloning experiment [4].
19. Upon conclusion of the ligation reaction, it is best practice to purify the ligated plasmid before performing bacterial transformation. This step can be executed by either using columns filled with gel filtration resin or by gel purification.
20. Bacterial transformation is an essential process in the luciferase assay workflow. It serves to verify the success of ligation and enables the production of large amounts of psiCHECK™-2 vectors containing the 3′ UTR insert. This process relies on bacteria's ability to absorb foreign DNA. The closed circular plasmid DNA enters the competent *E. coli* cell, crosses the bacterial membrane, and, once inside the cytoplasm, begins replication. Linearized plasmid DNA is more susceptible to intracellular nuclease degradation; therefore, circular DNA transforms more efficiently [8].
21. The most commonly used competent strains of *Escherichia coli* for this process are DH5α and TOP10.
22. The plate containing the psiCHECK™-2 vector with the insert should display a higher colony count than the control

plate, which contains only the linearized recipient plasmid. The control plate serves as a baseline for the expected background colony count.

23. After purifying DNA using the manufacturer's protocol for the Miniprep plasmid purification kit, the psiCHECK™-2 vector containing the 3′ untranslated region (UTR) of the gene of interest is ready for sequencing. This step ensures the accuracy and quality of the cloning protocol before transfection into eukaryotic cells. Following sequencing, it is recommended to prepare a large quantity of plasmid under sterile conditions to achieve high plasmid concentrations, minimizing contamination risk during transfection.
24. Cell transfection is a complex technique enabling the introduction of foreign DNA or RNA into eukaryotic cells. A widely used technique involves using cationic lipids, which facilitate the delivery of nucleic acids into cells [5].
25. Easily transfectable cell lines, such as HEK293 (an embryonic kidney cell line), are preferred for transfection.
26. Within the serum is negatively charged protein albumin, which in the presence of the liposome reagent, alters the transfection complex's charge, thus interfering with its capacity to undergo cellular membrane fusion and reducing transfection efficiency.
27. A serum-free medium is used for more delicate cells to ensure cells are adequately nourished even in the absence of FBS.
28. Antibiotics should be removed from the culture medium before transfection to prevent unintended antibiotic uptake, protect cells from toxicity, and improve overall transfection efficiency.
29. Liposomes, composed of hydrocarbon chain tails and positively charged head groups, facilitate nucleic acid delivery by binding to the phosphate backbone of nucleic acids. The cationic nature of liposomes facilitates their uptake by cells through endocytosis, driven by their interaction with the negatively charged cell membrane, allowing genetic material to enter the cell.
30. The luciferase assay is designed to assess the interaction between the miRNA(s) and the 3′ UTR of a target gene using a luciferase reporter system. In this phase, cells are transfected with a psiCHECK™-2 plasmid and a pre-miR precursor corresponding to the miRNA under investigation. Two additional control groups should be included: (1) cells transfected with an empty psiCHECK™-2 for the 3′ UTR target region, and (2) cells transfected with a pre-miR precursor for an exogenous miRNA. If the miRNA interacts with the 3′ UTR, it will

bind to the UTR, leading to luciferase mRNA degradation and inhibition of fusion protein translation. Conversely, if no interaction occurs, the luciferase mRNA will be transcribed and translated, producing yellow light. To ensure statistical validity, each transfection condition should be performed in triplicate.

31. It is recommended to use only the supplied passive lysis buffer, as reagents in the kit are specifically optimized for compatibility. The lysis buffer is formulated to minimize background cell auto-luminescence.
32. Luciferase Assay Reagent II (LAR II) contains the firefly luciferase substrate, which facilitates the generation and stabilization of yellow light emission.
33. If the equipment is configured for multisampling and plate reading, sequential measurements can be collected using the accompanying software. Otherwise, individual tubes must be replaced upon completion of each reading. When using a reagent injector, it should be cleaned thoroughly with ethanol and water to prevent contamination and properly primed with the reagent solutions.
34. The reagent significantly reduces the luminescent signals from firefly luciferase, decreasing intensity over 10,000-fold. It also contains coelenterate luciferin, the substrate for Renilla luciferase. By suppressing the firefly luciferase signal while providing the Renilla luciferase substrate, it allows for measurement of Renilla luciferase activity.
35. The coding region for Renilla luciferase is positioned upstream of the miRNA binding site in the psiCHECK™-2 vector. If the miRNA binds to the 3′ UTR, Renilla luciferase expression should be downregulated. To evaluate the miRNA's effect on the 3′ UTR, Renilla luciferase activity in samples transfected with the psiCHECK™-2 vector containing the 3′ UTR and the specific pre-miR precursor should be compared with negative control samples. This comparison assesses miRNA-induced degradation of the target gene's mRNA. The procedure should be performed at least three times, followed by statistical analysis to ensure the reliability and significance of the results.

Conflict of Interest

The authors have no conflicts of interest to declare that are relevant to the content of this chapter.

References

1. Janga SC, Vallabhaneni S (2011) MicroRNAs as post-transcriptional machines and their interplay with cellular networks. Adv Exp Med Biol 722:59–74
2. Solberg N, Krauss S (2013) Luciferase assay to study the activity of a cloned promoter DNA fragment. Methods Mol Biol 977:65–78
3. Zhang S, Cahalan MD (2007) Purifying plasmid DNA from bacterial colonies using the Qiagen Miniprep kit. J Vis Exp 6:247
4. Cranenburgh RM (2004) An equation for calculating the volumetric ratios required in a ligation reaction. Appl Microbiol Biotechnol 65:200–202
5. Liu D, Ren T, Gao X (2003) Cationic transfection lipids. Curr Med Chem 10(14):1307–1315
6. Kirtana Reddy K, Ramachandran B (2024) A reproducible and affordable method of conducting luciferase assay. Anal Biochem 692:115558
7. Lorenz TC (2012) Polymerase chain reaction: basic protocol plus troubleshooting and optimization strategies. J Vis Exp 22(63):3998
8. Johnston C, Martin B, Fichant G, Polard P, Claverys JP (2014) Bacterial transformation: distribution, shared mechanisms and divergent control. Nat Rev Microbiol 12(3):181–196

Check for updates

Chapter 17

Experimental Validation of MicroRNA Targets: Mutagenesis of Binding Regions Mutagenesis-Based Validation of MicroRNA Target Binding

Roma Karna, Soumya Sharma and Federica Calore

Abstract

MicroRNAs (miRNAs) are small, noncoding RNA molecules, typically 22 to 25 nucleotides long, that serve as critical regulators of gene expression at the post-transcriptional level. By binding to specific messenger RNAs (mRNAs), miRNAs can lead to translational repression or mRNA degradation, thereby modulating various cellular processes. While computational tools utilizing algorithmic models have been developed to predict miRNA–mRNA interactions, experimental validation remains essential. To ensure the reliability of these predictions, laboratory techniques have been continuously refined to enhance target verification accuracy. Among the established validation methods, the luciferase reporter assay is widely employed to evaluate miRNA–mRNA interactions. This approach involves a reporter plasmid containing the 3′ untranslated region (UTR) of the target gene, enabling the assessment of miRNA-mediated regulation through alterations in luciferase activity. Introducing targeted mutations within the 3′ UTR serves as a critical strategy to ascertain the specificity of miRNA binding by assessing changes in luciferase signal output. This chapter provides an in-depth overview of the experimental workflow for 3′ UTR mutagenesis, emphasizing optimized protocols and best practices for integrating this technique into miRNA functional studies. Additionally, it highlights methodological refinements that improve the precision and reproducibility of luciferase assay-based target validation.

Keywords MicroRNA target verification, 3' UTR mutagenesis, Gene regulation, PCR amplification, Luciferase reporter assay

1 Introduction

The targeted alteration of short nucleotide sequences critical for microRNA binding is a key approach for validating their interaction with mRNA targets. This strategy serves as an essential control for luciferase assays and provides indirect confirmation of experimental findings.

MicroRNAs bind to various regions of mRNAs, including the 3′ UTR, 5′ UTR, and coding sequences, through base pairing with target sites typically six to eight nucleotides in length. This interaction

Alessandro Laganà and Giovanni Nigita (eds.), *MicroRNA Target Identification*, Methods in Molecular Biology, vol. 3056, https://doi.org/10.1007/978-1-0716-5392-0_17,

recruits the RNA-induced silencing complex (RISC), leading to either mRNA degradation or translational repression, depending on the level of complementarity between the miRNA and its target [1].

Bioinformatics tools, such as TargetScan (http://www.targetscan.org/vert_72/), assist in identifying these binding sites within the mRNA sequence.

Among the five recognized classes of binding sites, 8-mer sites exhibit the strongest binding affinity, while 6-mer site are the least effective. This classification is based on complementarity within the *seed* region (nucleotides two to eight), often followed by an adenine (A).

Oligonucleotide-directed mutagenesis is a widely used technique for introducing specific mutations into DNA sequences. This method utilizes a synthetic oligonucleotide primer designed to incorporate the desired mutation during PCR amplification. The primer anneals to the wild-type DNA sequence, facilitating mutation incorporation. The extension of the mutated primer during PCR results in a modified DNA sequence carrying the desired mutation within the miRNA-binding region [1].

A streamlined adaptation of the Megaprimer method, originally developed by Kammann et al. [2], is commonly employed for mutagenesis. This approach involves two PCR steps using three distinct oligonucleotide primers. The first PCR step, conducted at a low annealing temperature, facilitates the incorporation of the mutation via a mutagenic primer. The second step, performed at a higher annealing temperature, allows standard primers to amplify the modified DNA while preventing residual mutagenic primer binding. In the initial PCR reaction, the mutagenic primer and a flanking primer generate a mutated DNA fragment, termed the "megaprimer." In the subsequent step, this megaprimer is used alongside the second flanking primer to complete amplification. The mutagenic primer, typically 16–25 nucleotides long, is designed for optimal annealing during the first PCR, while the other primers, generally 25–30 nucleotides long, require higher annealing temperatures to ensure specificity [2,3].

Recent advancements in polymerase enzymes have streamlined this method, improving efficiency and accessibility. The latest modification involves a single-step PCR protocol that requires only using just two primers—forward and reverse—that are designed to overlap and incorporate the desired mutation into the sequence (Fig. 1).

PCR-based site-directed mutagenesis can be performed using either a custom protocol or a commercially available kit. The homemade approach is cost-effective but lacks standardization, whereas commercial kits, while more expensive, offer a streamlined and reliable alternative (*see* **Note 1**).

This chapter outlines key steps in mutagenesis protocols, which, when combines with luciferase reporter assays and Western blotting,

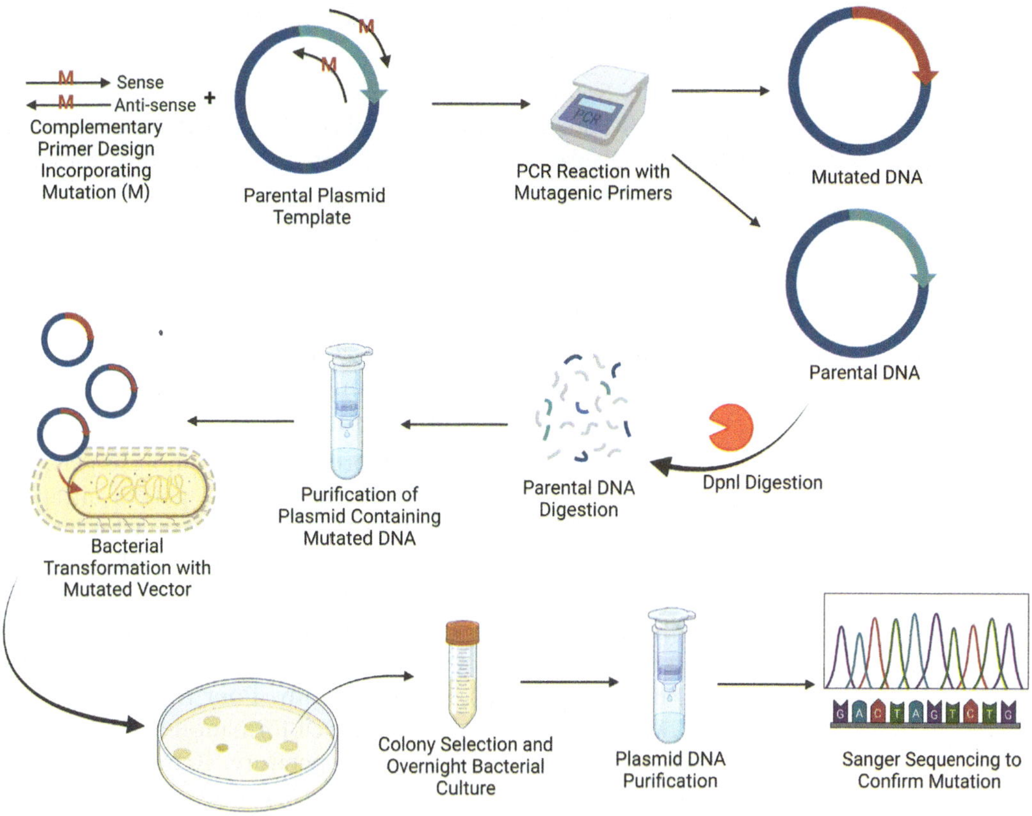

Fig. 1 Schematic representation PCR-based workflow for introducing a desired mutation into a DNA sequence. This figure was created with BioRender

provide a robust experimental framework for validating computationally predicted microRNA targets.

2 Materials

1. psiCHECK™-2 Vector: Pre-engineered to include the 3′ UTR sequence, ready for downstream applications.
2. Oligonucleotides: Dissolve lyophilized oligos in DNase- and RNase-free water to make a 100 μM stock. For experimental use, dilute the stock 1:10 with the same water to obtain a 10 μM working solution (*see* **Notes 2** and **3**).
3. PCR Buffer (10×): Composed of 15 mM $MgCl_2$, 500 mM KCl, and 100 mM Tris–HCl (pH 8.3 at 25°C) (*see* **Note 1**). To prepare, dissolve 0.142 g of $MgCl_2$ and 3.72 g of KCl in 10 mL of 1 M Tris–HCl (pH 8.3), then adjust the final

volume to 100 mL with DNase- and RNase-free water. Store aliquots at −20°C.

4. dNTP Solution (10 mM final concentration): Dilute the 100 mM stock solution (supplied with the PCR kit) tenfold using DNase- and RNase-free water. Keep at −20°C for long-term storage.
5. High Fidelity Taq Polymerase: Maintain at −20°C (*see* **Note 4**).
6. DpnI Restriction Enzyme: Store at −20°C.
7. CutSmart® Restriction Buffer: This buffer is composed of 50 mM potassium acetate, 20 mM Tris–acetate, 10 mM magnesium acetate, and 100 μg/mL BSA. Keep at −20°C.
8. Agarose Gel Preparation: To prepare electrophoresis gels, dissolve 0.6 g (for a 1% gel) or 1.2 g (for a 2% gel) of ultrapure agarose in 60 mL of 1 × TBE buffer (prepared by dissolving 10.8 g of Tris, 5.5 g of boric acid, and 4 mL of 0.5 M Na_2EDTA pH 8.0 in 1 L of water). Heat in a microwave until fully dissolved. Afterward, under a fume hood, add ethidium bromide (EtBr) to achieve a final concentration of 0.2–0.5 μ g/mL (*see* **Note 5**), and mix gently. Pour the solution into a casting tray with a comb for well formation and allow it to solidify for 20 min at room temperature. Remove the comb before electrophoresis.
9. DNA Ladder for Molecular Weight Estimation: Store at −20°C before first use, then keep at 4°C.
10. Loading Dye (10×): Contains 10 mM Tris–HCl (pH 7.6), 60 mM EDTA, 0.03% bromophenol blue, and 60% glycerol.
11. Gel Extraction Kit: Used for DNA purification from agarose gels.
12. PCR Purification Kit: For cleaning up PCR products.
13. Spin Columns with Gel Filtration Resin: For rapid nucleic acid purification and size selection.
14. Competent *E. coli* Cells: Store aliquots at −80°C. Avoid repeated freeze–thaw cycles to maintain efficiency.
15. LB Broth: Prepare by dissolving 10 g of Bacto tryptone, 10 g of NaCl, and 5 g of yeast extract in 800 mL of distilled water while stirring with a magnetic stirrer. Adjust pH to 7.5 using NaOH, bring the volume to 1 L, and sterilize by autoclaving. Once cooled, supplement with ampicillin to a final concentration of 100 mg/mL (*refer to* **Note 6**).
16. SOC Media: Prepare by combining 20 g of Bacto tryptone and 5 g of yeast extract in a solution containing 10 mM NaCl, 2.5 mM KCl, 10 mM $MgCl_2$, 10 mM $MgSO_4$, and 20 mM

glucose. Add distilled water to a final volume of 1 L and sterilize by autoclaving.

17. Agar Plates with Selective Antibiotic (Ampicillin): Dissolve 20 g of agar, 10 g of Bacto tryptone, 10 g of NaCl, and 5 g of yeast extract in 800 mL of distilled water with stirring. Adjust pH to 7.5 using NaOH, then bring the final volume to 1 L with distilled water and sterilize the mixture by autoclaving. After cooling, supplement with ampicillin (final concentration of 100 μg/mL) before agar solidification. Pour into Petri dishes (~25 mL per 100-mm plate) or re-melt the agar using a microwave if necessary before use.
18. Plasmid Mini Preparation Kit [4]: For small-scale plasmid DNA purification.
19. Centrifuge.
20. Microcentrifuge.
21. Thermal Cycler.
22. Transilluminator.
23. Heated Block.
24. Bacterial Shaker.
25. Microbiological Incubator.
26. Pipette Aid.
27. Pipettes and Filter Tips.
28. 15 mL Conical Tubes.
29. 1.5 mL Microcentrifuge Tubes.
30. Flasks.
31. Microwave.
32. Laboratory Scale.
33. NanoDrop Microvolume Spectrophotometer.

3 Methods

3.1 Mutagenesis-PCR

To introduce specific mutations into plasmid DNA, a refined PCR approach can enhance amplification efficiency and specificity [2, 3].

1. Use 25–50 ng of plasmid DNA as the template.
2. The total PCR volume is 20 μL. Prepare a reagent mix based on the number of samples plus one ($n+1$). For each reaction, combine:

   2 μL of 10× buffer,

   0.4 μL of each primer (10 μM),

0.4 μL of 10 mM dNTP mix,
0.4 μL of Taq polymerase.
Adjust the total volume with nuclease-free water.
Dispense the mix into PCR tubes, add DNA, mix gently, and briefly centrifuge. Keep tubes on ice before starting the reaction.

3. Set up the thermal cycler as follows (*see* **Notes 7** and **8**):
   Initial denaturation: 95°C for 3 min (1 cycle),
   Denaturation: 95°C for 30 s (40 cycles),
   Extension: 72°C for 1 min/kb (40 cycles),
   Final extension: 72°C for 7 min (1 cycle),
   Hold at 4°C.
4. Preheat the thermal cycler lid to 105°C before starting the reaction.
5. Load tubes into the thermal cycler and wait for amplification process to complete (*see* **Note 9**).
6. Following PCR, briefly centrifuge the tubes, then add to each sample 2 μL of 10× loading buffer, and prepare for gel electrophoresis.
7. Prepare the agarose gel in the electrophoresis chamber with 1× TBE. Load PCR samples and molecular weight ladder, and run at 120 V. Stop electrophoresis once the dye front reaches 75%–80% of the gel length.
8. Visualize the bands under a UV transilluminator.
9. Excise the correct fragment (*see* **Note 10**), weigh the gel slice, and purify DNA using a gel extraction kit as per instructions.

Advances in Mutagenesis Technologies

Recent advances have significantly improved the accuracy and efficiency of mutagenesis, particularly in miRNA target validation. CRISPR–Cas9-assisted mutagenesis now allows direct genome editing at the miRNA target sites, reducing off-target mutations compared to traditional PCR-based methods [5]. Additionally, digital droplet PCR (ddPCR) offers precise quantification of mutant sequences, eliminating the need for traditional gel electrophoresis [6].

Another innovation involves the use of nanopore sequencing for rapid validation, enabling real-time detection of introduced mutations without requiring time-intensive Sanger sequencing [7]. Additionally, in silico prediction tools powered by AI and machine learning can optimize mutagenesis design by predicting mutation efficiency and binding site disruptions before experimental validation [8]. These advancements collectively enhance the accuracy, scalability, reproducibility of mutagenesis-based functional studies.

3.2 Selection of Mutants Using DpnI Digestion

After confirming successful PCR amplification and purifying the mutated vector, DpnI digestion selectively removes the parental (non-mutated) plasmid, enriching the sample for the mutated construct (*see* **Note 11**) (Fig. 1).

1. Mix the gel-purified plasmid with:
   5 μL of the appropriate 10× buffer (*see* **Note 12**),
   1 μL of DpnI enzyme,
   ddH_2O to a final volume of 50 μL.
2. Incubate at 37°C for at least 2 h to ensure complete digestion.
3. Purify the digested plasmid using spin columns containing gel filtration resin, following the manufacturer's protocol (*see* **Note 13**).

3.3 Transformation of Bacteria

1. Thaw the required aliquots of competent bacterial cells on ice (*see* **Note 14**).
2. Prewarm a thermal block to 42°C.
3. Mix 50 μL of competent bacteria with the digested plasmid. Incubate on ice for 30 min, gently flicking the tube one to two times. Then, heat shock at 42°C for 30 s and return to ice for 2 min to allow plasmid uptake.
4. Add 250 μL of SOC medium to the bacterial mixture and incubate at 37°C with shaking (~220–250 rpm) for 1 h.
5. Plate the transformed bacteria on nutrient agar containing the selection antibiotic (*see* **Note 6**).
6. Incubate the plates at 37°C for 12–16 h to allow colony formation.
7. Pick five to ten colonies with a sterile tip and transfer each into a sterile 15 mL tube containing 5 mL of LB broth with antibiotic. Incubate overnight at 37°C with shaking (220–250 rpm).
8. Pellet the bacterial suspension by centrifugation at ~350-400 × *g* for 20 min.
9. Purify the plasmid DNA from the bacterial cells following the method outlined by Zhang et al. [4].
10. Verify the mutation by sequencing the target region.

4 Notes

1. Several commercial kits are available for PCR-based site-directed mutagenesis, significantly simplifying and accelerating the process. These kits follow a standardized protocol, with reagents optimized for maximum efficiency. They are widely

utilized for microRNA binding site mutagenesis, achieving an efficiency of approximately 80%. Mutagenic primers can be designed using the QuikChange Primer Design Program, accessible at www. agilent.com/genomics/qcpd.

2. Primer design is a critical factor in this protocol. Unlike the megaprimer approach, the primers must be sufficiently long to ensure annealing, even in the presence of mismatches introduced for mutagenesis. Key considerations for primer design include:

- GC content of more than 30%.
- The terminal bases should preferably be G or C.
- Primer length between 25 and 45 nucleotides.
- Melting temperature (T_m) ideally close or above 78°C.
- The T_m can be calculated using the formula below:

$$T_m = 81.5 + 0.41[\%(G+C)] - \left(\frac{675}{n}\right) - (\%\text{mismatch})$$

where n= total number of nucleotides.

3. There are two primary strategies for mutating microRNA binding sites:

   a. Deleting the bases forming the binding site.

   b. Replacing the wild-type sequence with a scrambled version.

   It is recommended in both methods to place the mutation sequence in the center of the primers in order to optimize their annealing.

4. Due to the length of the amplified region in mutagenesis PCR, it is recommended to use a high-fidelity DNA polymerase. The most accurate commercially available enzymes exhibit an error rate averaging 0.9% per kilobase.

5. Ethidium bromide is a known mutagen. Always wear personal protective equipment (PPE) when handling it.

6. Given the high melting temperature of the primers, a multi-stage PCR can be performed, where the annealing and extension steps occur at 72°C. The extension time must be sufficient to ensure complete vector amplification.

7. The psiCHECK™-2 vector carries the gene for ampicillin resistance.

8. Most vectors are several thousand bases long. The psiCHECK™-2 vector is 6273 base pairs in length. Since DNA polymerase processivity averages 1 kb/min—it is

recommended to equate the elongation time to 1 min/kb of plasmid length.

9. While the efficiency of this reaction is relatively low, only a small amount of mutated plasmid is required for DpnI digestion and subsequent transformation into competent bacterial cells.
10. The transilluminator utilizes a UV lamp for DNA detection. Always wear UV-blocking safety glasses when operating it.
11. The DpnI enzyme specifically targets methylated DNA, making it useful in mutagenesis protocols for digesting the template while selecting for newly synthesized, mutation-containing DNA. Its recognition site is 5′-Gm6ATC- 3′, requiring adenine methylation at the GATC sequence. DNA isolated from *E. coli* strains is naturally methylated by a DAM methylase (deoxyadenosine methylase), making it susceptible to DpnI cleavage.
12. Reaction buffers contain varying salt and BSA concentrations. Only a single buffer supports optimal endonuclease activity, which must maintain at least 75% activity for each enzyme used.
13. To verify successful plasmid digestion, a "control gel" should be run, comparing the digested plasmid to the uncut vector.
14. The DH5alpha and TOP10 strains of *E. coli* are the most commonly utilized for competent cell transformation.

Conflict of Interest

The authors have no conflicts of interest to declare that are relevant to the content of this chapter.

References

1. Janga SC, Vallabhaneni S (2011) MicroRNAs as post-transcriptional machines and their interplay with cellular networks. Adv Exp Med Biol 722:59–74
2. Kammann M, Laufs J, Schell J, Gronenborn B (1989) Rapid insertional mutagenesis of DNA by polymerase chain reaction (PCR). Nucleic Acids Res 17(13):5404
3. Sambrook J, Russell RW (2001) Molecular cloning: a laboratory manual, 3rd edn. Cold Spring Harbor Laboratory Press, N.Y
4. Zhang S, Cahalan MD (2007) Purifying plasmid DNA from bacterial colonies using the Qiagen Miniprep kit. J Vis Exp 6:e247
5. Chen Y et al (2024) MicroRNA-focused CRISPR/Cas9 screen identifies miR-142 as a key regulator of Epstein-Barr virus reactivation. PLoS Pathog 20(6):e1011970. https://doi.org/10.1371/journal.ppat.1011970
6. Huggett JF, Simon C, Carole AF (2015) Considerations for digital PCR as an accurate molecular diagnostic tool. Clin Chem 61(1):79–88. https://doi.org/10.1373/clinchem.2014.221366
7. Cumbo C et al (2019) Nanopore targeted sequencing for rapid gene mutations detection in acute myeloid leukemia. Genes 10(12):1026. https://doi.org/10.3390/genes10121026
8. Gjoni K, Katherine SP (2024) SuPreMo: a computational tool for streamlining in silico perturbation using sequence-based predictive models. Bioinformatics 40(6):btae340

The manufacturer's authorised representative in the EU is Springer
Nature Customer Service Centre GmbH, Europaplatz 3, 69115 Heidelberg,
Germany. If you have any concerns regarding our products, please
contact ProductSafety@springernature.com

Printed and bound by CPI Group (UK) Ltd, Croydon, CR0 4YY
18/09/2026
02219020-0005